Challenges Towards Ecological Sustainability in China

Xiaojun Yang · Shijun Jiang
Editors

Challenges Towards Ecological Sustainability in China

An Interdisciplinary Perspective

Springer

Editors
Xiaojun Yang
Department of Geography
Florida State University
Tallahassee, FL, USA

and

Institute of Groundwater
and Earth Sciences
Jinan University
Guangzhou, China

Shijun Jiang
Institute of Groundwater
and Earth Sciences
Jinan University
Guangzhou, China

ISBN 978-3-030-03483-2 ISBN 978-3-030-03484-9 (eBook)
https://doi.org/10.1007/978-3-030-03484-9

Library of Congress Control Number: 2018960729

This Springer imprint is published by the registered company Springer Nature Switzerland AG
The registered company address is: Gewerbestrasse 11, 6330 Cham, Switzerland

Foreword

The complex interplay between human population growth, nonrenewable natural resource depletion, and environmental degradation has been challenging the human society since the mid-twentieth century. While human fertility is generally declining across the world, the global population will continue to grow for many years to come, increasing pressure on the environment. While population numbers alone are often considered to be the main driver of environmental degradation, more evidence suggests that economic development and poor industrial and agricultural practices carry an increasing share of the blame in some regions. This is particularly true for China where four decades of rapid industrial and economic development have been accompanied by a dramatic change in consumption patterns, demanding more energy and escalating environmental stress. Fortunately, the Chinese society has come into consensus that unsustainable development must be changed, and economic and social progress must be adjusted to minimize the impacts upon the Earth's life support systems. The Chinese government has recently drafted massive plans to promote sustainable development, strengthen eco-environmental protection, and battle against all forms of pollution. There has been an urgent need to conduct intensive research on sustainability that can help implement these ambitious plans. Encouragingly, China's Natural Science Foundation has recently funded an increasing number of projects to support sustainability research. Despite the fast Chinese society's transition, the incomparable demand for sustainability, and the extensive research conducted by Chinese scholars on the subject, there are only few English books focusing on the experience from China, the world's largest developing country. Most published books have targeted the case studies drawn from developed countries in Europe and North America.

Xiaojun Yang and Shijun Jiang's *Challenges Towards Ecological Sustainability in China* fills the gap identified above and provides a timely and welcome addition to the field. The book is appealing due to a number of reasons. First, it brings together active scholars from several fields including geography, ecology, environmental science, and economics, providing an interdisciplinary perspective on ecological sustainability in China, particularly in the Great Pearl River Delta. This region is well known as the birthplace of China's economic reform since 1978 and the home of the world's largest urban area in both size and population. Second, it covers a selection of important issues challenging the transition towards ecological sustainability in China, such as urban growth and landscape changes, surface and groundwater pollution, water resources management, flood hazards and risk, ecological security, and green development. Third, it includes a strong methodology component focusing on the spatially explicit approach that has its root in the field of geography but has been increasingly adopted in sustainability science. Last, it incorporates a policy dimension that is critical to help link knowledge with action in pursuit of a transition towards sustainability.

Both editors are native Chinese, but received their postgraduate training and earned their terminal degree (Ph.D.) from the United States. They have demonstrated outstanding scholarship records in the field of Earth Sciences. I have known Dr. Xiaojun Yang, Lead Editor of this volume, for more than one decade during the time we worked together for the "Ecosystem Processes and Services" project funded through the International Partnership Program for Creative Research Teams Initiative by the Chinese Academy of Sciences (2009–2014). He has published in numerous academic journals and edited several books with John Wiley, Taylor Francis, and Springer. His *Urban Remote Sensing* was translated into Chinese and won a Best Imported Book Award in 2016. Dr. Yang is one of few scholars who is capable of leading an interdisciplinary effort to put together a volume on ecological sustainability issues in China because of not only his educational background in both natural and social sciences but also his passions in pursuit of research on the subject.

Finally, I would like to take this opportunity to congratulate the successful completion of this book project. Designed for a wide variety of end users, this book will appeal to students, researchers, and professionals in the fields of sustainability science, geography, ecology, and environmental science and policy. Although this book is mostly authored by Chinese scholars with case studies predominately drawn from China, the knowledge gained from this country with the world's fastest growing economy can be applied to other areas in the world. I highly recommend this book!

Beijing, China
July 2018

Bojie Fu, Ph.D.
Professor and Academician
Chinese Academy of Sciences

Preface

As the world's largest developing country, China has witnessed significant economic growth since 1978 when it initiated the reform and open door policy. However, the country has been facing grave challenges because of diminishing natural resources, continued environmental degradation, and increasing social and economic marginalization. While the society has been undergoing a transition from seeking strong economic growth into pursuing environmental sustainability, the Chinese central government has recently drafted massive plans to reinforce eco-environmental protection and battle against pollution. There has been an urgent need to conduct basic and applied research on sustainability that can help implement these ambitious plans. Despite the undergoing Chinese society transition, the increased demand for sustainability, and the extensive research on the subject, there are only few books targeting the experiences in China. Most of the published books are concentrated on the case studies from developed countries in Europe and North America.

Within the above context, a book on ecological sustainability in China is timely. Designed for academic and business sectors, this book brings together leading scholars in geography, ecology, environmental science, and economics, providing an interdisciplinary perspective on various issues challenging ecological sustainability in China, especially in the Southern China, where rapid urbanization over the past four decades has caused economic and ecological strains on land and people. While the range of issues is constantly increasing, this book covers some major ones, such as land use and land cover change, urban growth and landscape changes, urbanization impacts upon the thermal environment, surface and groundwater pollution, land use and non-point pollution in coastal watersheds, urban and coastal flood hazards and risk, water resources utilization and management, ecological security, green development for resource-based regions, and sustainable landscape design. Additionally, the book incorporates a strong component of methodology targeting the spatially explicit approach that is supported by geospatial technologies such as remote sensing and geographic information systems. Finally, this book includes a policy dimension that is critical for linking knowledge with action to pursue a transition towards ecological sustainability in the world's largest developing country. Although this book is mostly authored by Chinese scholars with case studies predominately drawn from China, the knowledge gained from this country with the world's fastest growing economy should be applicable to other areas globally.

This book is the outcome of intensive research conducted by multidisciplinary scholars and is intended for students, academics, and professionals in the fields of sustainability science, geography, ecology, and environmental science and policy. We are very grateful to all authors who contributed their work and revised their manuscripts multiple times according to our requests. Special thanks are due to Drs. Ting Liu and Jinxin Yang for their help in the manuscript review and proofreading and to Ms. Hong Su for her administrative support to the Jinan Forum on Geography and Ecological Sustainability held during February 7–9, 2017, Guangzhou, China. The Lead Editor is very appreciative of the support from Jinan University during his several recent visits to China. Last, we would like to extend our sincere gratitude to Petra van Steenbergen for her help and assistance leading to the completion of this book project.

Guangzhou, China
July 2018

Xiaojun Yang
Shijun Jiang

Contents

Editors and Contributors

About the Editors

Dr. Xiaojun Yang Editor of this volume, is a tenured Full Professor of Geography in the College of Social Sciences and Public Policy at the Florida State University, USA. He has also had a few visiting academic positions with the Chinese Academy of Sciences and the Jinan University, China. His research interests include the development of remote sensing and geospatial technologies with applications in the urban and environmental domains. He has authored or co-authored six English books and over 100 articles in these areas.

Dr. Shijun Jiang Co-editor of this volume, is a Full Professor of Paleoecology at the Institute of Groundwater and Earth Sciences, Jinan University, China. His research focuses on the use of fossils and geochemical measurements to examine the biotic responses to environmental changes and the dynamic processes of aquatic ecosystems over various timescales. He has authored or co-authored more than 40 peer-reviewed journal articles.

Contributors

Wei Bian State Key Laboratory of Desert and Oasis Ecology, Xinjiang Institute of Ecology and Geography, Chinese Academy of Sciences, Urumqi, Xinjiang, China

Jianyao Chen School of Geography and Planning, Sun Yat-sen University, Guangzhou, Guangdong, China

Jiayu Chen International Institute for Earth System Science (ESSI), Nanjing University, Nanjiang, Jiangsu, China

Liding Chen State Key Laboratory of Urban and Regional Ecology, Research Center for Eco-Environmental Sciences, Chinese Academy of Sciences, Beijing, People's Republic of China; University of Chinese Academy of Sciences, Beijing, China

Qingyi Chen School of Geography and Planning, Guangxi Teachers Education University, Nanning, Guangxi, China

Shuisen Chen Guangdong Open Laboratory of Geospatial Information Technology and Applications, Guangdong Key Laboratory of Remote Sensing and GIS Technology Application, Guangdong Engineering Technology Center for Remote Sensing Big Data Applications, Guangzhou Institute of Geography, Guangzhou, Guangdong, China

Weiqi Chen Department of Geography and Anthropology, Louisiana State University, Baton Rouge, LA, USA

Yaning Chen State Key Laboratory of Desert and Oasis Ecology, Xinjiang Institute of Ecology and Geography, Chinese Academy of Sciences, Urumqi, Xinjiang, China

Jianquan Cheng Key Laboratory of Environment Change and Resources Use in Beibu Gulf, Ministry of Education, Guangxi Teachers Education University, Nanning, China

Jianhui Cong School of Economics and Management, Shanxi University, Taiyuan, Shanxi, China; Shanxi Research Center for Green Development, Taiyuan, Shanxi, China

Ayu Ervinia College of the Environment & Ecology, Xiamen University, Xiamen, Fujian, People's Republic of China

Gonghuan Fang State Key Laboratory of Desert and Oasis Ecology, Xinjiang Institute of Ecology and Geography, Chinese Academy of Sciences, Urumqi, Xinjiang, China

Lei Gao School of Geography and Planning, Sun Yat-sen University, Guangzhou, Guangdong, China

Jinliang Huang College of the Environment & Ecology, Xiamen University, Xiamen, Fujian, People's Republic of China

Yaling Huang College of the Environment & Ecology, Xiamen University, Xiamen, Fujian, People's Republic of China

Fei Jiang International Institute for Earth System Science (ESSI), Nanjing University, Nanjiang, Jiangsu, China

Shijun Jiang Institute of Groundwater and Earth Sciences, Key Laboratory of Eutrophication and Red Tide Prevention of Guangdong Higher Education Institutes, Jinan University, Guangzhou, Guangdong, China

Richard William Jordan Department of Earth & Environmental Sciences, Faculty of Science, Yamagata University, Yamagata, Japan

Fanhua Kong International Institute for Earth System Science (ESSI), Nanjing University, Nanjiang, Jiangsu, China

Zhi Li State Key Laboratory of Desert and Oasis Ecology, Xinjiang Institute of Ecology and Geography, Chinese Academy of Sciences, Urumqi, Xinjiang, China

Zhishan Li Guangdong Provincial Academy of Environmental Science, Guangzhou, Guangdong, China

Yuanda Lei Institute of Groundwater and Earth Sciences, Key Laboratory of Eutrophication and Red Tide Prevention of Guangdong Higher Education Institutes, Jinan University, Guangzhou, Guangdong, China

Jia Liu Guangdong Open Laboratory of Geospatial Information Technology and Applications, Guangdong Key Laboratory of Remote Sensing and GIS Technology Application, Guangdong Engineering Technology Center for Remote Sensing Big Data Applications, Guangzhou Institute of Geography, Guangzhou, Guangdong, China

Ting Liu Department of Geography and Environmental Studies, Northeastern Illinois University, Chicago, IL, USA

Nianxiu Qin School of Geography and Planning, Guangxi Teachers Education University, Nanning, Guangxi, China; Key Laboratory of Environment Change and Resources Use in Beibu Gulf, Ministry of Education, Guangxi Teachers Education University, Nanning, China

Ranhao Sun State Key Laboratory of Urban and Regional Ecology, Research Center for Eco-Environmental Sciences, Chinese Academy of Sciences, Beijing, People's Republic of China

Junneng Wang School of Geography and Planning, Guangxi Teachers Education University, Nanning, Guangxi, China; Key Laboratory of Environment Change and Resources Use in Beibu Gulf, Ministry of Education, Guangxi Teachers Education University, Nanning, China

Xianwei Wang School of Geography and Planning, Sun Yat-sen University, Guangzhou, Guangdong, China; Guangdong Key Laboratory for Urbanization and Geo-simulation, Guangzhou, Guangdong, China

Yasu Wang Institute of Groundwater and Earth Sciences, Key Laboratory of Eutrophication and Red Tide Prevention of Guangdong Higher Education Institutes, Jinan University, Guangzhou, Guangdong, China

Changguang Wu Department of Landscape Architecture, College of Horticulture and Forest Science, Huazhong Agricultural University, Wuhan, Hubei, China

Rongbo Xiao Guangdong Provincial Academy of Environmental Science, Guangzhou, Guangdong, China; School of Environmental Science and Engineering, Guangdong University of Technology, Guangzhou, Guangdong, China

Lichun Xie School of Geography and Tourism, Guangdong University of Finance & Economics, Guangzhou, Guangdong, China

Jun Yang School of Economics and Management, Shanxi University, Taiyuan, Shanxi, China; Shanxi Research Center for Green Development, Taiyuan, Shanxi, China

Lei Yang Research Center for Eco-Environmental Sciences, Chinese Academy of Sciences, Beijing, China

Xiaojun Yang Department of Geography, Florida State University, Tallahassee, FL, USA; Institute of Groundwater and Earth Sciences, Jinan University, Guangzhou, Guangdong, China

Haiwei Yin Department of Urban Planning and Design, Nanjing University, Nanjing, Jiangsu, China

Qin Zhang School of Economics and Management, Shanxi University, Taiyuan, Shanxi, China; Shanxi Research Center for Green Development, Taiyuan, Shanxi, China

Zhou Zhou School of Economics and Management, Shanxi University, Taiyuan, Shanxi, China; Shanxi Research Center for Green Development, Taiyuan, Shanxi, China

Part I
Conceptual and Technical Issues

Part I (Chapters "Challenges Towards Ecological Sustainability in China: An Introduction"–"Diatoms as an Evaluation Tool for the Ecological and Environmental Conditions of Rivers and Streams in China: A Retrospective Study") provides an overview or a review on several major conceptual and technical issues. Chapter "Challenges Towards Ecological Sustainability in China: An Introduction" discusses the rationale and motivation behind this book project and provides an overview on some core concepts on sustainability, major methods for sustainability research, and the book content. It also identified several major areas deserving further research. Chapter "Regional Eco-security: Concept, Principles and Pattern Design" discusses the objectives, principles, and frameworks in regional eco-security pattern design emphasizing the practice towards ecosystem restoration and landscape optimization. Chapter "How to Design an Urban Ecological Landscape: Sustainability, Efficiency, and Harmony?" discusses several core principles for urban ecological landscape design aiming to minimize negative outcomes caused by urbanization. Chapter "Monitoring Urban Growth and Land Changes in Beijing, China's Capital City by Remote Sensing: Progress and Challenges" reviews several essential conceptual and technical issues for remote sensing-based monitoring of urban growth in Beijing, arguably the most frequently researched megacity in the world. Finally, Chapter "Diatoms as an Evaluation Tool for the Ecological and Environmental Conditions of Rivers and Streams in China: A Retrospective Study" reviews the history and applications of using diatoms for environmental assessment of rivers and streams in China.

Challenges Towards Ecological Sustainability in China: An Introduction

Xiaojun Yang and Shijun Jiang

Abstract As a rising global superpower, China is facing grave challenges for economic and social prosperity because of diminishing natural resources and deteriorating environmental health. There has been an urgent need to conduct extensive research on sustainability that can help implement the eco-environmental protection and battle against pollution plans drafted by China's central government. Within this context, a book on some major ecological sustainability issues in China is timely. This chapter discusses the motivation behind this book project, provides an overview on the historical development of sustainability and the nature of sustainability science, introduces some major methods appropriate for ecological sustainability research, and previews the book structure. It concludes by highlighting several areas deserving further research.

Keywords Ecological sustainability · Sustainability science
Spatially explicit approaches · Socio-environmental systems · China

1 Introduction

The global population has grown from 1 billion in 1800 to 7.6 billion in 2018, and expects to reach 10 billion by mid-2050 according to the United Nations and U.S. Census Bureau. Such rapid population growth has prompted concerns about

X. Yang (✉)
Department of Geography, Florida State University, Tallahassee, FL 32306, USA
e-mail: xyang@fsu.edu

S. Jiang
Institute of Groundwater and Earth Sciences, Key Laboratory of Eutrophication and Red Tide Prevention of Guangdong Higher Education Institutes, Jinan University, Guangzhou, Guangdong 510632, China

X. Yang
Institute of Groundwater and Earth Sciences, Jinan University, Guangzhou, Guangdong 510632, China

X. Yang and S. Jiang (eds.), *Challenges Towards Ecological Sustainability in China*, https://doi.org/10.1007/978-3-030-03484-9_1

diminishing natural resources, continued environmental degradation, and increasing social and economic marginalization (Daily and Ehrlich 1992; Clay and Reardon 1998; Bass and Dalal-Clayton 2012; Chiras and Reganold 2014). There has been an urgent need to create and maintain the state of equilibrium in which humans and nature can co-exist harmonically, which permits fulfilling the ecological, economic, political, and cultural requirements for the present and future generations (Kates et al. 2001). Sustainable development or sustainability has therefore become a vital imperative of international, national, and local policies (Sachs 2015).

The literature concerning sustainability or sustainable development has been growing very quickly, with contributors across an increasing number of disciplines and from different regions of the world. While most of the literature is scattered across different publication venues, there have been some dedicated books with sustainability or sustainable development as the general subject. But they are concentrated on the cases and experiences largely drawn from developed countries in Europe and North America (e.g., Blewitt 2015; James 2015; Matson et al. 2016; NASEM 2016). This is understandable because affluent countries usually with higher levels of industrialization may worry more than poorer countries about the continued economic prosperity given the decreasing natural resources and the deteriorating environment.

This volume focuses on China for several major reasons. First, China as the largest developing country has witnessed significant economic growth since 1978 when the country initiated the reform and open to the outside world policy. What the Chinese people desire most has evolved during the past four decades of economic reform, which can be represented with a few "slogans" found throughout social media. For example, people went to cities and joined armies for better living conditions in 1970s; went to colleges and night schools to earn a diploma in 1980s; went to the United States and Western Europe for better personal opportunities in 1990s; went to private and foreign companies for higher salaries in 2000s; went to work with governments for better job security in 2010s; and finally go to oases and any places with a preserved natural environment free from toxins, hormones, and pigments in 2020s. Apparently, the Chinese society has been undergoing a transition from looking for higher monetary returns and better job security into pursuing an environmentally sustainable country, and it would be interesting to examine the issues challenging the move towards sustainability in China.

Second, accompanying with the Chinese society transition, the demand for sustainable development from the governmental sector has become quite high, particularly when President Jinping Xi began his second term in 2017. China's central government has recently been reorganized with a new department called "Ministry of Ecology and Environment" that is responsible for ecological monitoring and environmental protection. Xi and his central government have formulated various initiatives and plans to promote the long-term ecological civilization emphasizing economic and social development being achieved with the man-nature harmony. On 16 June 2018, China's State Council issued a memorandum aiming to strengthen eco-environment protection and battle against pollution (State Council of China 2018). Under this general framework, there is an urgent need to conduct intensive research on sustainability that can help implement the plans and materialize the ambitious goals of sustainable development set by the central government.

Last, despite the undergoing Chinese society transition, the increasing demand for sustainability, and the extensive research on the subject, there are only a few published books targeting the cases and experiences in China. For example, Tilt (2009) discussed an ethnographic study on Futian within China's Sichuan Province, a rural community suffering from pollution. Shi (2010) presented methodological perspectives and practical suggestions for the comprehensive analysis of Chinese ecological agriculture as inputs to improve agricultural policy-making for sustainability practices. Andressen et al. (2013) compiled a volume discussing some broader issues challenging China's continued sustainable development mostly in economic, political and social domains, with a few chapters on the environment. Miller et al. (2014) discussed the relevance of the culture and religion issues in order to understand China's environmental problems and the society's transition towards sustainability. Apparently, most of these books cover either a specific, in-depth study or very broad areas surrounding sustainability, and there is no book focusing on the ecological dimension of sustainability in China.

Within the above context, a book on ecological sustainability in China is timely. Designed for academic and industrial sectors, this book brings together leading scholars in geography, ecology, environmental science, and economics, providing an interdisciplinary perspective on various issues challenging the ecological sustainability in China, especially in the Southern China, where rapid urbanization over the past four decades has caused economic and ecological strains on land and people. While the range of issues is constantly increasing, this book covers some major ones, such as land use and land cover change, urban growth and landscape changes, urbanization impacts upon the thermal environment, surface and groundwater pollution, land use and non-point pollution in coastal watersheds, urban and coastal flood hazards and risk, water resources utilization and management, ecological security, green development for resource-based regions, and sustainable landscape design. Additionally, the book incorporates a strong component of methodology targeting the spatially explicit approach that is supported by geospatial information technologies, especially remote sensing and geographic information systems. Finally, this book includes a policy dimension that is critical for linking knowledge with action to pursue a transition towards ecological sustainability in the world's largest developing country.

Unlike many edited volumes with contributions from a single event, this book is written by a combination of two carefully selected author pools: scholars who present a paper at the Jinan Forum on Geography and Ecological Sustainability held during 7–9 February 2017, Guangzhou, China; and some other active researchers generally identified by their recent conference presentations. A total of 38 authors from China, Japan, and U.S.A. contribute to this book. The author affiliations include Guangdong University of Finance and Economics, Guangdong Provincial Academy of Environmental Science, Guangzhou Institute of Geography, Guangxi Teachers Education University, Huazhong Agricultural University, Jinan University, Nanjing University, Shanxi University, State Key Lab Laboratory of Urban and Regional

Ecology (Chinese Academy of Sciences), State Key Laboratory of Desert and Oasis Ecology (CAS), Sun Yat-sen University, and Xiamen University in China; Yamagata University in Japan; Florida State University, Louisiana State University, and Northeastern Illinois University in the United States. Although this book is mostly authored by Chinese scholars with case studies predominately drawn from China, the knowledge gained from this country with the world's fastest growing economy should be valuable and applicable to other areas globally.

The sections to be followed will provide an overview on the concepts of sustainability and sustainable development, discuss the nature of sustainability science, introduce several methods that may be appropriate for sustainability research, preview the book structure, and highlight several areas deserving further research.

2 Sustainability, Sustainable Development, and Sustainability Science

Sustainability and sustainable development are among the keywords most frequently sought through major search engines because the topic belongs to the most popular areas for research and policy deliberations. Sustainability can be considered as the action in maintaining the continuation of productivity without using up or destroying natural resources (Kahle and Gurel-Atay 2014). It is an old idea because for centuries societies have believed that human demands should not outstrip nature's supply in the long run, as indicated by the long-lasting ideas about fallowing fields for agricultural activities and protecting water sources for multiple purposes (Matson et al. 2016). Sustainability is a term increasingly being used in business, policy-making, and academia.

Sustainable development can be considered as development fulfilling human demands for both the present and future generations. It is a relatively new concept although it also has its root in much earlier ideas about sustainable forest management advocated in Europe during the 17th and 18th centuries (e.g., Evelyn 1662; Von Carlowitz 1732) and the developing environmental movement after publishing Rachel Carson's *Silent Spring* in 1962 (Carson 1962). The term was firstly coined in 1980 by the International Union for the Conservation of Nature and then formally articulated in a report by United Union's World Commission on Environment and Development in 1987 (also called the Brundtland Report after the Commission chairwoman's last name) (Brundtland Commission 1987). The term has frequently appeared in high-level discussions of United Nations, World Bank, and Non-Governmental Organizations (e.g., World Wildlife Foundation-WWF) (Matson et al. 2016).

Shaker (2015) suggested that the term "sustainability" can be considered as the target goal of human-nature equilibrium while "sustainable development" is for the holistic approach and dynamic processes leading to sustainability. While the two terms are often used on various occasions by different groups of people, they share a very important commonality that the continuation of our current and future prosperity depends upon balancing economic development with environmental protection and natural resource conservation. Therefore, these two terms are often used interchangeably (Matson et al. 2016).

While the term "sustainable development" has prompted a global consensus that the Earth's carrying capability is limited due to the availability of natural resources and economic and social progress must be adjusted to minimize the impacts upon the Earth's life support systems, there are some possible conceptual and practical limitations with this term (Baker 2006). The term emphasizes an anthropocentric viewpoint, assuming consumption of natural resources is the most important contribution to human welfare. But placing human being as just one of many species would make more sense. Under this framework, human beings can be viewed as stewards of nature having the responsibility to love and care for nature (New World Encyclopedia contributors 2015). It has been also argued that overall sustainable development does not actually exist in an industrialized world that is heavily dependent upon the utilization of non-renewable resources because any positive extraction rate will ultimately exhaust the finite stock (Turner 1988). Another critic argued that the Brundtland Commission manipulated the concept of sustainable development as a public relations slogan since business will still be a usual strategy for world development (Perez-Carmona 2013).

Sustainability science is a field emerging since late 1990s, seeking to develop an understanding of the interactions between human and natural systems. According to Kates et al. (2001), "sustainability science is focused on examining the interactions between human, environmental, and engineered systems to understand and contribute to solutions for complex challenges that threaten the future of humanity and the integrity of the life support systems of the planet, such as climate change, biodiversity loss, pollution and land and water degradation". Kates et al. (2001) also identified a set of core research questions of sustainability science (p. 642). Unlike many basic or applied fields, sustainability science is defined by the problems it targets through collaboration from interdisciplinary scholars across natural and social sciences. It brings together academics and professionals and global and local perspectives, creating a dynamic bridge between knowledge and action (Kates et al. 2001).

As an emerging field with a history of less than 20 years, sustainability science has witnessed rapid growth, as shown by the mushrooming of NGOs, conferences, publications, degree programs, and grant programs within recent years. An increasing number of NGOs have been formed to promote the sustainable development goals defined by the United Nations (e.g., SDGs 2015), and some most influential NGOs include World Business Council for Sustainable Development (https://www.wbcsd.org/), Science and Technology Alliance for Global Sustainability (http://stalliance.org/), Nature Conservancy (https://www.nature.org/), World Wildlife Fund (https://www.worldwildlife.org/), among others. There have been many dedicated high-profile conferences (e.g., Earth Summit 1992, 2002 and 2012) and professional meetings (e.g., Sustainability Submit). Sustainability theme has been included in a wide variety of related conferences. For example, the annual meetings of the Association of American Geographers (AAG) have included multiple oral paper sessions on sustainability or sustainable development; UN World Data Forum 2017 included several sessions related to sustainability: harnessing the power of data for sustainable development; moving forward—harnessing the power of data for the 2030 Agenda for Sustainable Development; and geospatial and remote sensing data for sustainable development applications. Also an increasing number of dedicated monographs on sustainability have been published within recent years (e.g., Blewitt 2015; James 2015; Sachs 2015; Matson et al. 2016), along with at least three major academic journals on the subject, namely, *Sustainability Science* (https://link.springer.com/journal/11625), *Sustainability* (http://www.mdpi.com/journal/sustainability), and *Nature Sustainability* (https://www.nature.com/natsustain/). A variety of undergraduate and graduate degree programs (BA, BS, MA, MS, and Ph.D.) in Sustainability have been established (https://www.sustainabilitydegrees.com/degrees/). Finally, US National Science Foundation (NSF) has established the Environmental Sustainability program to support engineering research in promoting sustainable engineered systems, and some similar grant programs can be found elsewhere (e.g., China's Natural Science Foundation).

3 Research Methods

Given the nature of sustainability science, there may be no specifically preferred methods for research. Based on literature review, however, Kates (2010) compiled a group of methods that can be used to address various research questions in sustainability science (see Kates et al. 2001, p. 642), which include place-focused studies; observations and monitoring; analytic methods; and models. Since sustainable development is place-based by nature, both long-term and short-term studies can help answer different research questions. Long-term place-focused studies can help move beyond snapshots of place and time captured by most case studies and into an in-depth understanding of the dynamics of complex social-environmental systems (e.g., Turner et al. 2016). It is worth mentioning that the NSF has funded to establish the U.S. Long-Term Ecological Research (LTER) network since 1980, which now includes 28 sites covering various types of ecosystems (Vanderbilt and Gaiser 2017). Additionally, it has been expanded into the International Long-Term Ecological Research (ILTER) network since 1993, which now includes 40 country-level members and over 600 sites. The ILTER network has played a critical role in advancing our understanding of global ecosystems and informing solutions to various environmental problems (Mass et al. 2016). On the other hand, short-term place-focused case studies offer opportunities to put them together for knowledge generalization that can help answer big questions in sustainability science although challenges exist due to incompatible research designs and inconsistent methodologies among individual case studies (Kates 2010).

Several methods can be used for observations and monitoring, which include indicators, geospatial information techniques, decision support systems, and community participation (Kates 2010). An increasing number of quantitative indicators have been developed to assess the progress towards sustainability, although their popularity varies greatly. These indicators can be grouped into three broad categories, i.e., ecological indicators for evaluating the impacts on a specific natural system, social indicators for measuring impacts across the social and economic domain, and environmental indicators for quantifying environmental health and ecosystem vitality (NASEM 2016). Geospatial information techniques are being increasingly used by sustainability scientists, which mainly include remote sensing for environmental data acquisition to support environmental and social indicator development and geographic information systems for integrating biophysical and socio-economic data to support ecosystem assessment (Yang et al. 2013). Decision support systems are a collection of data, tools, and information products that support business or organizational decision-making activities by combining multiple potentially conflicting criteria from different users, practitioners, and scientists (Craig 2018). Finally, participatory approaches allow all end users to be involved in knowledge formation, dissemination, and utilization leading to decision-making and implementation (Kates 2010).

Several methods offer the capability for analysis that can help identify causes and consequences of a sustainability transition and assess the long-term transition, which include portfolios of tools for analysis, driving force analysis, scenario development, and assessment (Kates 2010). Analyzing causes and consequences may end with a wrong conclusion when research designs and specific methods are not appropriately handled. To deal with this potential issue, Young et al. (2006) argued a portfolio of approaches such as statistical analyses, pattern comparisons and meta-analyses of case studies, narratives, and systems analysis may be more helpful. Driving force analysis aims to analyze causes and consequences. For this type of analysis, Plieninger et al. (2016) suggested that several critical issues should be appropriately handled, which include temporal and spatial scales of analyses, separating underlying drivers from proximate drivers, defining and measuring proximate drivers, and more robust tools and methods for causality assessment. Scenario analysis is a forward-looking method for imaging alternative possible futures for human-environment systems under a set of different biophysical and socioeconomic conditions as well as a powerful tool for asking 'what if' questions to explore the consequences of uncertainty (Beach and Clark 2015). Finally, analytic methods can be used to assess hazards or risk, human-environment problems, and transnational problems (Kates 2010).

Different types of models, non-spatial or spatially explicit, can be used to examine human-environment interactions, model long-term trends in the sustainability transition, and identify appropriate solutions for a sustainability problem (Kates 2010). Some commonly used spatially explicit models include statistical models, machine learning models, Markov chain, cellular automata, and agent-based models (ABM). Built upon various theories and technologies, these models are ordered according to the theoretical transition from aggregate to individual modeling frameworks (Liu and Yang 2015). The best model to be used depends upon specific sustainability applications. For modeling the sustainability transition, a dynamic modeling framework can help capture the necessary temporal behavior changes. Statistical models and machine learning are most useful for problem identification and exploration at an early stage. For modeling the complexity of human-environment systems, however, these two approaches are giving way to cellular automata and agent-based models that simulate emergent properties from the bottom up. The agent-based modeling also provides the opportunity to examine multilevel interactions and feedbacks within a human-environment system (Liu and Yang 2015).

4 Overview of the Book

This book includes 14 chapters, falling within three major sections. The first section consists of five chapters discussing some conceptual and technical issues. The second section includes four chapters examining several issues challenging towards ecological sustainability in the Great Pearl River Delta, home of more than 60 million people and the birthplace of China's economic reform since 1978, while the last section includes five chapters discussing various issues in other areas of China.

This book contains at least three types of chapters by writing styles and contents. The first type is mainly an overview or a review on one or more major conceptual or technical issues, which includes five chapters: an overview on some core concepts on sustainability, major methods for sustainability research, and the book content (chapter "Challenges Towards Ecological Sustainability in China: An Introduction"); an overview on the objectives, principles, and frameworks in regional eco-security pattern design (chapter "Regional Eco-security: Concept, Principles and Pattern Design"); an overview on several core principles in urban ecological landscape design (chapter "How to Design an Urban Ecological Landscape: Sustainability, Efficiency, and Harmony?"); a review on the progress of several essential issues being related to the research design and implementation for remote sensing-based monitoring of urban growth and landscape changes in Beijing, China's Capital city (chapter "Monitoring Urban Growth and Land Changes in Beijing, China's Capital City by Remote Sensing: Progress and Challenges"); and a review on the history and applications of using diatoms for environmental assessment in rivers and streams in China (chapter "Diatoms as an Evaluation Tool for the Ecological and Environmental Conditions of Rivers and Streams in China: A Retrospective Study"). The second type of chapters examines various issues challenging towards ecological sustainability in China through a spatially explicit approach using geospatial information technologies, such as vegetation health (chapter "Hyperspectral Remote Sensing of Vegetation Health at the Baiyun Mountain National Forest Park, China"), urbanization impacts upon the eco-environment (Chapters "Comparison of Urbanization and its Eco-environmental Effects in three Large Pearl River Delta Metropolises, China" and "Potential Impacts of Urban Sprawl on the Thermal Environment in the Nanjing Metropolitan Area Based on the SLEUTH and WRF Models"), land use-water quality relationship in coastal watersheds (chapter "Linking Land Use with Water Pollution in Coastal Watersheds of China"), and spatial linkage of an urban agglomeration (chapter "The Spatial Network of Beibu Gulf Urban Agglomeration and Its Challenges for Coordinated Integral Development"). The last type of chapters discusses some issues in China with hybrid methods, such as nutrient and trace metal issues in the Pearl River Delta (chapter "Nutrient and Trace Metal Issues in the Pearl River Delta, China"), flood hazards and risk in the Pearl River Delta (chapter "Flooding Hazards and Risk Analysis in the Pearl River Delta, China"), water and ecological security at the core area of China's Silk Road (chapter "Water and Ecological Security at the Heart of China's Silk Road Economic Belt"), and strategies to promote green development in China's Shanxi Province that is well known as a coal resource-based region (chapter "Strategies to Promote Green Development in Coal Resource-based Regions with China's Shanxi Province as an Example"). The following paragraphs will provide a chapter-by-chapter overview.

Chapter "Challenges Towards Ecological Sustainability in China: An Introduction" is an introduction to the entire book focusing on the rationale and motivation behind this book project. It provides an overview on the historical development of sustainability and the nature of sustainability science, introduces several major methods that can be used to support ecological sustainability research, and previews the book structure. It concludes by highlighting several major areas deserving additional research.

Ecological security can be defined as the degree of human security not being affected by ecological and environmental issues (Xiao and Chen 2002). It has recently received much attention in China. However, few people really understand the nature and scope of the subject, and there is a lack of relevant practices being successfully implemented. Chapter "Regional Eco-security: Concept, Principles and Pattern Design" discusses the objectives, principles, and frameworks in regional eco-security pattern design emphasizing the practice towards ecosystem restoration and landscape optimization. The chapter also includes several regional eco-security pattern design case studies.

Urbanization has become a world-wide phenomenon, developed and developing countries alike. While urban growth has often been seen as a positive sign of a regional economy, poorly planned development has provoked concerns over environmental degradation (Yang 2002). Chapter "How to Design an Urban Ecological Landscape: Sustainability, Efficiency, and Harmony?" discusses several core principles for urban ecological landscape design aiming to minimize negative outcomes caused by urbanization. It introduces the definition of ecological landscape, discusses its characteristics, and identifies a set of guidelines for ecological landscape design based on a questionnaire survey.

Urban growth has been a subject in numerous studies mostly through the use of remote sensing (Yang 2011). However, these studies varied greatly in their research design and implementation. Chapter "Monitoring Urban Growth and Land Changes in Beijing, China's Capital City by Remote Sensing: Progress and Challenges" reviews several essential conceptual and technical issues for remote sensing-based monitoring of urban growth in Beijing, arguably the most frequently researched megacity in the world. Several major challenges have been identified, along with some future research directions. Such a longitudinal study can not only help improve research design but also assist formulating effective strategies to deal with major challenges towards ecological sustainability in global megacities.

As a primary producer at the base of aquatic ecosystems, diatoms, particularly their community dynamics and diversity patterns, are sensitive to ecological and environmental changes (Smol and Stoermer 2010). Chapter "Diatoms as an Evaluation Tool for the Ecological and Environmental Conditions of Rivers and Streams in China: A Retrospective Study" reviews the history and applications of using diatoms for environmental assessment of rivers and streams in China. The use of diatoms for assessing aquatic ecosystem conditions in China has witnessed an increase since 1980, particularly after 2000. Several diatom-based indicators, such as biomass, diversity, species dominance, and autecology, have been used for several decades in China, while the diatom biotic indices and multimetric indices have been tested across the country and widely considered to be more effective. This indicates that diatoms are promising biological indicators in aquatic eco-environmental monitoring programs that can help improve world-wide water resource management.

Native vegetation around the world is under threat due to increasing anthropogenic activities, which may result in ecosystem service degradation (Lawley et al. 2016). Chapter "Hyperspectral Remote Sensing of Vegetation Health at the Baiyun Mountain National Forest Park, China" discusses a remote sensing approach to assess vegetation health in the Baiyun Mountain National Forest Park, which is perhaps the most popular recreational site in Guangzhou, the largest megacity in southern China with a population of more than 13 million. Two forest health related indicators, anthocyanin and carotenoid, were evaluated by using hyperspectral imagery and a spectral model of leaf pigment reflectance. The remote sensing-derived outcome was validated through in situ sample analyses of canopy leaves. The result shows that the concentrations of anthocyanin and carotenoid, vegetation stress indicators, can be quantified using the reflectance index derived from hyperspectral imagery.

The Pearl River Delta (PRD) is considered as the front-runner in China's economic reform since 1978 and the most representative region in terms of rapid urbanization in the country and even in the world (Enright et al. 2005). Chapter "Comparison of Urbanization and its Eco-environmental Effects in Three Large Pearl River Delta Metropolises, China" examines the similarities and differences of three PRD megacities, i.e., Guangzhou, Shenzhen and Hong Kong, according to their natural and socio-economic conditions. An index system was identified and then used to examine the urbanization process and evaluate the eco-environment quality change since 1980. Such a comparative analysis of long-term changes in multiple megacities within a certain region can help improve our understanding of the impacts of urbanization upon the eco-environment system.

Urbanization belongs to the most significant factors affecting the environment in the Pearl River Delta (Ouyang et al. 2006). Chapter "Nutrient and Trace Metal Issues in the Pearl River Delta, China" discusses urbanization and industrialization impacts on the surface water and groundwater systems in the PRD region through several case studies targeting nutrients and trace metals pollution in different environmental settings. It specifically examined the surface-water nitrogen (N) and phosphorous (P) budget and its regional variations, and linked them with appropriate biophysical and socio-economic conditions. Based on water samples, it also examined the nitrogen and phosphorus pollution in the groundwater system of the Tangjiawan Town, Zhuhai, which is believed to be caused by domestic wastewaters. It finally analyzed the trace metal pollution in the Shima River, Dongguan, which was mainly associated with watershed-wide industrial and agricultural activities.

China's Pearl River Delta is the home of more than 60 million people and the world's largest urban area by both size and population (Van Mead 2015). However, its complex river networks and low-lying terrain make the region vulnerable to floods. Chapter "Flooding Hazards and Risk Analysis in the Pearl River Delta, China" discusses flooding hazards and risk in the PRD region and also describes some defense strategies and important infrastructures there. It includes several sections introducing PRD's physiographic settings; heavy precipitation, fluvial floods, and key defense infrastructures in the Beijiang watershed; urban pluvial floods in Guangzhou including major flood types, storm features, and flood hazards; and storm surge and coastal floods including flood types, storm surge hazards, and several extreme typhoon and storm surge events occurred in the 21st century. The chapter concludes by highlighting the need of a more thorough and quantitative analysis that can help better understand location-specific flood hazards and risk in the PRD region.

The urban heat island (UHI) effect has been observed in many world cities, large or small, which threatens the livability and sustainability (Rizwan et al. 2008). Chapter "Potential Impacts of Urban Sprawl on the Thermal Environment in the Nanjing Metropolitan Area Based on the SLEUTH and WRF Models" discusses the potential impacts of urban growth on the thermal environment in China's Nanjing metropolitan area, home of more than 8 million people by using a dynamic spatial model and the Weather Research and Forecasting (WRF) model. The results can help urban planners and managers better understand the potential impacts of urban growth on the thermal environment, which can further help take necessary measures to ensure a livable and sustainable city.

Land changes through urbanization, vegetation removal, and extensive agricultural practices have greatly increased nitrogen cycling that contributes to the degradation of global coastal ecosystems (Yang 2012). Chapter "Linking Land Use with Water Pollution in Coastal Watersheds of China" examined the land use/cover-water pollution relationship with two coastal watersheds in China's Fujian Province as the study site. Satellite remote sensing revealed accelerating watershed-wide land transformations during the past three decades, and water pollution was more serious in urban watersheds than in agricultural and natural ones. Correlation and regression analyses indicate that anthropogenic inputs and agricultural activities contributed to the watershed-wide variation of water quality. Moreover, watershed modeling techniques were used to calculate diffuse nitrogen and phosphorus emissions, determine the contaminant sources, and develop nutrient management strategies. The chapter further discusses some implications of the land-water relationship research for water resource management.

In late 2013, China announced an ambitious initiative called "One Belt, One Road (OBOR)" that combines land powers and maritime powers to bolster China's existing oceanic hegemony in Asia (Tsui et al. 2017). For the land-based Silk Road economic belt, eco-environmental management faces severe challenges due to the limited water resources and fragile ecosystems for many countries.

Chapter "Water and Ecological Security at the Heart of China's Silk Road Economic Belt" discusses recent research progress on water security and ecosystem stability in the central Asia, the core region of the Silk Road economic belt. An overall decrease of the solid-state water resources and increasing water system uncertainty were observed, which was due to global climate changes. Meanwhile, intensifying human activities such as unplanned agricultural expansion exacerbated the water shortage problem. Therefore, this core region has been facing severe eco-environmental problems. The chapter concludes by highlighting the research need on future water resources and possible ecological responses to ensure a healthy eco-environment that is critical in constructing the Silk Road economic belt.

With several decades of strong economic development, a hierarchical urban agglomeration system begins to emerge in China, which represents not only the ultimate spatial form of urbanization but also national and regional growth centers (Fang and Yu 2017). This is particularly relevant to China's ambitious "One Belt, One Road" initiative. Chapter "The Spatial Network of Beibu Gulf Urban Agglomeration and Its Challenges for Coordinated Integral Development" discusses the urban networks and their inter-city interactions of the Beibu Gulf Urban Agglomeration (BGUA), a newly planned urban agglomeration in Southern China that is expected to become a strategic focus to support future regional economic development. The results revealed limited coordination between cities within the agglomeration due to their varying levels of urbanization as well as weak inter-city interactions due to its insufficient competition with other cities. Overall, the BGUA can be viewed as a type of "weak-nuclear" model, and the central cities (i.e., Nanning, Zhanjiang, and Haikou) have not done enough to promote regional economy. The chapter concludes by suggesting strategies towards enhancing the networking effect and optimizing the inter-city coordination.

For a quite long period, resource-based economy has dominated several provinces in northern China. However, many resource-based regions in China nowadays face unprecedented challenges due to resource depletion, environmental pollution, ecological degradation, and negative socio-economic repercussions (Li et al. 2018). Promoting sustainable development of resources-based regions is important to protect the supply of resources and maintain healthy economic development (Li et al. 2016). Chapter "Strategies to Promote Green Development in Coal Resource-based Regions with China's Shanxi Province as an Example" discusses some strategies to promote the green development in coal resource-based regions with China's Shanxi province as an example. It first defined the essential dimensions of the green development pattern and discussed several major obstacles toward the green development in Shanxi. It suggested that coal resource-based regions should promote the coal supply-side structural reform, improve the environmental capacity and resource values, and promote the quality of life for local people.

5 Summary

In this chapter, we have discussed the rationale and motivation behind our current book project. Then, we have provided an overview on the concepts of sustainability and sustainable development and introduced the emerging sustainability science including relevant methods and techniques that can be used to support ecological sustainability research. Moreover, we have previewed the book structure through a chapter-by-chapter overview.

While exciting progress has been made in ecological sustainability research in China, as discussed in this book, there are several main conceptual or technical areas that deserve further attentions. First, given the nature of sustainability science and the actual situation in China, more efforts are needed to encourage the dialogue between natural and social scientists in order to shape a place-focused understanding of the human-environment interactions. Second, scientists being involved in sustainability research in China should be equipped with not only essential knowledge across natural and social sciences but also solid technical skills for observations, monitoring, analysis and modeling of socio-environment systems. Third, more efforts are needed to promote the collaboration between scientists, corporations, governments, and communities in order to materialize the sustainable development goals as defined by China's central government. Fourth, more efforts are needed to promote better utilization of existing tools (especially geospatial information technologies) and processes linking knowledge with action to pursue a transition to sustainability (Kates et al. 2001). Last, while most existing higher education programs in sustainability science have been established in the United States and Europe, China should make more efforts to develop similar programs that will help train the next generation of sustainability scientists for the world's largest developing country.

References

Andressen C, Mubarak AR, Wang X (eds) (2013) Sustainable development in China. Routledge

Baker S (2006) Sustainable development. Routledge, New York

Bass S, Dalal-Clayton B (2012) Sustainable development strategies: a resource book. Routledge

Beach DM, Clark DA (2015) Scenario planning during rapid ecological change: lessons and perspectives from workshops with southwest Yukon wildlife managers. Ecol. Soc. 20(1). http://dx.doi.org/10.5751/ES-07379-200161 (Last accessed on 26 June 2018)

Blewitt J (2015) Understanding sustainable development, 2nd edn. Routledge, London

Brundtland Commission (1987) Report of the world commission on environment and development. United Nations. https://www.un.org/documents/ga/res/42/ares42-187.htm (Last accessed on 22 June 2018)

Carson RL (1962) Silent spring. Houghton Mifflin Harcourt (reproduced on 2002)

Chiras DD, Reganold JP (2014) Natural resource conservation: management for a sustainable future. Pearson/Prentice Hall

Clay DC, Reardon T (1998) Population and sustainability: understanding population, environment, and development linkages. Sustain Agric Rural Dev 117

Craig NAP (2018) A GIS-based multi-criteria decision analysis to select roadside wildflower planting sites for ground-nesting bees in Leon County, Florida. M.S. Thesis, Florida State University

Daily GC, Ehrlich PR (1992) Population, sustainability, and Earth's carrying capacity. Bioscience 42(10):761–771

Enright MJ, Scott EE, Chang KM (2005) Regional powerhouse: the greater Pearl River Delta and the rise of China. Wiley, Chichester

Evelyn J (1662) Sylva, or a discourse of forest-trees and the propagation of timber. In: Paper presented to the Royal Society (on 16 February 1662). http://www.gutenberg.org/files/20778/20778-h/20778-h.htm (Last accessed on 22 June 2018)

Fang C, Yu D (2017) Urban agglomeration: an evolving concept of an emerging phenomenon. Landsc Urban Plan 162:126–136

James P (2015) Urban sustainability in theory and practice: circles of sustainability. Routledge, London

Kahle LR, Gurel-Atay E (eds) (2014) Communicating sustainability for the green economy. M.E. Sharpe, New York

Kates RW (ed) (2010) Readings in sustainability science and technology. In: CID working paper no. 213. Center for International Development, Harvard University. Harvard University, Cambridge, MA, December 2010. http://www.hks.harvard.edu/centers/cid/publications/faculty-working-papers/cid-working-paperno.-213 (Last accessed on 18 June 2018)

Kates RW, Clark WC, Corell R, Hall JM, Jaeger CC, Lowe I, McCarthy JJ, Schellnhuber HJ, Bolin B, Dickson NM, Faucheux S, Gallopin GC, Grübler A, Huntley B, Jäger J, Jodha NS, Kasperson RE, Mabogunje A, Matson P, Mooney H, Moore B III, O'Riordan T, Svedin U (2001) Sustainability science. Science 292(5517):641–642

Liu T, Yang X (2015) Land change modeling: status and challenges. In: Li J, Yang X (eds) Monitoring and modelling global changes: a geomatics perspective. Springer

Lawley V, Lewis M, Clarke K, Ostendorf B (2016) Site-based and remote sensing methods for monitoring indicators of vegetation condition: An Australian review. Ecol Ind 60:1273–1283

Li L, Lei Y, Pan D, Si C (2016) Research on sustainable development of resource-based cities based on the DEA approach: a case study of Jiaozuo, China. Math Probl Eng. http://dx.doi.org/10.1155/2016/5024837

Li L, Lei Y, Wu S, He C, Yan D (2018) Study on the coordinated development of economy, environment and resource in coal-based areas in Shanxi Province in China: based on the multi-objective optimization model. Resour Policy 55:80–86

Maass M, Balvanera P, Bourgeron P, Equihua M, Baudry J, Dick J, Forsius M, Halada L, Krauze K, Nakaoka M, Orenstein DE, Parr TW, Redman CL, Rozzi R, Santos-Reis M, Swemmer AM, Vădineanu A (2016) Changes in biodiversity and trade-offs among ecosystem services, stakeholders, and components of well-being: the contribution of the International Long-Term Ecological Research network (ILTER) to Programme on Ecosystem Change and Society (PECS). Ecol Soc 21(3):31. https://doi.org/10.5751/ES-08587-210331

Matson P, Clark WC, Andersson K (2016) Pursuing sustainability: a guide to the science and practice. Princeton University Press

Miller J, Yu DS, van der Veer P (eds) (2014) Religion and ecological sustainability in China. Routledge

NASEM (National Academies of Sciences, Engineering, and Medicine) (2016) Transitioning toward sustainability: advancing the scientific foundation: proceedings of a workshop. The National Academies Press, Washington, DC

New World Encyclopedia contributors (2015) Sustainable development. New World Encyclopedia, http://www.newworldencyclopedia.org/p/index.php?title=Sustainable_development&oldid=991812 (Last accessed on 22 June 2018)

Ouyang T, Zhu Z, Kuang Y (2006) Assessing impact of urbanization on river water quality in the Pearl River Delta Economic Zone, China. Environ Monit Assess 120(1–3):313–325

Perez-Carmona A (2013) Growth: a discussion of the margins of economic and ecological thought. In Meuleman L (ed) Transgovernance: advancing sustainability governance. Springer

Plieninger T, Draux H, Fagerholm N, Bieling C, Bürgi M, Kizos T, Kuemmerlef T, Primdahla J, Verburg PH (2016) The driving forces of landscape change in Europe: a systematic review of the evidence. Land Use Policy 57:204–214
Rizwan AM, Dennis LY, Chunho LIU (2008) A review on the generation, determination and mitigation of Urban Heat Island. J Environ Sci 20(1):120–128
Sachs JD (2015) The age of sustainable development. Columbia University Press, New York
SDGs UN (2015) United Nations sustainable development goals. https://www.un.org/sustainabledevelopment/sustainable-development-goals/ (Last accessed on 23 June 2018)
Shaker RR (2015) The spatial distribution of development in Europe and its underlying sustainability correlations. Appl Geogr 63:304–314
Shi (2010) Sustainable ecological agriculture in China: bridging the gap between theory and practice. Cambria Press
Smol JP, Stoermer EF (eds) (2010) The diatoms: applications for the environmental and earth sciences. Cambridge University Press
State Council of China (2018) Memorandum on comprehensively strengthening eco-environment protection and battling against pollution. http://www.mep.gov.cn/xxgk/hjyw/201806/t20180625_443663.shtml (Last accessed on 25 June 2018)
Tilt B (2009) The struggle for sustainability in rural China: environmental values and civil society. Columbia University Press
Tsui S, Wong E, Chi LK, Wen T (2017) One belt, one road: China's strategy for a new global financial order. Mon Rev 68(8):36–45
Turner BL II, Geoghegan J, Lawrence D, Radel C, Schmook B, Vance C, Manson S, Keys E, Foster D, Klepeis P, Vester H, Rogan J, Chowdhury RR, Schneider L, Dickson R, Ogenva-Himmelberger Y (2016) Land system science and the social–environmental system: the case of Southern Yucatán Peninsular Region (SYPR) project. Curr Opin Environ Sustain 19:18–29
Turner RK (1988) Sustainability, resource conservation and pollution control: an overview. In: Turner RK (ed) Sustainable environmental management: principles and practice. Belhaven Press, London
Van Mead N (2005) China's pearl river delta overtakes Tokyo as world's largest megacity. https://www.theguardian.com/cities/2015/jan/28/china-pearl-river-delta-overtake-tokyo-world-largest-megacity-urban-area (Last accessed on 13 June 2018)
Vanderbilt K, Gaiser E (2017) The international long term ecological research network: a platform for collaboration. Ecosphere 8(2):1–7
Von Carlowitz HC (1732) Sylvicultura oeconomica. Andesite Press (reproduced in 2015)
Xiao D, Chen W (2002) On the basic concepts and contents of ecological security. J Appl Ecol 13(3):354–358 (in Chinese)
Yang X (2002) Satellite monitoring of urban spatial growth in the Atlanta metropolitan area. Photogramm Eng Remote Sens 68(7):725–734
Yang X (ed) (2011) Urban remote sensing: monitoring, synthesis and modeling in the urban environment. Wiley
Yang X (2012) An assessment of landscape characteristics affecting estuarine nitrogen loading in an urban watershed. J Environ Manage 94(1):50–60
Yang X, Fu B, Chen L (2013) Remote sensing and geospatial analysis for landscape pattern characterization. In: Fu B, Jones BK (eds) Landscape ecology for sustainable environment and culture. Springer
Young OR, Lambin EF, Alcock F, Haberl H, Karlsson SI, McConnell WJ, Myint T, Pahl-Wostl C, Polsky C, Ramakrishnan PS, Schroeder H, Scouvart M, Verburg PH (2006) A portfolio approach to analyzing complex human-environment interactions: institutions and land change. Ecol Soc 11(2):31. https://www.ecologyandsociety.org/vol11/iss2/art31/ (Last accessed on 26 June 2018)

Dr. Xiaojun Yang Editor of this volume, is a tenured Full Professor of Geography in the College of Social Sciences and Public Policy at the Florida State University, USA. He has also had a few visiting academic positions with the Chinese Academy of Sciences and the Jinan University, China. His research interests include the development of remote sensing and geospatial technologies with applications in the urban and environmental domains. Yang has authored or co-authored six English books and over 100 articles in these areas.

Dr. Shijun Jiang Co-Editor of this volume, is a Full Professor of Paleoecology at the Institute of Groundwater and Earth Sciences, Jinan University, China. His research focuses on the use of fossils and geochemical measurements to examine the biotic responses to environmental changes and the dynamic processes of aquatic ecosystems over various time scales. Jiang has authored or co-authored more than 40 peer-reviewed journal articles.

Regional Eco-security: Concept, Principles and Pattern Design

Liding Chen, Ranhao Sun and Lei Yang

Abstract As a popular topic, eco-security has recently received much attention in China. However, some unexpected outcomes related to eco-security often happened in China since few people understand its real meaning and little practices were implemented. In fact, the issues associated with eco-security involve all aspects of human societies. The nature of eco-security is to balance the relationship between nature and human and between ecosystem restoration and environmental background. Eco-security is an ambitious target on socio-economic development and human surviving in the long term. In this chapter, we discuss the objectives, principles, and frameworks in regional eco-security pattern design. Contrary to the conceptual eco-security, regional eco-security pattern design emphasizes the practice on ecosystem restoration and landscape optimization. It is required to set specific objectives in a foreseeable period and to make every effort in resolving key issues faced by humans. In regional eco-security pattern design, biodiversity should be properly maintained or restored, the integrity of ecosystem and ecological processes needs to be kept, the local environmental issues have to be resolved, and the human needs on daily life, production and recreation in a foreseeable period have to be achieved. Finally, we analyze some practical cases on regional eco-security pattern design.

Keywords Eco-security · Regional eco-security pattern · Pattern design
Sustainability

L. Chen (✉) · R. Sun · L. Yang
Research Center for Eco-Environmental Sciences, Chinese Academy
of Sciences, Beijing 100085, China
e-mail: liding@rcees.ac.cn

L. Chen
University of Chinese Academy of Sciences, Beijing 100035, China

X. Yang and S. Jiang (eds.), *Challenges Towards Ecological Sustainability in China*, https://doi.org/10.1007/978-3-030-03484-9_2

1 Introduction

As a popular topic, eco-security has been widely recognized in China by different levels of governments and the general public. What does it mean by the term "eco-security" and how to implement and ensure regional eco-security are still not quite clear even though the term 'eco-security' has been widely used by the society. As population increases and human activities expand, formalizing a framework towards regional eco-security and sustainability is important given that more natural capital diminishes. Eco-security is not only related to the environmental issues at a large scale, such as soil and water loss, desertification and land degradation, but also closely related to the occurrence of regional environmental geological hazards, such as landslides, debris flow, earthquakes, and so on. Eco-security is becoming a rather broad topic in China, and is often linked to the daily life of people. Many issues are ascribed to the scope of eco-security damages, such as urban flooding, urban heat island, fog and haze, etc. At large scales, the soil and water loss as a critical issue relevant to eco-security can affect the national eco-security, and the desertification can often bring harms to the Eco-security. Another important issue concerned by the publics is the urban flooding. The frequent occurrence of urban flooding in China has drawn much attention from the publics, and it has therefore become a hot topic. About 62% of the cities in China have been affected by rapid flooding according to the statistical data of the Ministry of Housing and Urban-Rural Development of China, and its occurrence frequency is becoming higher. Such an unanticipated incident happened on 21 July 2012 in Beijing had resulted in serious losses both in human life and properties. Another incident occurred in June of 2016 in Wuhan, another mega-city in central China, which caused widespread damages. Frequent urban flooding brings troubles on the daily life of local people, and also leads to serious threats to the urban eco-security and human living environment.

Although eco-security has become a hot topic for the publics, little firm and effective measures are proposed to realize ideal eco-security at the regional or the state level. The current research on eco-security mainly has been focused on theoretical and comprehensive assessment, and few studies on practical eco-security pattern design (Huang et al. 2017; Liang et al. 2010; Kang et al. 2007; Li et al. 2010a, b; Meng et al. 2011; Ren and Liu 2013). Currently, several abnormal phenomena can be observed in China. First, eco-security is becoming a public hot word but little serious actions were taken in China. The leaders from the central to the local government, as well as the publics from different communities talk about eco-security often, but few of them understand the real meaning of the term and have no ideas on how to make eco-security to become true. Second, the relationship of eco-security planning between the state level and the department level is often confused. Many projects and thematic planning on eco-security have been conducted, for example, water eco-security planning, food eco-security planning, state land eco-security planning, environmental eco-security planning, among others. But whether each specific eco-security planning is fully fallen in the scope of the master eco-security planning at the state level has been little investigated. Last, which scale would be appropriate for eco-security planning or eco-security pattern design is not quite clear. Studies on eco-security conducted in China were made at different scales, such as the state-level, the province-level, the county-level or the town-level. For each level, different issues and topics have been explored. However, it is not clear which level would be the best one for eco-security studies and planning, and what issues are to be distinguished and what topics are to be focused in different levels.

2 Concept of Eco-security and Its Characteristics

2.1 Concept of Eco-security

Generally, security is closely related to human surviving and sustainable development in the safe condition that is free from any threat, risk and disturbance from the environment that may affect human daily life (Xiao et al. 2002; Gao et al. 2006). Effective eco-security is a state without threats on human capacity to adapt to the environmental change in future, and natural capital required for meeting the needs of human daily life, for production, for health-care and for recreation is satisfactory. Eco-security includes natural eco-security, economic eco-security and social eco-security conducted by different departments. An eco-security situation implies that human activities and surviving will not be influenced as eco-environmental changes happen, or the environmental capacity to meet human needs in daily life, production and recreation is not attenuated.

The intrinsic nature of eco-security includes two aspects: the ecological balance and harmonious relationship is to be reached between different on-site environmental elements; and the trade-off and balance between human needs on ecosystem services and provision of ecosystem services from natural capitals have to be considered. For the first aspect, there are several considerations. First, the ecological balance and suitability of human activities to the environmental settings should be considered. Any human-introduced biological measures or engineering measures have to be carefully examined before they are employed for ecosystem restoration and eco-security pattern building. Otherwise, the human intervention may lead to deleterious effects to the regional eco-security. Second, harmonious relationship is required on ecological processes from a large spatial scale, i.e., the ecosystem restoration in a region or catchment should be reasonable, and the ecological processes at the spatial scale should not be disturbed. The impacts of human interference on the integrity of ecological processes in spatial scales have to been addressed when the regional eco-security pattern is made, and at the same time, an optimized arrangement of different ecosystems on spatial scales is to be attained. Last, the adaptability and sustainability of the eco-security pattern to environmental change is to be addressed, i.e., the potential effects of the ecological restorations should be at the minimum in the future and the eco-security pattern should be adapted to the environmental change in future.

For the second aspect, only when the human needs on ecosystem services are fully satisfied, or the value of ecosystem services offered from nature is much higher than the human demand, the relationship between human and nature will be in balance or harmonious, and it is the expected eco-security pattern for humans.

2.2 *Characteristics of Eco-security*

Eco-security is virtually characterized with several traits (Chen et al. 2006; Xiao et al. 2002). First, it is a comprehensive concept that realizing the ideal situation of eco-security has to consider all the factors associated with the ecological, the social and the economic aspects. It is impossible or difficult to reach an ideal eco-security if the ecological, the social and the economic factors are addressed separately. Second, the eco-security pattern is to be developed according to the features of different regions since different issues may be faced in different regions. Thus, different requirements on building eco-security pattern are be considered for resolving different issues, and the solutions to realizing regional sustainability can vary in regions. Additionally, since eco-security is closely related to regional ecological processes, it is required to build the eco-security pattern from regional/basin scale aiming to keep the integrity of ecosystem and the connectivity of ecological process unimpaired. Third, the eco-security pattern developed is often temporary since human needs on ecosystem services may change with their understanding on eco-security and improvement of daily life as economic development increase. The eco-security pattern developed by humans through solving some specific issues may be useful at the specific periods and it needs to be modified when new challenges appear. Last, the eco-security pattern design should be adaptable to the environmental change in the near future. A safe socio-economic-ecological complex system may become unsafe when a new human disturbance is introduced, or abrupt environmental hazards happen. In contrary, a degraded ecosystem or unsafe ecosystem may become healthy after wise human interference, restoration or land closuring. The above-mentioned characteristics of eco-security pattern indicate the complexity of eco-security and its difficulty to realize in long term.

2.3 *Comparison on the Concepts Associated with Eco-security*

Eco-security: the ecological balance and harmonious relationship between human and nature is the key component of eco-security in theory. However, the issues faced by human have to be resolved in the first place before we could consider an ideal eco-security pattern, otherwise the so-called eco-security is of low quality, or can't reach the level as we expect. Thus, the human surviving and socio-economic development have to be emphasized before we can find a best way on ecosystem conservation and human utilization of the natural capital. In fact, the term eco-security didn't appear until the crisis occurred on the socio-economic development and natural conservation last century that was resulted from the over exploitation and damage of the natural capitals. In a word, eco-security is an abstract concept, an ideal and bold target on socio-economic development, and it is too difficult to become truth because of its comprehensive and complicated features.

Eco-security pattern design: Compared to the abstract and ideal eco-security, eco-security pattern design focuses on concrete measures and actions for realizing a safe environment for human surviving. Building eco-security patterns need to address some specific issues humans are facing, and a rather practical goal should be emphasized. A complex and abstract eco-security pattern is hard to imagine and therefore difficult to implement. However, the integrity of ecosystem and ecological processes need to be considered when establishing an eco-security pattern, and appropriate restoration measures and landscape optimization at the regional/basin scale should be attempted. All the human-introduced measures to be used in building an eco-security pattern should be compatible with the processes in the whole region or basin which may vary by the issues to tackle. One of the most important issues in building an eco-security pattern is the choice of restoration measures and allocation of landscape types at the regional scale.

Eco-security warrantee: Eco-security warrantee has become a common sense in eco-security pattern design. It is also a comprehensive and complex concept. But it is more concerned about the policies related to management and decision-making on eco-security. The measures on financial supports, laws formalization, policy making for implementing eco-security at the national or regional level is particularly emphasized when we talk about the eco-security warrantee. Additionally, a long-term strategic planning on ecosystem restoration and management should be made for regional eco-security and sustainability, and practical restoration projects or landscape restoration measures are required to make the planning into action. At the same time, the early-warning on eco-security and contingency plans to solve the unexpected incidents are to be made (Li et al. 2010a, b; Ma et al. 2017).

State master eco-security planning, department eco-security planning and regional eco-security planning: In fact, the state master eco-security planning is the fundamental rule for all the departments or administrations to make their own practical eco-security planning. The state master eco-security planning is to be put into force by concrete tasks or projects that are to be implemented by different departments or administrations. It is unwise to address the eco-security issues from the point of individual department or region. The state master eco-security planning is to be complied with the sub-eco-security planning of department, and to realize state master eco-security planning demands each department to solve its own problems properly. The state master eco-security planning can be made based on spatial scale of administration or basin, and the specific eco-security planning has to be incorporated into the state master eco-security planning as the general background. However, the state master eco-security planning requires a detailed plan on the practical restoration projects and optimization of landscape pattern in regional scale.

Regional eco-security and sustainability: Both regional sustainable development and eco-security have the similar ultimate goal, i.e., the sustainable development of human society. The eco-security model is to be made according to specific regions or specific development stages, cautions should be taken when it is extended to other regions. Regional sustainable development is an abstract and ideal social development target in the long-term. It is also largely dependent on human needs since the understanding on sustainable development and requirements from different groups or different regions is changeable. It is almost impossible for humans to reach an absolute regional eco-security pattern or regional sustainable development, at least by present knowledge. Therefore, to make an explicit and practical target at certain stages is important for human. To resolve the existing problems properly we faced in each stage is much helpful for regional eco-security pattern design and sustainable development.

3 Objectives, Principles and Framework of Eco-security Pattern Design

As discussed above, the regional eco-security pattern design is a concrete engineering or practice that is to be implemented. It is different from the concept of pure eco-security.

3.1 *Objectives of Eco-security Pattern Design*

The ultimate goal of eco-security pattern design is to meet human needs on daily life, production and recreations at the minimum cost of natural capitals, and thus the following objectives have to be reached when a regional eco-security pattern is developed (Ma et al. 2004; Li et al. 2004; Chen et al. 2006).

First, the regional biodiversity should be maintained or restored by eco-security pattern design. Biodiversity is proven to be the foundation of human surviving, and a rich bio-diversity may provide better ecosystem services to humans, including foods, water, living environment, recreation places, etc., other than a poor bio-diversity. This means that human surviving and sustainable development can be kept only in case of high quality ecosystem and rich biodiversity. Based on the estimation of Constanz et al. (1997), the ecosystem services offered by the global ecosystems was about $\$3.3 \times 10^{13}$, much higher than the total global GDP $\$1.8 \times 10^{13}$. Therefore, the primary target of regional eco-security pattern design is to protect or restore biodiversity that is crucial for human sustainable development.

Second, the integrity of ecosystem structure and ecological processes is to be assured by regional eco-security design. The emergence of eco-security is closely related to human surviving and social and economic development, its realization, however, is completely dependent on the quality and the function of natural ecosystems. Therefore, the basic requirement is to protect the integrity of ecosystem structure as well as the processes when the regional eco-security pattern is to be established, otherwise the foundation of supporting human surviving on goods provision will be destroyed. The spatial units that the ecological processes occurred have to be addressed, and how to maintain the ecosystem functions is to be concerned.

Thirds, the local environmental issues have to be resolved by eco-security pattern design. The ultimate goal of eco-security design is to alleviate the effects of environmental hazards on human activities, and thus the important work in eco-security pattern design is to control the environmental issues humans faced. However, there are usually many issues to be dealt with in a highly-disturbed region, and it is a tough and arduous task to solve all the issues through eco-security pattern design in a short term. The best way is to resolve the most pronounced ones that may lead to huge damage to human society if it were not eliminated properly. In this case, some pertinent measures to control the key environmental issues by eco-security pattern design is particularly significant. Even though the environmental issues can't be controlled thoroughly, it should be mitigated or improved through eco-security pattern design and restoration. In another word, the requirements of human society in a rather long term should be satisfied through eco-security pattern design, even though not totally.

Last, the human needs on daily life, production and recreation in a foreseeable period should be achieved through regional eco-security design. Eco-security is a very complex problem that is closely related to many affecting factors associated with social, economic and natural dimensions, but also dependent on human understanding on eco-security and human needs on ecosystem services. It seems that human needs in the far future may not be satisfied through regional eco-security pattern design since it is hard to version by the people living in this era. It is practical to satisfy the human needs on ecosystem services in a certain period by controlling the environmental hazards and enhancing ecosystem services.

3.2 Principles of Eco-security Pattern Design

When eco-security pattern is to be developed, several principles have to be abided by Ma et al. (2004) and Chen et al. (2006). First, ecosystem integrity should be kept when building an eco-security pattern. The regional ecosystem integrity has to be followed when designing an eco-security pattern since it is a regional issue. The regional eco-security will be affected when any change happened in one of the ecosystems or patches in a region. Thus, the eco-security pattern design has to be based on the regional or basin scale, and the integrity of ecosystem structure, ecosystem processes and ecosystem functions has to be followed thoroughly. Furthermore, the measures on ecosystem restoration and landscape optimization should be examined carefully from the regional or basin scale.

Second, the multi-scale features of eco-security pattern have to be addressed. Eco-security involves multiple scales, including on-site vegetation restoration and ecosystem construction that should not give rise to adverse effects on the soil, water resource. Landscape allocation and ecosystem optimization at the catchment scale should be helpful for improving the integrity of ecological processes and ecosystem services. Thus, it is difficult to find an ideal and universal model on eco-security pattern design in nature.

Third, harmonious relationships between different factors and ecosystems have to be established. Since eco-security is mainly focused on the harmonious relationship between human and nature, the suitability and adaptability of any measures for eco-security design has to be evaluated, which include the harmony between human activities and natural conservation, the harmony between measures on ecosystem restoration and environmental factors such as soil, water and landscape. Additionally, the harmonious relationship of the factors on site and landscape allocation in spatial scale should be addressed, including the adaptability of the measures to be used for Eco-security pattern design.

Last, differences in regional ecosystems have to be considered when building an eco-security pattern. Even though eco-security is comprehensive in term of concept, some specific issues and the principal problem have to be addressed in priority when an eco-security pattern is to be established. All the measures used for the eco-security pattern design should be totally based on the features of the regional ecosystems, and different eco-security patterns may be required according to the differences in regional ecosystems. Of course, the requirements of human activities and socio-economic development have to be embraced in the eco-security pattern design.

3.3 Framework for Eco-security Pattern Design

The following steps have to be followed in eco-security pattern design: ecosystem assessment and issue analysis, ecosystem service assessment, ecological risk and key issues identification, landscape and eco-security pattern design (Fig. 1).

Ecosystem assessment and issue analysis: Ecosystem assessment is mainly devoted to analyze the features of the ecosystem composition, functions and environmental background such as landform, soil, vegetation, etc. Based on ecosystem assessment, the existing issues are to be identified and the potential solutions should be addressed. Generally, thematic mapping such as landform mapping, soil mapping, vegetation mapping, land-use mapping, and some derived thematic mapping (e.g., rainfall, surface run-off, evapotranspiration, aridity, and geological hazards) is to be conducted. Then, the ecosystem susceptibility and environmental capacity have to be defined based on the ecosystem features. At the same time, the social and economic data on population, economic development, and urban development are to be collected. Finally, the contradictions between environmental conservation and human needs on socio-economical development are to be examined, and the ultimate development goal and long-term strategy have to be formulated.

Ecosystem service assessment: Ecosystem service assessment, as a widely concerned topic, has been paid much attention by many countries and employed as

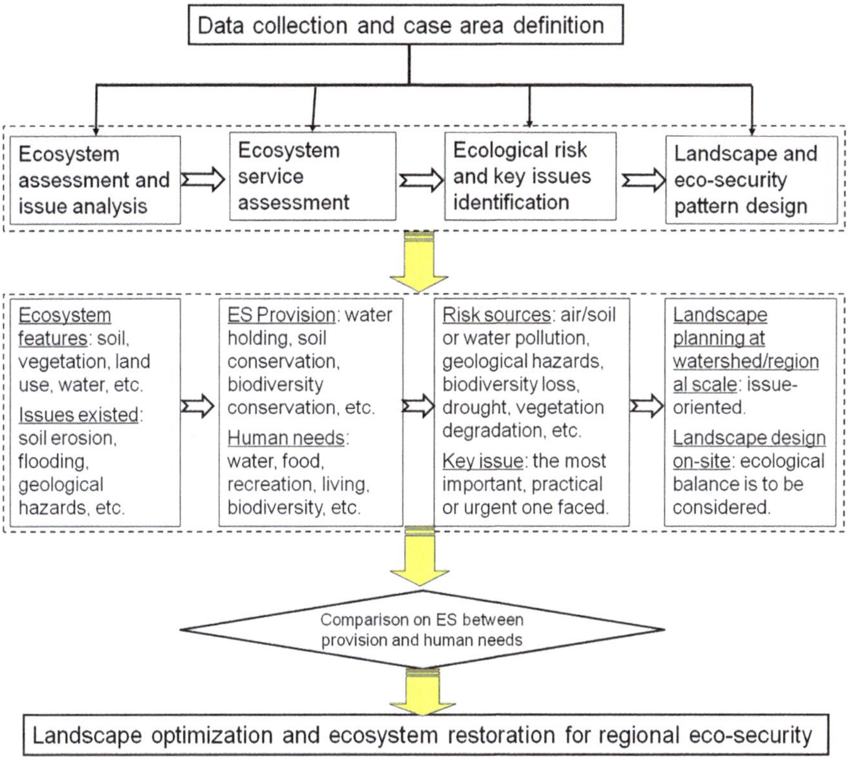

Fig. 1 The flowchart on eco-security pattern design

a baseline in decision-making to balance the socio-economical development and ecosystem conservation. Since regional pattern design is virtually to address the trade-off between the provision of ecosystem services from natural ecosystems and the human needs for socio-economic development. Thus, the benefit value of natural ecosystems offered to human has to be estimated thoroughly before conducting the regional eco-security pattern design. At the same time, the human needs on ecosystem services have to be analyzed, and the shortage on ecosystem service provision has to be discriminated.

Ecological risk and key issues identification: After ecosystem analysis and ecosystem service assessment, the potential risk that may affect the regional eco-security pattern has to be analyzed, including the frequency and the degree of air/soil or water pollution, the occurrence of geological hazards, the loss of biodiversity, the risk of drought and flooding, the degree of vegetation degradation, etc. Which one is the most important for the regional eco-security and which one is the key issue to be resolved in the near future will have to be determined. In most cases, it is difficult to resolve all the issues which may be closely related to the eco-security pattern since a high uncertainty remained on the occurrences of such issues. Also, since the ecosystems and human needs on ecosystem services are both changeable that result in the measures used for building eco-security pattern is to be adjusted according to the new situation.

Landscape planning and eco-security pattern design: At this stage, the landscape planning at the regional scale and vegetation restoration on-site has to be made. The landscape planning should be made based on the results of ecosystem assessment and risk analysis, and the ecosystem integrity for realizing maximum ecosystem services has to be considered. The measures of vegetation restoration on-site include both biological measures and engineering measures that should be planned according to the environmental factors, i.e., rainfall availability, terrain variation, soil nutrient and soil water, climate, etc. The final landscape planning and eco-security pattern will be reached based on resolving the key and practical issues faced to human.

4 Practices of Eco-security Pattern Design in China

As discussed above, there are so many factors to be addressed in eco-security pattern design, e.g., the ecosystem integrity, the ecosystem differences in different scales, the harmonious relationship of basic elements on-site and the key regional issue to be faced. However, the real eco-security pattern has to consider the features of the regional ecosystems and the human needs on regional ecosystem services. In practice, there are many studies having been conducted in China aiming to establish an eco-security pattern for human surviving although many of them may not be thoroughly perfect. In this section, we will discuss some case studies relating to eco-security pattern design in China.

4.1 The Terrace Landscape of Hani Minority in China

Hani terrace landscape, as one cultural landscape in the list of Globally Important Agricultural Heritage Systems (GIAHS), is mainly composed of three sub-systems along hill-slope, i.e., forest landscape, village and terrace farming system (Fig. 2). It is a typical eco-security pattern developed on hill slopes characterized with an integral system of landscape pattern and ecosystem services. The general features of this eco-security pattern are as follows: a parcel of forest is located and protected carefully on the upper hill-slope, a village landscape as human living and daily activities is located at the middle part of the hill-slope, and the farm landscapes with terrace engineering are below the village. The landscape pattern with forest, village and terrace farming along the contour line on the hill-slope constitutes an eco-security pattern that is welcomed by the local people from Hani minority (Jiao et al. 2012).

In this eco-security pattern, there is few waste water and solids emitted to the surroundings because the forest landscape located at the upper hill-slope can offer clean water to the local people for daily use and production while stopping soil and

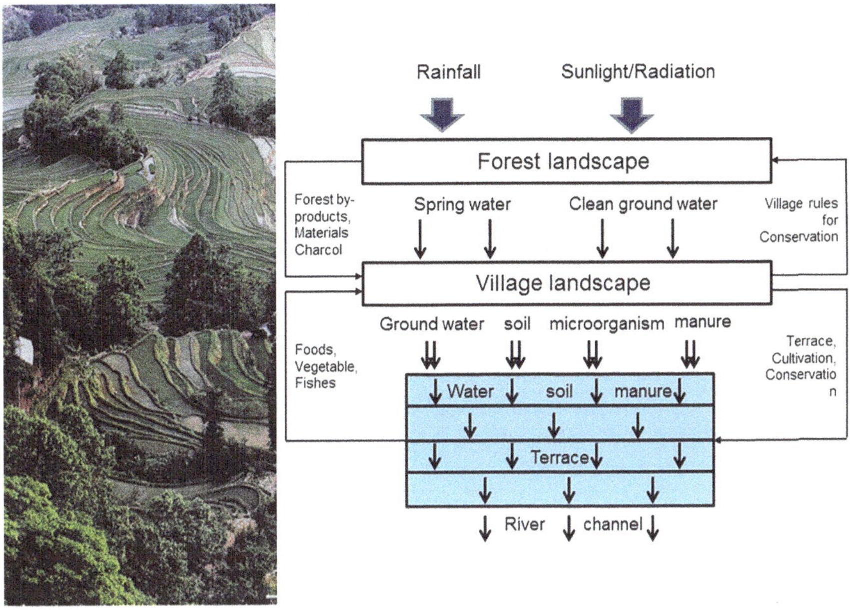

Fig. 2 An environment-friendly landscape pattern of the Hani terrace system (based on Feng et al. 2008; Zhang et al. 2016)

water losses, as well as some specific forest by-products be captured. As the major living place, the village in the middle hill-slope accommodates the local people for daily activities and the drinking water is from the upper forest landscape and their waste water and rubbish are drained to the lower terrace farmland where the agricultural goods are produced. The waste water and rubbish generated from daily life of local people can be used as nutrients to support crop/vegetable growth. A catena with benign circulation of nutrients along the hill-slope, and safe and harmonious landscape pattern are developed. The human needs on clean water can be captured from the upper forest landscape and the human-induced waste can be disposed by returning them to the terrace farming landscape in the lower hill-slope. In such a landscape pattern, little waste solids and water are emitted to the surrounding environment while the human needs on water and agricultural goods are satisfied.

4.2 *The Comprehensive Treatment on Soil and Water Conservation at the Watershed Scale in the Loess Hilly Areas, China*

Soil and water losses in the loess hilly area are the most important issue to affect the regional sustainable development in the loess plateau. Many efforts and funds have been made by the Chinese central government to support ecological construction aiming to reduce soil and water losses and to enhance ecosystem services. After long-term research and practices on vegetation restoration and ecosystem management, a model on the comprehensive control of soil and water losses at the watershed scale was established (Zhu 1998). The general features of this model are that specific measures according to the terrain units and landscape factors are employed from the watershed scale (Fig. 3). This landscape pattern becomes a typical eco-security pattern that was dedicated to control the soil and water loss, and further to improve ecosystem services. The general pattern of this model is to restore woodlands (including shrubs) in the upper steep slope areas, to develop agroforestry or agricultural productions in the gentle middle slope areas by introducing soil conservancy measures, and to adopt engineering measures such as stone/willow check dams in the lower gully areas. This model is characterized by a complete ecological continuum on ecosystem function with soil and water conservation in the upper area, goods provisions and accommodations in the middle parts and sediment stopping in the lower parts in the watershed.

Details about this model are: (1) Given the steep slope and serious soil and water erosion in the upper slopes, it may cause serious soil erosion if intense human activities were introduced. Planting trees, shrubs or land closure is the best choice that may keep the steep hill-slope from soil and water erosion by increasing vegetation

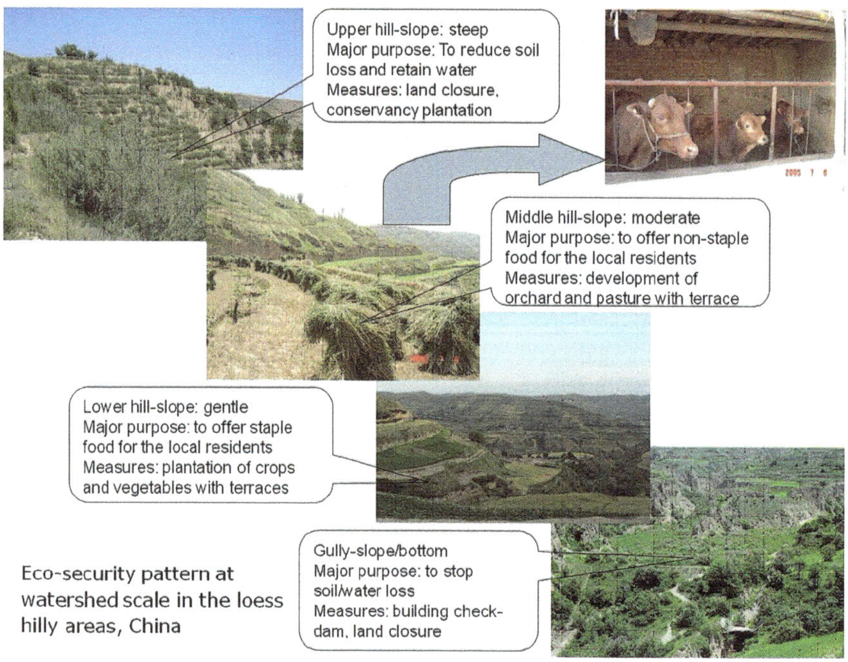

Fig. 3 Eco-security pattern developed at watershed scale in the loess hilly area, China (based on Zhu 1998; Yu et al. 2011)

coverage. As a result, the adverse effects of upper slope due to soil erosion will be decreased and the human activities in the lower parts of the watershed are protected. (2) Human activities are mainly distributed in the middle parts of the watershed, the middle-lower hill-slope with moderate to gentle slope where a high-productivity farming ecosystem or agro-forestry ecosystem is developed by remolding the natural slope into terrace or flat strip. By this system, the basic needs of local people in daily life and survivor will be produced, and the water used in daily life is from the rainfall collected by cellars. (3) The valley bottom will be the last defense against soil and water erosion in watershed scale. However, only the engineering measures such as stone/willow check dam and check dams can be used in these areas because of the steep gully slope where is used for ecosystem function improvement by increasing vegetation coverage. In this model, the soil and water erosion will be reduced to the minimum while the needs of local people are properly satisfied from the watershed scale, and a beneficial cycle will be developed. A relative safe ecosystem pattern is developed, a typical Eco-security aiming to control soil and water erosion.

4.3 The Multi-pond System in Southern China

The multi-pond system is a very popular landscape in southeastern China since it may offer benefits to the local people. Although a plentiful of rainfall and developed drainage system is available in southern China, the seasonal shortage of water resources happened often that may lead to some troubles to the local people in both daily life and agricultural production. Man-made ponds or wetlands are built beside the farmlands or villages to capture surface run-off for meeting human needs in daily life, and thus a multi-pond system constituted of natural, man-made ponds and river channels is developed. As a beautiful landscape existed in southern China, the multi-pond system can offer many ecosystem services to the local people. In general, the multi-pond system can collect the surface run-off in the rainy seasons that can be used for breeding fish and irrigating farmland. At the same time, the multi-pond system can retain the non-point source pollutants and protect the targeted river or waterbody from pollution. In dry seasons, the water collected in the ponds can be used for agricultural irrigation and for improving the efficiency of water utilization, and the sludge in the ponds can be taken out as manure to support crop growth by returning them into farmlands, and finally to purify water. The ecosystem services of the multi-pond system have been studied by many Chinese scholars (Mao et al. 2004, 2006; Wang et al. 2005; Yin et al. 1993). The general values of multi-pond system on ecosystem services include: (1) water provision to the farmland by irrigation in dry seasons; (2) flooding alleviation and regulation by reducing peak run-off in rainy seasons; (3) water purification and protection of the targeted water body from pollutions by removing the non-point source pollutants from the surface run-off; (4) provision of aquatic-goods such as fish to meet human needs in daily life; (5) enhancing nutrient cycle by returning the pond-sludge of ponds to farmlands;

(6) improving land productivity; and (7) provision of recreation places for local people such as go boating, go-fishing, sightseeing, and so on.

Apart from the ecosystem services of the multi-pond system mentioned above, a safe landscape pattern, i.e., eco-security pattern, is also developed at the regional scale. This eco-security pattern is devoted to modifying the utilization pattern of water resources at the catchment scale by reducing peak run-off through the pond systems in the rainy seasons and offering water resources to farmland for irrigation in the dry seasons. Additionally, an eco-security pattern on water quality is generated by improving the nutrient cycle with the sludge returning to farmlands to support crop growth, and a healthy ecosystem at a large scale is developed. However, the pressure and threats due to increasing human activities have resulted in high risks to the existing eco-security pattern. How to define the size, the configuration, the distribution of ponds in a catchment and how to manage the multi-pond system for sustainable use will have to be addressed. Furthermore, how to keep the sustainability of the eco-security pattern on both flooding mitigation and non-point source pollution reduction needs further research.

4.4 Landscape Patterns with Network Optimization of Nature Reserves

Establishing nature reserves is important for biodiversity conservation (Cuonga et al. 2017; Liu et al. 2017; Schumaker et al. 2014). However, how to establish a reasonable and effective nature reserve to improve the conservation efficiency on targeted species is particularly significant. Usually, the following works have to be conducted when a new nature reserve is to be established for a safe pattern for the endangered animals (Fig. 4) (Chen et al. 2000).

(1) Evaluation of habitat suitability: to make an evaluation on the habitat suitability to the targeted species is fundamental for establishing a nature reserve. The weight of each affecting factor to the suitability has to be assigned according to the importance of each factor to the targeted species before the final habitat suitability is defined.
(2) Identification of core patches in the nature reserve. Usually, a large patch may accommodate more animals than a small one does, and will be the core habitat patches other than the small ones which are usually less helpful for wildlife conservation, and sometimes become sinks of the target species (Chen et al. 2000; Wu et al. 1993). The identification and protection of the core habitat patches in a nature reserve is rather important. To become the core patches, the quality of habitat, i.e., the habitat suitability should be better enough besides its size is large enough to bear enough animals, the minimum population number.
(3) Buffer design: the traditional way of buffer design is based on the rule that buffers are installed around the core patches with certain width. A consequence of this approach is that the core patches can't be totally embraced by the buffers.

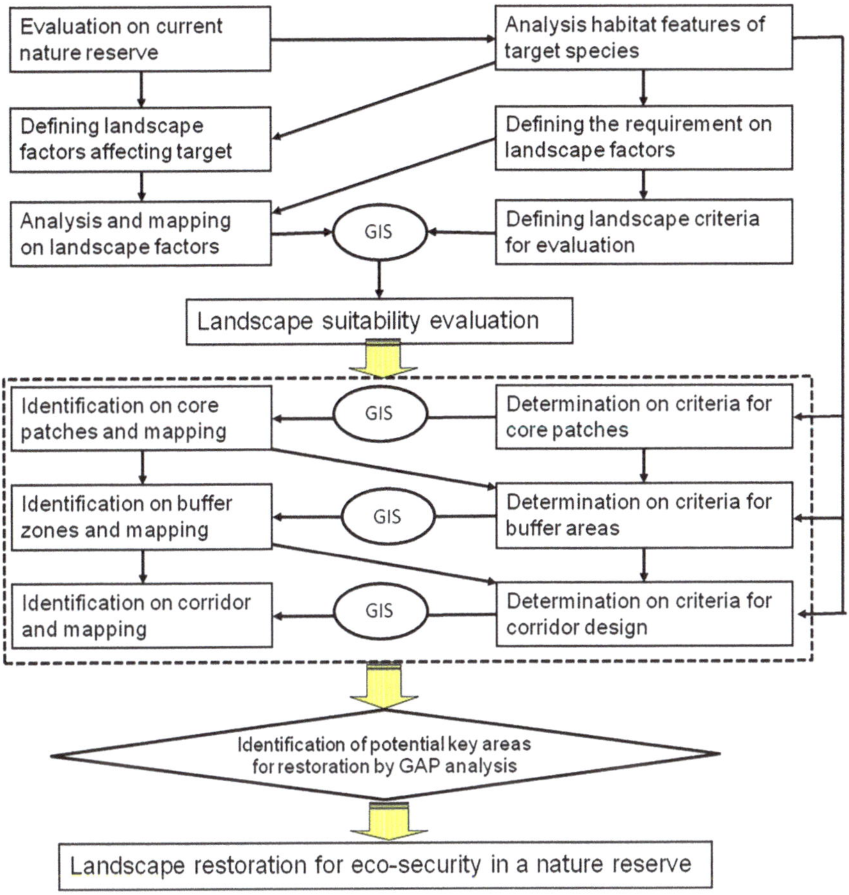

Fig. 4 Flowchart for eco-security pattern in a nature reserve

The best way is that all the core patches in a nature reserve should be put together to design the buffers for wildlife conservation. In this case, the animals living in different core patches can move freely, and little human disturbance is exerted to the animals under the protection of buffers. Therefore, the following two criteria have to be considered in the buffer design: the outside boundary of the buffers to each core patch must be bigger than a threshold such as the dimeter of circle with same size to the home range of the target animals, and all the core patches should be incorporated in a continuous rather than discrete buffer.

(4) Protection of existing habitat corridors and re-construction of potential habitat corridors. It is much helpful for gene-exchange and wildlife conservation if suitable and enough habitat corridors are built to connect the separate core patches. The core patches have already be covered by the buffers that are beneficial for animal movement in the nature reserve, however these measures only

constrict the types and intensity of human activities in the nature reserve, but not forbidden totally. Some long and narrow conduits existed in the core patches are of high significance for wildlife conservation because the core patches may become more fragmented once they were destroyed. Thereby, these existing habitat conduits require to be protected strictly. Additionally, some potential habitat corridors are to be rebuilt for a safe and better living places to the target species. The potential habitat corridors are some long and narrow conduits which are currently unsuitable for animal to move freely but they can become the habitat corridors after specific landscape restoration. In identification of potential habitat corridors, the following factors are to be examined. One is that those landscape factors hard to change should have a high suitability, and some other factors which are flexible such as the vegetation have a low quality. Another one is that the width of the conduit should be no less than a threshold which is equal to the dimeter of a circle having the same size to the home range of the target animals.

(5) The identification of key areas for ecosystem restoration and habitat reconstruction. In theory, the size of the core patches may be enlarged after vegetation restoration in some special places of a nature reserve, but it may not if the vegetation restoration occurred in some other places. The former ones playing a pivotal role in enlarging the core patches are proven to be the key areas for ecosystem restoration in case of financial shortage. However, the latter ones don't contribute much to enlarge the core patches, it can be restored if enough fund is available than being restored in the first place.

The nature reserve plays a significant role in biodiversity conservation. However, sometimes it may cause adverse effects on biodiversity conservation due to inappropriate distribution of nature reserves, particularly in the strong human-affected regions. Establishing a reasonable natural reserve network at a large scale is often required. Many Chinese landscape scholars have conducted studies addressing such issues (Guan et al. 2015; Jiang et al. 2007; Xu et al. 2010). For example, Xu et al. (2010) proposed an optimal network of natural reserves which is devoted to protecting Giant Panda better in the Qinling mountainous areas through habitat suitability evaluation and accessibility analysis. Before the optimization on the natural reserve network, there are about 17 natural reserves established there and 20 separate populations living there, which have resulted in some adverse effects on Giant Panda conservation. How to optimize the networks of the total nature reserves and how to improve the conservancy efficiency on Giant panda are essential for Giant panda conservation. Hence, developing an eco-security pattern for Giant panda conservation in a large regional scale is required by establishing an optimized network on nature reserves (Xu et al. 2010). Based on the network planning, the following three measures are suggested: (1) to enlarge the core areas by adjusting the function areas to the existing nature reserves; (2) to build two new nature reserves and (3) to increase landscape connectivity by establishing three habitat corridors between the natural reserves (Fig. 5). Finally, a safe landscape pattern for Giant panda living in the concerned regions is developed. Whatever the optimization on nature reserve

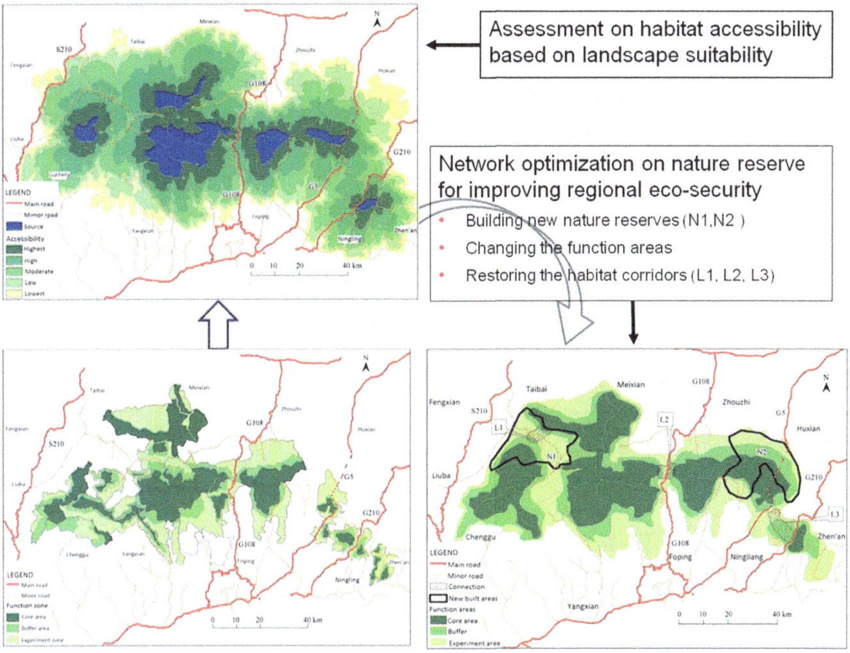

Fig. 5 Spatial optimization on nature reserve networks for Giant Panda conservation in Qinling mountainous areas (based on Xu et al. 2010)

networks or detailed landscape pattern design in an individual nature reserve, the ultimate goal is to establish an eco-security pattern for the target species conservation.

5 Concluding Remarks

Eco-security has already become a popular topic for the publics in China. It has been given a lot of attentions both in the construction of large projects and in the daily life. However, the unanticipated outcomes still happened often in China, such as badly damage to railways, highway, and pipeline due to abrupt geological hazards, bringing troubles on safe operations of traffic systems and energy supply. The occurrence of unwanted urban flooding, the urban heat island and haze may bring troubles to urban traffic that may disturb the daily life of urban residents. Eco-security dealt with the environmental issues in large scale rather than on-site or small scale, but its effects may involve in all sides of human societies including current and potential, direct and indirect. The nature of eco-security is to balance relationship between nature and human, between the ecosystem restoration and surrounding environment, between current human need and provision of ecosystem services. It is an abstract concept and affected by many factors associated with social, economic and natural aspects. In other words, eco-security is a bold target on socio-economic development and human surviving in long-term that is too hard to realize. However, eco-security pattern design, contrary to the conceptual eco-security, is an artificial practice on ecosystem restoration and landscape optimization in a period. To set a specific goal in solving the key problems faced by humans is especially important in eco-security pattern design, and the integrity of regional ecosystem and ecological processes has to be addressed carefully.

Acknowledgements This research was partly supported by the National Key Research and Development Program of China (Project No. 2016YFC0503000).

References

Chen LD, Fu BJ, Liu XH (2000) Landscape pattern design and wildlife conservation in nature reserve. J Nat Res 15(2):164–169
Chen LD, Guo SH, Jiang CL (2006) Ecosystem assessment and ecological security in the nearby regions along West-To-East Pipeline. Science Press, Beijing
Costanz R, d'Arge R, de Groot R et al (1997) The value of the world's ecosystem services and natural capital. Nature 387:253–260
Cuong CV, Dart P, Dudley N et al (2017) Factors influencing successful implementation of biosphere reserves in Vietnam: challenges, opportunities and lessons learnt. Environ Sci Policy 67:16–26
Feng JC, Shi S, He SJ (2008) Hani terrace ecosystem in Yunnan Province. J Minzu University China (Natural Sciences Edition) 17(Suppl.):146–152
Gao CB, Chen XG, Wei CH et al (2006) Regional ecological security: the concept and assessment theoretical basis. Ecol Environ 15(1):169–174
Guan TP, Wang F, Li S et al (2015) Nature reserve requirements for landscape-dependent ungulates: the case of endangered takin (*Budorcas taxicolor*) in Southwestern China. Biol Conserv 182:63–71
Huang H, Chen B, Ma ZY et al (2017) Assessing the ecological security of the estuary in view of the ecological services. Ocean Coast Mang 137:12–23
Jiang Y, Swallow AK, Paton PWC (2007) Designing a spatially-explicit nature reserve network based on ecological functions. Biol Conserv 140:236–249
Jiao YM, Li XZ, Liang LH et al (2012) Indigenous ecological knowledge and natural resource management in the cultural landscape of China's Hani Terraces. Ecol Res 27:247–263
Kang XW, Liu XH, Zhang S et al (2007) Regional eco-security assessment of southwest Beijing. Chin J Appl Ecol 18(12):2846–2852
Li XY, Ma KM, Fu BJ et al (2004) The regional pattern for ecological security (RPES): designing principles and method. Acta Ecol Sin 24(5):1055–1062
Li YF, Sun X, Zhu XD et al (2010a) An early warning method of landscape ecological security in rapid urbanizing coastal areas and its application in Xiamen, China. Ecol Model 221:2251–2260
Li ZC, Liu DL, Sun YF et al (2010b) Assessment method of regional ecological security based on pressure-state-response model. Acta Ecol Sin 30(23):6495–6503
Liang P, Du LM, Xue GJ (2010) Ecological security assessment of Beijing based on PSR model. Proc Environ Sci 2:832–841
Liu HQ, Li WD, Lv GH (2017) The design of nature reserves in the face of habitat loss. Ecol Model 358:50–58
Ma KM, Fu BJ, Li XY et al (2004) The regional pattern for ecological security (RPES): the concept and theoretical basis. Acta Ecol Sin 24(4):761–768

Ma SW, Xie DT, Zhang XC et al (2017) Measures of land ecological security early warning and its spatial-temporal evolution in the ecologically sensitive area of the Three Gorge reservoir area. Acta Ecol Sin 37(24):8227–8240
Mao ZP, Peng WQ, Yin CQ et al (2004) Spatial variability of non-point source pollutants within a multi-pond system. J Agro-Environ Sci 23(3):530–535
Mao ZP, Yin CQ, Shan BQ et al (2006) Spatial variability of agricultural pollutions in complicated landscape system with multiple ponds. J Hydraul Eng 37(6):727–733
Meng JJ, Zhao CH, Liu MD (2011) Regional ecological security assessment based on land-use change. J Nat Resour 26(4):578–590
Ren ZY, Liu YX (2013) Exploring the regional ecological security evaluation methods based on values. Geogr Res 32(10):1771–1781
Schumaker NH, Brookes A, Dunk JR et al (2014) Mapping sources, sinks, and connectivity using a simulation model of northern spotted owls. Landsc Ecol 29:579–592
Wang XH, Yin CQ, Shan BQ (2005) Control of runoff and retention of diffuse P-pollutants by s ink landscape structures of agricultural watershed. Acta Sci Circum 25(3):293–299
Wu J, Vankat JL, Barlas Y (1993) Effects of patch connectivity and arrangement on animal metapopulation dynamics. Ecol Model 65:221–254
Xiao DN, Chen WB, Guo FL (2002) On the basic concepts and contents of ecological security. Chin J Appl Ecol 13(3):354–358
Xu WH, Luo C, Ouyang ZY et al (2010) Designing regional nature reserves group: the case study of Qinling mountain range. China. Acta Ecol Sinica 30(6):1648–1654
Yin C, Zhao M, Jin W et al (1993) A multipond system as a protective zone for the management of lakes in China. Hydrobiologia 251:321–329
Yu HB, Chen LD, Cai GJ (2011) Comprehensive control on soil erosion and water loss in the loess hilly areas. Science Press, Beijing, China
Zhang YX, Min QW, Jiao WJ et al (2016) Values and conservation of Honghe Hani rice terraces system as a GIAHS site. J Resour Ecol 7(3):197–204
Zhu XM (1998) Theory and practices of "28 words of grand plan on land management in the loess plateau". Bull Chin Acad Sci 3:232–236

Dr. Liding Chen is a Full Research Professor at the Research Centre for Eco-Environmental Sciences under the Chinese Academy of Sciences and Director of the State Key Laboratory of Urban and Regional Ecology. His research interests include landscape patterns and ecological processes, land use/cover changes and their environmental consequences, landscape planning and spatial modeling. He has authored or co-authored 9 books and more than 300 articles in these areas.

Dr. Ranhao Sun is currently an Associate Research Professor at the Research Center for Eco-Environmental Sciences under the Chinese Academy of Sciences. His main areas of expertise include landscape ecology, physical geography, and geographic information systems. Sun has authored or co-authored five books and over 90 articles in these areas.

Dr. Lei Yang is an Associate Research Professor at the Research Center for Eco-Environmental Sciences of the Chinese Academy of Sciences. His research interests focus on simulating landscape dynamics and ecological processes, linking ecosystem processes and ecosystem services in the critical zone, and exploring and understanding vegetation dynamics and ecohydrological processes in arid and semi-arid areas. Yang has published more than 50 peer-reviewed papers.

How to Design an Urban Ecological Landscape: Sustainability, Efficiency, and Harmony?

Ranhao Sun and Liding Chen

Abstract Urban areas are rapidly expanding across the world, causing many negative environmental effects. A challenge for reaching a sustainable urban socio-ecological system is to understand how urban landscapes could be designed and managed to minimize adverse impacts. Urban ecological landscapes are products of ecological planning and design considering landscape patterns and their related ecological processes across various spatial scales. In this chapter, we firstly introduce the definition of ecological landscape that can be considered as an anthropogenic or managed landscape. Second, we discuss the characteristics of ecological landscapes highlighting their three major aspects, i.e., sustainability, efficiency, and harmony. Last, we develop some guidelines for ecological landscape design that is based on a questionnaire survey we conducted in order to understand the public opinions regarding ecological landscapes. The questionnaires covered various issues such as overall understandings, knowledge of connotation and features, difficulties and challenges, and evaluation guidelines and criterions of urban ecological landscapes. The findings from this work can help develop a conceptual framework for urban landscape studies and also serve as guidance for urban landscape design.

Keywords Landscape design · Landscape sustainability · Ecosystem service
Urbanization

1 Dilemma of Urban Ecological Landscapes

Cities become human's major settlements in recent decades. Human civilization can be viewed as a process of people moving into cities and urban development. The urban population expects to account for 69% of the global population by 2050, and will be more than five times the population at the beginning of the 20th century

R. Sun (✉) · L. Chen
State Key Laboratory of Urban and Regional Ecology, Research Center for Eco-Environmental Sciences, Chinese Academy of Sciences, Beijing 100085, People's Republic of China
e-mail: rhsun@rcees.ac.cn

X. Yang and S. Jiang (eds.), *Challenges Towards Ecological Sustainability in China*, https://doi.org/10.1007/978-3-030-03484-9_3

(United Nations 2010). Urbanization is a historical process in which traditional rural societies dominated by agriculture gradually switch into a modern urban society emphasizing industries and services. The switches involve occupations, industrial structure, landscape, and social organization. Reasonable urbanization will improve humans' standard of living, enhance social development, and reduce the pressure of human activities on the environment through various strategies such as infrastructure construction and the building of a green environment. Unreasonable urbanization, however, will result in traffic congestion, resource shortages, and a decline in the quality of living (Makhzoumi 2000; Termorshuizen et al. 2007; Lovell and Johnston 2009). With rapid social and economic development, urbanization particularly increases its pace in developing countries (Rosenzweig et al. 2010), and the ecological construction of urban landscapes has become a concern for the society. Local governments increase investment in human, material, and financial resources to improve the urban landscape (Zhao et al. 2011). Therefore, establishing guiding principles and assessment indices for the urban ecological landscape is important and imperative (Mayer et al. 2010).

In general, the urban ecological landscape includes all landscapes that are able to provide natural ecological functions or ecosystem services, such as urban forests, grasslands, green public arcades, and water bodies. A city is a society-economy-nature compound ecological system. The construction of an ecological landscape should consider the reasonable use and allocation of human's mental and physical health, resources, and energies, as well as social and economic benefits. The construction of an ecological landscape should also follow certain principles considering fundamental ecological functions and being compatible with relevant assessment standards. Currently, a number of international and national ecological landscapes are being planned or constructed. However, due to the lack of consistent guiding principles and assessment standards, many constructed urban ecological landscapes are not "ecological" for the following reasons. First, the ecological integrity of the urban landscape has been destroyed to varying degrees. Some urban landscapes are scattered in cities and designed only for beautification. The patch-matrix-corridor relationships have not been taken into consideration in the landscape planning. Some urban landscapes transform ecological systems without reference to scientific knowledge, such as the over-planting of aquatic plants and the over-cultivation of aquatic animals. These unreasonable transformations have resulted in the breeding of mosquitoes and flies and the deterioration of water quality (Lund et al. 2013). Second, the ecological benefits of urban landscapes are relatively low. The current design of urban landscapes and public facilities focuses only on the appearances and does not fully consider ecological functions and benefits. These constructions are generally large, expensive, and pretentious, but lack a sense of harmony between human and nature. Some urban landscapes with high energy, water, and land consumption are under construction. The development direction with a focus on perception instead of ecological values would destroy urban landscapes (Georgescu et al. 2014). Last, the current urban landscapes do not indicate highly artistic or cultural values. Each country has its own unique cultural traditions. Specifically, the accumulation of historical cultures constructs the flesh and blood of a city (Kuitert 2013). However, most urban landscapes, such as holiday flower beds, landscape avenues, residential communities, city squares, and historical sites, are non-ecological designs that prompt unsustainable "landscape manufacturing". It seems that urban landscape design is only equipped with concepts but without appropriate approaches. Fake ecological and humanistic landscape phenomena sometimes exist, and many landscapes are duplicated or imitated without much originality.

2 What Is an Ecological Landscape?

The definition of an ecological landscape varies across different disciplines such as forestry, landscape ecology, architecture, or urban planning. In the discipline of forestry, an ecological landscape is an umbrella term used to represent all vegetation landscapes or green landscapes with a focus on vegetation. In Landscape Ecology, an ecological landscape refers to landscapes with biological and humanistic characteristics with certain ecological functions or services, which are different from generally visual landscapes (Forman 1995; Fu et al. 2011). In Urban or Regional Planning and Environmental Science, an ecological landscape is the one being ecologically designed and sustainable, and with a harmony between humans and nature (Gao and Tian 2007; Chen et al. 2004; Wang and Li 2006). In Architecture, the ecological landscape concept is often combined with traditional Chinese "FengShui" theory and emphasizes the harmony between humans and nature and environmental protection. According to Zhao et al. (2011), "Feng Shui" is a venerable idea in China, using climate to find desired locations according to the type and configuration of the surrounding landscape. In general, ecological landscape design should integrate the input from landscape ecology aiming to reinforce the natural and cultural spirit of a living place (Makhzoumi and Pungetti 1999; Makhzoumi 2000; Lovell and Johnston 2009).

The international research on ecological landscape emphasizes natural design. For example, the ecological design originated from Howard's garden city idea in the 19th century to Mcharg's nature-oriented design in the 20th century. The core idea is the compatibility between human activities, regional resources and environmental characteristics based on ecological suitability analyses. In the late 20th century, Forman (1995) proposed the ecological design and ecological system management thought emphasizing the spatial pattern of landscapes and the control and impact of the process to achieve the goal of sustainable planning. Chinese scholars, such as Shijun Ma, Liangyong Wu, Rusong Wang, and Kongjian Yu, proposed ecological thoughts of landscape design emphasizing the harmony between humans and nature. Different from traditional nature-oriented ecological planning, the Chinese ecologists considers human activities as an integral part of the whole landscape instead of an independent part outside the landscape.

Therefore, we interpret urban ecological landscapes as spatial landscape elements or combinations that are planned and designed based on landscape ecology, landscape architecture, and urban planning. The ecological landscapes don't have to be the natural landscapes. Sometimes, they may be the product of ecological landscape planning, which are man-made landscapes and managerial landscapes with healthy ecology and sustainable standards. To achieve the goal of sustainable development, urban ecological landscapes emphasize the whole values of ecological functions and ecosystem services. The ecological design emphasizes the ecological processes whereby landscapes are able to maintain and protect the ecosystem and reduce pollution and resource use. The ecological landscapes are constructed to ensure the compatibility and coordination among existing landscapes, and to maximize the reduction of human activities in the local, regional, and global environments. As recognized by many scholars, it is insignificant to cheaply assign an object as "ecological". Ecological landscapes are not a zero-sum judgment. The opposite is not "non-ecological landscapes". The essential nature of ecological landscapes is the degree of ecological health and integration.

There have some specific studies in relation to the theories and practices of ecological landscapes. For sustainable landscape design, research has focused more on biodiversity and environmental protection (Makhzoumi and Pungetti 1999; Termorshuizen et al. 2007). Green communities have received much attention. For example, some researchers compiled 22 indices in five aspects, i.e., ecological protection and restoration strategies, green materials and products, water resource use and wastewater treatment, rainwater management, and the heat island effect (Cassidy 2003; Calkins 2005). Chinese researchers have developed quantitative assessment indices of ecological community construction (Zhou et al. 2005; Tian et al. 2007). Recently, Zhou et al. proposed 25 indices to assess ecological communities, concerning the environment, architecture, ecological environment infrastructure, residents' environmental protection awareness, and service management (Zhou et al. 2011). Cheng et al. proposed other indices for ecological landscape assessment, such as atmospheric environmental quality, physical environmental quality, architectural environmental quality, and ecological environmental quality, and 13 factors for the an assessment system of the ecological environmental quality of residential communities (Cheng et al. 2011). In addition to the community scale, similar research has also been conducted on the district and city scales in recent years. For example, Wang and Ye (2004) developed assessment indices of ecological cities based on three aspects: ecological industry metabolism, ecological landscape integrity, and ecological culture. The Sino-Singapore Tianjin Eco-city is a typical example of ecological urban construction. A total of 22 controllable indices and four guiding indices were used in the construction to realize a harmonious coexistence among humans, economic activities, and the environment. At the nation-wide scale, relevant governmental departments have attempted to promote a scientific and quantitative assessment of urban ecological landscapes (HIPCM 2001). China's Ministry of Construction released the Evaluation Standard for Green Building focusing on water, land, energy, and material conservation of the living environment. The Environmental Protection Department released Regional Green Community Assessment Indices and Assessment Criteria consisting of 20 specific indices with a focus on environmental quality, environmental construction, organization, management, and public participation.

The current research on urban ecological landscapes has some limitations. In particular, the lack of unified and quantitative assessment principles and standards restricts the promotion and application of urban ecological landscapes. Termorshuizen et al. (2007) analyzed 38 urban landscape planning schemes in the Netherlands and revealed that some schemes involved environmental protection, ecological system structure and functions. However, most of these indices are qualitative rather than quantitative (Termorshuizen et al. 2007). Szenasy's survey on urban planners revealed that 93% of the respondents had interests in ecological landscapes, but 70% of them were not able to implement ecological landscape design due to their lack of substantial understanding of ecological landscapes (Szenasy 2002). The International Interior Design Association (IIDA) conducted a survey of 100 interior designers, and found that 83% of respondents believed that it was an obligation to provide customers with sustainable solutions but only 37% of projects integrated quantitative ecological landscape schemes. The most difficulty respondents felt was the lack of scientific guiding principles, sufficient time, and no demonstration cases that ecological landscapes can bring economic benefits. Cassidy's survey revealed that the cost and market recognition are two major barriers for the promotion of ecological landscapes. The first Green Building conference sponsored by the US Green Building Council summarized the difficulties faced by ecological landscape practices, including a higher raw cost (78%), uncertain input cost and long-term economic benefits (47%), inadaptability of new technologies (39%), and lack of market recognition (24%). Smida's survey on ecological landscape design materials revealed that 68% of respondents partially used green materials. Difficulties of using green materials experienced by respondents included low demand by customers (37%), low accessibility of materials (21%), high cost (14%), and uncertain determination of green materials (23%) (Smida 2003).

3 Characteristics of Ecological Landscapes

According to our own case studies and literature survey, some characteristics of urban ecological landscapes can be summarized, which include dynamics and systematicness (Makhzoumi 2000), health (Cao et al. 2002), sustainability (Termorshuizen et al. 2007), greenness (Calkins 2005), multi-functionality (McCuskey et al. 1994; Givoni 1997; Lovell and Johnston 2009), culture (Nassauer 2004), harmony, livability, environmental protection, and energy conservation (Yu et al. 2001).

Urban ecological landscapes have sustainable characteristics, including ecological, social, and economic sustainability. First, urban ecological landscapes should follow systematic and dynamic principles of ecosystems, maintain the structure and process of ecosystems, and provide biodiversity and environmental sustainability protection (Makhzoumi and Pungetti 1999). Second, urban ecological landscapes should consider the current/past and local/regional landscape patterns and processes, restrictions of the social reality, and the natural and humanistic background

(Makhzoumi 2000). Therefore, ecological landscape construction needs to integrate interdisciplinary methods, such as landscape ecology, landscape architecture, urban planning, architecture, and system science. In addition to a thorough comprehension of landscapes based on an analysis and description of landscape ecology, an intuitive and creative problem-solving capability using interdisciplinary methods is also required to provide cost-effective solutions for the sustainable development of future landscapes. Dynamics, health, and a systematic approach are characteristics that must be addressed for the sustainability of ecological landscapes.

Urban ecological landscapes should be efficient to maintain ecological and economic benefits. First, they should meet the 4R principles (Weddle 1967; McHarg 1971), namely, reduce, reuse, recycle, and renew. Second, ecological and economic benefits are based on the condition of landscape health. Landscape health is an important factor for the assessment of benefits, including impacts on and response to environmental background, landscape structure and functional changes, as well as an integration of social-economic-natural compound systems. The ecological and economic benefits have the following characteristics: (1) providing ecological goods and services to meet present and future needs; (2) no or minimum external subsidies of landscapes for the maintenance of ecosystem functions; and (3) no negative or minimum effect on adjacent landscapes and ecosystems. Therefore, the assessment of ecological landscapes should be based on the landscape maintenance and renewal capability instead of the extent of naturalness. In such a way, landscapes will be truly efficient to maintain ecological and economic benefits. Ecological and economic benefits are one core value of ecological landscapes and reflect the greenness, health, environmental protection, energy conservation, and multi-functionality characteristics of ecological landscapes.

Urban ecological landscapes should exhibit harmonies among the shape, structure, and function, as well as between history and the present, the local area and the region, and nature and humanity. First, ecological landscapes should adhere to the harmony between historical and current landscapes, respect cultural characteristics present in landscapes, the evolution process, and historical tradition, and enhance natural and humanistic values of landscapes (Makhzoumi and Pungetti 1999). Second, ecological landscapes have both subjective and objective natures. Humanistic elements play an important role. Therefore, in addition to meeting people's needs, ecological landscapes also pursue a harmony and co-existence relationship among all lives (Zhao et al. 2011). Last, ecological landscapes should emphasize the relationship between spots and the regional background and have an understanding of natural, historical, social, and cultural factors associated with the background. These help with the design of landscape units being more compatible with the environment, culture, and aesthetic appreciation.

4 How to Design an Ecological Landscape?

Taking into consideration the importance of public participation in the construction of urban ecological landscapes, the preliminary selection of assessment indices under study should involve public participation. We conducted a public survey using household questionnaires from 1 July to 1 September 2011 (Fig. 1). Survey participants included students, teaching and research staff, and enterprise staff. Most respondents lived in large, medium, or small cities, and there was a generally even distribution of the education level. A total of 170 valid questionnaires were collected. All questions were designed to be short, concise and easy to understand so that more respondents would enjoy participating in the questionnaire. The questionnaires focused on four aspects: ecological landscape connation, guiding principles, difficulties, and assessment indices. There was a total of 19 questions including two single-answer and 17 multiple-answer questions. Specifically, a total of four questions pertained to the information of the respondents; four questions were related to public opinions on the understanding and demands for ecological landscapes; three questions were related to the content and guiding principles of ecological landscapes; and a total of eight questions were related to assessment indices of ecological landscapes.

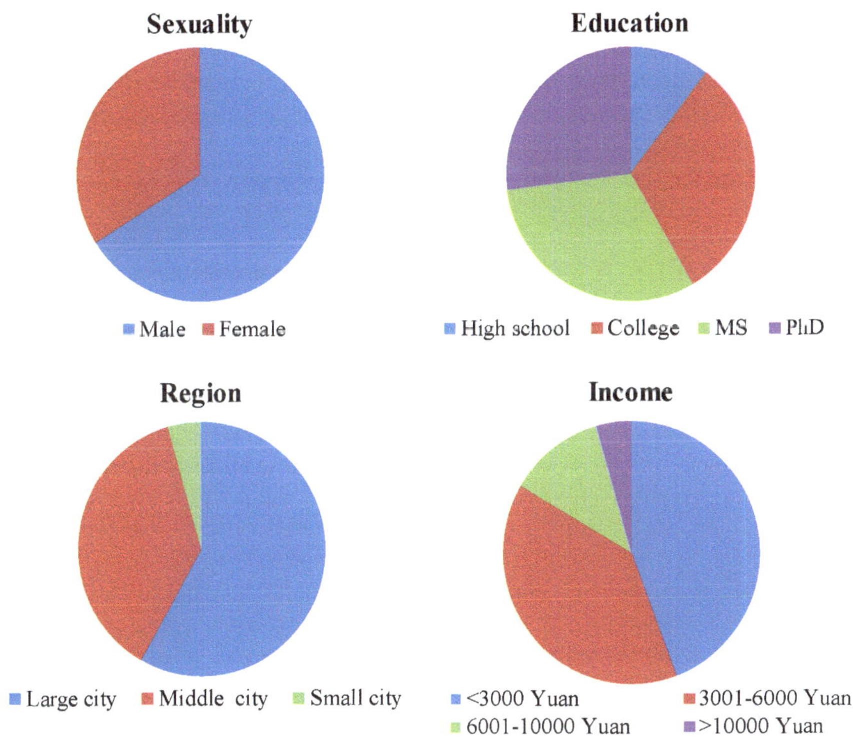

Fig. 1 Participant characteristics

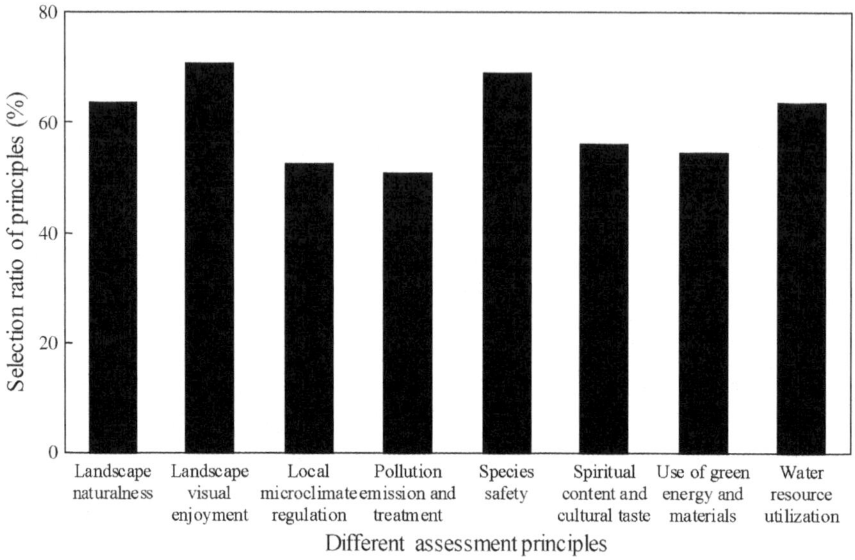

Fig. 2 Selection ratios of different principles based on public questionnaires

Our survey results revealed that the public had a high recognition of the eight assessment principles. "Agree" answers for all principles exceeded 50%. The specific principles with the highest to lowest "agreement" were landscape visual enjoyment, species safety, landscape naturalness, water resource use efficiency, spiritual content, cultural taste, use of green energy and materials, local microclimate regulation, and pollution emission and treatment (Fig. 2). Table 1 summarizes the overall recognition of respondents in terms of the indices, as well as the recognition ratios of different groups of respondents. In particular, respondents' residential locations had a linear relationship with 13 indices that accounted for 30.23% of all indices. Respondents' education level had a linear relationship with 10 indices that accounted for 23.26% of all indices. Respondents' incomes had a linear relationship with only three indices that accounted for 6.98% of all indices. The above results indicated that respondents' location (cities) and intellectual levels had a more significant effect on the choices of the assessment indices, whereas respondents' incomes had a small effect.

As shown in Table 1, four indices, namely, "seasonal coordination of plants", "rainwater recycling facilities and utilization", "species richness and collocation of trees", and "species safety", had the highest weights. These indices are common assessment criteria in domestic industrial technique guidelines. Indices such as "green roof utilization", "pavement color and texture", and "green construction materials" are also relatively important in the assessment of green communities. However, they had lower weights in our study. The results indicated that further promotion and application of these indices are necessary. For example, modern permeable construction materials are an important technique for reducing urban floods and saving water resources. Some cities have implemented them at a small scale.

Table 1 Impacts of personal characteristics of respondents on the recognition of assessment indices of urban ecological landscapes

Assessment principle (selection ratio, %)	Assessment index	Selection ratio (%)	Recognition based on personal characteristics		
			Education level (from the lowest to highest)	Residential location (from large to small cities)	Income (from the highest to lowest)
Landscape naturalness (63.64)	Ratio of natural landscapes	66.92	0	0	0
	Ecological disturbance extent	62.41	0	0	0
	Landscape green space ratio	48.12	0	0	0
	Average green space ratio per capita	41.35	0	0	+1
	Proportion of local species	47.37	+1	−1	0
Species safety (69.09)	Proportion of alien species	63.91	+1	0	0
	Pest and disease hazards of species	66.92	+1	0	0
	Flowering and fruiting species ratio (attracting butterflies and birds to increase species diversity)	38.35	−1	0	0
Water resource utilization (63.64)	Proportion of low water-consuming plants	42.11	0	0	0
	Utilization ratio of a poorly and highly efficient irrigation system	58.65	0	0	0

(continued)

Table 1 (continued)

Assessment principle (selection ratio, %)	Assessment index	Selection ratio (%)	Recognition based on personal characteristics		
			Education level (from the lowest to highest)	Residential location (from large to small cities)	Income (from the highest to lowest)
	Rainwater recycling facilities and utilization ratio	79.70	0	−1	0
	Water percolation ratio of pavement materials	60.15	0	+1	0
	Hard pavement ratio	33.08	0	−1	0
	Natural conditions such as terrain and topography	41.35	0	+1	0
Pollution emission and treatment (50.91)	Radioactive content of natural stone materials	39.85	0	0	0
	Volatile substance, heavy metal, and formaldehyde content of paints and binders	58.65	0	0	0
	Heavy metal content of ceramic products	39.10	0	0	0
	Formaldehyde content of wooden products	44.36	0	0	0

(continued)

Table 1 (continued)

Assessment principle (selection ratio, %)	Assessment index	Selection ratio (%)	Recognition based on personal characteristics		
			Education level (from the lowest to highest)	Residential location (from large to small cities)	Income (from the highest to lowest)
	Proportion of on-site sewage treatment facilities	72.93	+1	0	0
	Utilization ratio of toxic materials (pesticides, paints, etc.)	75.19	0	0	0
	Control measures of non-source pollution sites	60.90	0	+1	0
Use of green energy and materials (54.55)	Utilization ratio of renewable sources	77.44	0	0	0
	3R material ratio	62.41	+1	0	0
	Recovery ratio of materials	56.39	0	0	−1
	Energy consumption and intensive production extent of materials	63.16	0	−1	0
	Local material ratio	36.84	0	0	0
Local microclimate regulation (52.73)	Landscape type, height, density, shape, connectivity, and location (pattern, etc.)	78.95	0	0	0
	Shade area	49.62	0	0	0
	Pavement material and color	39.10	0	0	0

(continued)

Table 1 (continued)

Assessment principle (selection ratio, %)	Assessment index	Selection ratio (%)	Recognition based on personal characteristics		
			Education level (from the lowest to highest)	Residential location (from large to small cities)	Income (from the highest to lowest)
	Use ratio of green roofs	42.11	0	0	0
Landscape visual enjoyment (70.91)	Linear coordination of landscapes	54.14	0	+1	0
	Seasonal coordination of plants	72.18	+1	0	0
	Species richness and collocation of trees	67.67	+1	0	0
	Ratio of vertical green space area in the total green space	54.14	0	+1	0
	Accessibility of landscapes	50.38	0	0	0
	Residents' participation extent	40.60	0	0	0
	Occupancy ratio of landscape per capita	45.11	0	+1	0
	Structural optimization	53.38	0	0	0
Spiritual content and cultural taste (56.36)	Historical nature	59.40	0	0	+1
	Cultural nature	78.20	+1	0	0
	Inheritance	62.41	0	+1	0
	Symbolism	32.33	+1	+1	0
	Position consistency	37.59	0	+1	0

Note −1: low; 0: neutral; +1: high

The construction of Beijing Olympic Park used a large number of permeable bricks. In the pavement repair project of Chang'an Street, Beijing, both sides of the green isolation belt from Xidan to Dongdan, as well as sidewalks, used permeable bricks as the main pavement-laying materials. More exciting, the ongoing transformation of old districts in Beijing has started to integrate these new sources and technologies. Green roofs and pavement materials, as elements of ecological landscapes, have been gradually used. Further publicity and policy promotion are still necessary.

5 Conclusions

This study examined the essential content and main characteristics of urban ecological landscapes. The core values include sustainability, ecological and economic benefits, and harmony. The eight guiding principles for the construction of ecological landscapes are landscape naturalness, species safety and diversity, rainwater utilization, pollution emission and treatment, use of green energy and materials, local microclimate regulation, visual enjoyment of landscapes, spiritual content, and cultural taste. Our questionnaire results showed that the public had a high recognition of ecological landscapes, and a deep awareness of problems in the current ecological landscapes. They had a lower recognition of the structure and function of ecosystems, energy resources and materials. Further education and promotion are necessary to improve the communication channels with the public and ultimately improve the recognition of the theory and technologies of ecological landscapes. Respondents' education level and residential location had an important effect on the recognition of ecological landscapes whereas their incomes had a minor effect. The recognition results of the public with respect to proposed assessment indices provide an important implication for the further determination of index weights.

First, the theoretical system of urban ecological landscapes should be strengthened through multiple fields. So far, the integrated theoretical system of urban ecological landscapes has not been established. The basic concepts, main functions, constitutive elements, and basic characteristics of urban ecological landscapes need to be further clarified. The theories and technologies in other relevant fields, such as modern landscape ecology, geography, urban planning, tourism, systematics, psychology, economics, and sociology, provide a solid theoretical foundation for the planning and design of urban ecological landscapes. Future research of urban ecological landscapes will involve an interdisciplinary and diversified system. Since ecological landscapes involve multiple fields, the establishment of an assessment system for an integrated and human–nature harmonious urban ecological landscape should consider multiple aspects, such as technology, management, engineering, public's participation, and social development. Appropriate landscape elements and indices should be selected using both qualitative and quantitative methods. A series of technical guidelines including data acquisition, quantification, and weight determination should be established.

Second, we suggest conducting research of multi-dimensional patterns of urban ecological landscapes in an innovative way. Human's habitat space can be divided by multiple layers from the global to the local level. Urban ecological landscapes are constituent elements of the globe and region and can have multiple spatial scales. Only if continuously absorbing materials, energy, and information from the outside can urban ecological landscapes be further developed. Future urban ecological landscapes must be compatible with global changes and regional sustainable development. As an interdisciplinary field, urban ecological landscapes involve different data acquisition and analysis methods. With the rapid development of modern remote-sensing technologies and mathematical models, research on urban ecological landscapes can use more powerful data and technological support. In future research, new and modern technologies and quantitative measurement methods should be integrated, such as complexity theory, life cycle analysis methods, and ecological footprint analysis. In particular, the new Geo-informatic Tupu theory can be applied in the research of urban ecological landscapes to realize a combination of "graph" and "number" and to more accurately describe the temporal and spatial patterns of urban ecological landscapes.

Last, we note the major limitations associated with the promotion of urban ecological landscapes. In addition to an improvement of the theory and practice of urban ecological landscapes, the promotion and application of urban ecological landscapes are restricted by some key factors, such as the payback, experimental feasibility, and observation feasibility of these types of landscapes. The payback indicates the ecological and economic benefits of ecological measures. It is a prerequisite for the promotion of urban ecological landscapes. The experimental feasibility demonstrates the implementation feasibility of urban ecological landscapes. For example, the planting of simple species as part of landscape planning is more deserving of an experiment than a large-scale pipeline transformation. Similar schemes can also be replicated. The observation feasibility is an important measure of assessing the benefits of ecological landscapes. The construction of the three aspects supports the implementation of urban ecological landscapes from the initial planning to the medium-term implementation and final assessment. This will help promote and apply urban ecological landscapes.

Acknowledgements The work was supported by the National Natural Science Foundation of China (Grant # 41471150).

References

Calkins M (2005) Strategy use and challenges of ecological design in landscape architecture. Landsc Urban Plan 73(1):29–48

Cao Y, Hasi BG, Song DM (2002) A review on the concept, characteristic and assessment of landscape health. Chin J Appl Ecol 13(11):1511–1515

Cassidy R (2003) White paper on sustainability. Build Des Constr 1–48

Chen S, Wang J, Zhan ZY (2004) Study on ecological landscape and its integration with the city form. Prog Geogr 23(5):67–77
Cheng Z, Sun Y, Yang LY (2011) The eco-environmental evaluation of communities in Changzhou. Environ Sci Manag 36(1):170–174
Forman RTT (1995) Land Mosaics: the ecology of landscapes and regions. Cambridge University Press, Cambridge
Fu BJ, Chen LD, Ma KM, Wang YL (2011) Landscape ecology principles and applications, 2nd edn. Science Press, Beijing
Gao JX, Tian MR (2007) Discussion on ecological community, the sustainable development mode for urban communities. China Dev 7(4):6–10
Georgescu M, Morefield PE, Bierwagen BG, Weaver CP (2014) Urban adaptation can roll back warming of emerging megapolitan regions. Proc Natl Acad Sci USA 118:2909–2914
Givoni B (1997) Climate considerations in building and urban design. Wiley, New York
Housing Industrialization Promoting Centre of Ministry (HIPCM) (2001) Outlines and technical principles for green ecological residential quarter construction. Hous Sci 6(9):3–10
Kuitert W (2013) Urban landscape systems understood by geo-history map overlay. J Landsc Arch 8:54–63
Lovell ST, Johnston DM (2009) Creating multifunctional landscapes: how can the field of ecology inform the design of the landscape? Front Ecol Environ 7(4):212–220
Lund A, McMillan J, Kelly R, Jabbarzadeh S, Mead DG, Burkot TR, Kitron U, Vazquez-Prokopec GM (2013) Long term impacts of combined sewer overflow remediation on water quality and population dynamics of Culex quinquefasciatus, the main urban West Nile virus vector in Atlanta, GA. Environ Res 129:20–26
Makhzoumi J (2000) Landscape ecology as a foundation for landscape architecture: application in Malta. Landsc Urban Plan 50:167–177
Makhzoumi J, Pungetti G (1999) Ecological landscape design and planning: the mediterranean context. E & FN Spon, London
Mayer P, Grimm N, Lepczyk C, Pickett S, Pouyat R, Warren P (2010) Urban ecosystems research joins mainstream ecology. Nature 467(7312):153
McCuskey SA, Conger AW, Hillestad HO (1994) Design and implementation of functional wetland mitigation: case studies in Ohio and South Carolina. Water Air Soil Pollut 77(3–4):513–532
McHarg IL (1971) Design with nature. Doubleday, New York
Nassauer JI (2004) Monitoring the success of metropolitan wetland restorations: cultural sustainability and ecological function. Wetlands 24(4):756–765
Rosenzweig C, Solecki W, Hammer SA, Mehrotra S (2010) Cities lead the way in climate-change action. Nature 467(7318):909–911
Smida J (2003) The reality of green: readers speak out on building with sustainable materials and techniques. Design/Build Business
Szenasy SS (2002) Metropolis magazine survey, teaching green. International Contemporary Furniture Fair, New York
Termorshuizen JW, Opdam P, van den Brink A (2007) Incorporating ecological sustainability into landscape planning. Landsc Urban Plan 79(3–4):374–384
Tian MR, Gao JX, Zhang B, Qiao Q (2007) Study on assessment index system of ecological community. Res Environ Sci 20(3):87–92
United Nations (2010) Department of Economic and Social Affairs (UNDESA). World urbanization prospects: the 2009 revision. http://esa.un.org/unpd/wup/index.htm. Last accessed 1 Feb 2018
Wang RS, Li F (2006) Urban ecological management. J Chin Urban For 4(2):8–13
Wang RS, Ye YP (2004) Eco-city development in China. Ambio 33(6):319–320
Weddle A (1967) Techniques of landscape architecture. William Heineman Ltd, London
Yu KJ, Li DH, Ji QP (2001) Ecological design for landscape and city: concepts and principles. Chin Landsc Arch 17(6):3–9
Zhao CJ, Fu GB, Liu XM, Fu F (2011) Urban planning indicators, morphology and climate indicators: a case study for a north-south transect of Beijing, China. Build Environ 46(5):1174–1183

Zhou CB, Dai X, Wang RS, Huang JL (2011) Indicators for evaluating sustainable communities: a review. Acta Ecol Sin 31(16):4749–4759

Zhou JF, Zeng GM, Jiao S, Yang F, Zhu H, Li Q, Xiong Y, Tang L (2005) On uncertainties of indicating system for eco-environmental evaluation of residential communities. J Saf Environ 5(2):24–27

Dr. Ranhao Sun is currently an Associate Research Professor at the Research Center for Eco-Environmental Sciences under the Chinese Academy of Sciences. His main areas of expertise include landscape ecology, physical geography, and geographic information systems. Sun has authored or co-authored 6 books and over 100 articles in these areas.

Dr. Liding Chen is a Full Research Professor at the Research Centre for Eco-Environmental Sciences under the Chinese Academy of Sciences and Director of the State Key Laboratory of Urban and Regional Ecology. His research interests include landscape patterns and ecological processes, land use/cover changes and their environmental consequences, landscape planning and spatial modeling. He has authored or co-authored 9 books and more than 300 articles in these areas.

Monitoring Urban Growth and Land Changes in Beijing, China's Capital City by Remote Sensing: Progress and Challenges

Ting Liu and Xiaojun Yang

Abstract Over the past two and a half decades, urban growth has been a subject in numerous studies mostly through the use of remote sensing technology. Although many cities, large or small, have been targeted, Beijing as China's capital city has probably been more frequently researched than any other metropolises in the world. This chapter aims to examine some major advances in remote sensing-based urban growth studies with Beijing as the focus. For this purpose, we surveyed peer-reviewed English literature paying attention on some journal articles reporting the subject. Specifically, we examined the progress on several issues related to the research design and implementation, namely, spatial extent or temporal scale, data sources, and quantified dimensions. Based on the literature review, we further identified several major challenges and discussed some future research directions. We believe our longitudinal study focusing on major English literature examining the urbanization pattern in Beijing through remote sensing can not only help better research design but also assist formulating effective strategies and polices to deal with major challenges towards ecological sustainability in large metropolises.

Keywords Urban growth · Remote sensing · Beijing · China
Ecological sustainability

T. Liu (✉)
Department of Geography and Environmental Studies, Northeastern Illinois University, Chicago, IL 60625, USA
e-mail: tliu1@neiu.edu

X. Yang
Department of Geography, Florida State University, Tallahassee, FL 32306, USA

X. Yang and S. Jiang (eds.), *Challenges Towards Ecological Sustainability in China*, https://doi.org/10.1007/978-3-030-03484-9_4

1 Introduction

With the rapid population growth and economic development, urbanization in China has taken off at an unprecedented scale (Seto et al. 2011). In 1960, only 16% of people in China lived in urban areas. The urban share of the population had grown to 57% by 2016 and is projected to reach 70% by 2030 (The World Bank 2018). Unregulated urban development could threaten the sustainability of the human and natural systems over the long term (Grimm et al. 2008; Seto et al. 2011). Chinese cities face many challenges in order to accommodate the world's largest population while maintaining the current economic growth pace (The World Bank 2018).

As the capital city and a top megacity in China, Beijing has undergone a rapid growth during the past four decades (Seto et al. 2011; Li et al. 2015; Xu et al. 2016). The current urban form in Beijing is a result from the complex interplay of various physical, socioeconomic, and policy factors (Wu et al. 2006; Xie et al. 2007; Li et al. 2013b; Akiyama 2017). Beijing is physiographically situated at the northern part of the North China Plain, with 62% mountainous landscape dominant on the northern, the northeastern, and the western region of the municipality. The topographic feature functions as the physical constraints for urban development in Beijing, with most urban and agricultural land concentrated in the plain area. As a historical capital and the cultural and political center of China, Beijing's urban history dates back to more than 3000 years ago, and has undergone substantial transformation especially since China initiated the open door policy and implemented the economic reform in the late 1970s. Suburbanization in Beijing began in the late 1980s and the form of urban expansion is expressed by the construction of more ring roads. The degree of urban growth in Beijing has caused a myriad of ecological and societal pressures, which attracted attentions from both the academic community and the society. Previous research has extensively documented the negative impacts of urbanization in Beijing, including air and water pollution (Shao et al. 2006), altered local weather patterns (Guo et al. 2006; Li et al. 2012; Song et al. 2014), congestion (Wen et al. 2014), unemployment (Xue and Zhong 2003), among others. Developing systematic understanding of the urbanization dynamics in Beijing is crucial for both academic research and policy making.

Advances in airborne and satellite remote sensing technology have enabled the monitoring of urbanization across multiple spatial and temporal scales, providing crucial information for assessing land use efficacy and the potential environmental and societal impacts of urbanization. Despite the numerous research efforts made for monitoring the spatiotemporal urban patterns in Beijing through remote sensing (e.g., Wu et al. 2006, 2015; Li et al. 2013b, 2015; Song et al. 2015; Zhang et al. 2017), there is a lack of consensus and comparable results that can be useful for theory development or decision-making. On the one hand, technique-driven studies usually aimed at improving image classification and change detection accuracy without detailed examination of the results and their implications, which provide limited information that can help address specific challenges in Beijing (e.g., Li et al. 2015; Yu et al. 2016; Hu et al. 2017). On the other hand, applied-oriented studies have generally focused on land change analysis using standard techniques, which somewhat duplicated with existing studies that often failed to reveal emerging urban growth dynamics (e.g., Huang et al. 2008; Wu et al. 2015; Zhang et al. 2016).

The objective of this chapter is to examine the status and progress of remote sensing-based urban growth studies focusing on Beijing and identify the challenges and future research directions towards a better understanding of its urbanization patterns. Specifically, we surveyed English literature paying attention on peer-reviewed journal articles published since 2000 that were on the monitoring of urban growth and land changes in Beijing by remote sensing. We searched both the ISI Web of Science and the Scopus database using keywords focusing on remote sensor data, urbanization, and Beijing. We further refined the search results by selecting research articles focusing on the urban growth patterns and processes using remote sensing, as opposed to their driving factors and consequences. This chapter discusses the challenges and issues in monitoring urban expansion using remote sensing and related geospatial techniques, with Beijing as the case study area. The following sections present our major findings on several major issues including spatial extent or temporal scale, data sources, and quantified dimensions. Based on the literature review, we further identified several major challenges and discussed some future research directions.

2 Scales

2.1 *Spatial Scale: Extent and Grain*

We adopted the definition of scale from landscape ecology to consider the effects of both extent and grain in landscape pattern analysis (Turner et al. 1989; Turner and Gardner 2015). While extent refers to the overall size of the study area, grain refers to the resolution of the data. Here we have included research conducted at various spatial scales. While majority of the studies chose the municipality of Beijing as the study area, we also surveyed regional studies that compared urbanization patterns and explored regional interactions in the Jing-Jin-Ji area (Tan et al. 2005; Dong et al. 2008; Wu et al. 2015; Zhang et al. 2016), as well as smaller extent studies that focused on a particular district/region within the municipality (An et al. 2007; Huang et al. 2008). While the image resolution determines the "grain", it is also important to factor in the level of image analysis, e.g., sub-pixel classifier, which can potentially go beyond the limits of image resolution.

Research conducted at varying spatial extents has revealed different dynamics regarding the urbanization patterns and regional interactions. For example, Zhang et al. (2016) revealed the interaction of Beijing with its nearby urban areas in a comparative study of urban expansion in the Beijing-Tianjin-Tangshan (so called the Jing-Jin-Ji) region. They have found out that the large scale southwestward urban expansion in Tangshan was largely influenced by the expansion of Beijing and Tianjin which caused a substantial centroid shift for the developed areas. Wu et al. (2015) suggested that Shijiazhuang was under similar directional influence from the expansion of Beijing and Tianjin and the overall growth of the three cities was modeled considering the regional development of the Jing-Jin-Ji Urban agglomeration.

The "grain" aspect of spatial scale in urban remote sensing depends both on the image resolution and the analysis level. Urban areas are highly variable in spectral and spatial characteristics, which usually cause the presence of large amount of "mixed" pixels (Myint 2006; Franke et al. 2009; Powell 2011). Sub-pixel analysis derives percentage land cover information that provides useful information for quantifying urbanization patterns, especially with moderate resolution image data. Research efforts were made but usually focused on the technical aspects of image analysis with little analysis of the actual urbanization patterns in Beijing (Hu et al. 2017).

While the image resolution and analysis levels basically determine the "grain" of the analyses, the level of details presented can be affected by the spatial extent. This may produce results on urbanization patterns that appear to be inconsistent. For example, Huang et al. (2008) has identified multiple urban expansion cores in Beijing during the past three decades by focusing on the changes in the four center districts. However, Wu et al. (2015) suggested a mononuclear concentric expansion of Beijing during the same time periods in a regional comparative study that included the nearby urban areas, i.e., Tianjin and Shijiazhuang. Indeed, Wu et al. (2015) drew the conclusion based on a directional analysis from the original urban core located at the city center, which led to the omission of smaller sub-centers of urban expansion.

The detected patterns of urbanization vary by spatial extent and grain of the analysis. When presenting and analyzing urbanization patterns using remote sensing, researchers need to be explicit regarding the analysis scales or conduct multiscale analysis if possible. With the establishment of the Xiong'an New Area in 2017, more research efforts are demanded at the regional level, which evaluate the regional integration of the Jing-Jin-Ji Urban Agglomeration as a result of strategic plans.

2.2 Temporal Scale: Duration and Frequency

At the temporal dimension, both duration and frequency affect the change analysis results. In the case of Beijing, many researchers have focused on change analysis during the economic reform era since the late 1970s or early 1980s (e.g., Wu et al. 2006; Li et al. 2015), while the analysis frequency varies. Overall, a monotonic growth trend was observed during this time period. However, the observations at varying temporal frequency could produce inconsistent results for characterizing the fluctuation of urban growth in Beijing. For example, researchers have reported slightly different results as for the highest growth period in the past several decades due to the use of varying time intervals in change detection (Li et al. 2015; Wu et al. 2015; Zhang et al. 2017).

As a city undergoing constant transformation due to policy reform and globalization, long term and high frequency monitoring of urbanization became crucial for sustainable land use planning. While such information is valuable for urban growth modeling and understanding time lag effects of policy factors, the feasibility of deriving such information is challenged by data availability and technical difficulties. Li et al. (2015) utilized temporal consistency check for urban land change detection in Beijing at an annual frequency over a 30-year time period. However, their research focused on the performance of classification and change detection algorithms without presenting detailed analysis of the results. Their methodology and procedures can also be time consuming to replicate.

Since land regulation and policies play an important role in the urbanization patterns of Chinese cities, many researchers attempted to define their analysis frequency based on the time frame of major land regulation polices. For example, Wu et al. (2006) mapped urban expansion from 1986 to 2001 at a 5-year interval and they have pointed out a slowdown of urban expansion between 1996 and 2001, which was attributed to the implementation of the "Ordinance for the Protection of Primary Agricultural Land" in 1994 for controlling the extremely rapid land development. While this approach could help determine the key turning point for monitoring urban land changes, it failed to capture the complex interplay of various environmental and socioeconomic factors that affect urban growth. Other factors such as the regional and global economic environment may also affect the magnitude of urbanization. Zhang et al. (2016) revealed a slowdown of urban expansion observed in the late 1990s due to the Asian financial crisis.

Acquiring long term and high frequency land change information of Beijing remains a challenge. At the research design level, researchers need to carefully select temporal frequency for change detection that account for the major shifts in policies and the regional and global socioeconomic dynamics. At the technical level, researchers may develop automated change detection techniques and tools for deriving reliable high frequency land change information in the urban environments.

3 Data Sources

By now, a wide variety of remote sensor data sources have been used in the monitoring of urban growth patterns in Beijing (Table 1). The Landsat system is among the most commonly used sources for urban remote sensing thanks to its global coverage and long term inventory since the 1970s (Wu et al. 2006; Jia et al. 2014; Li et al. 2015). In the meantime, various image analysis and change detection techniques have been developed for improving categorical precision and spatial and temporal consistency and accuracy (Jia et al. 2014; Li et al. 2015; Yu et al. 2016).

Table 1 Summary of the image data types and processing techniques used according to our literature survey

Image types	Image processing techniques	Examples
Moderate spatial resolution imagery	Improved change detection techniques	Zhang et al. (2002) Chen et al. (2003) Liu and Zhou (2004) Li et al. (2015) Yu et al. (2016)
	Image fusion	Jia et al. (2014)
	Sub-pixel analysis	Hu et al. (2017)
	Object-oriented methods	An et al. (2007) Tian et al. (2014)
High spatial resolution imagery	Object-oriented methods	Zhang et al. (2017)
	Visual interpretation	Zheng et al. (2017)
Hyperspectral imagery	Supervised classification	Li et al. (2016)
Nighttime satellite imagery	Object-oriented methods	Cai et al. (2017)
	Light intensity based measures	Yang et al. (2017)

3.1 Remote Sensor Data

In addition to the widely used Landsat system, researchers have attempted to extract urban land information in Beijing from various image data sources, including high spatial resolution data, hyperspectral imagery and nighttime imagery (Table 1). However, the possibilities of using active remote sensing, such as LiDAR and Radar data are less explored, which can both provide valuable information for characterizing the urban environments (Dong et al. 1997; Priestnall et al. 2000; Brenner and Roessing 2008; Li et al. 2013a; Rottensteiner 2010).

With the successful launch of Chinese high resolution satellites in the past several years, researchers made efforts to evaluate the applicability of these new sensors. Li et al. (2016) compared the hyperspectral satellite TianGong-1 (TG-1) with EO-1 Hyperion. TG-1 is a fine spatial resolution hyperspectral satellites launched on September 29, 2011, with a spatial resolution of 10 m, compared to the 30 m resolution of the EO-1 Hyperion. They have concluded that TG-1 outperforms EO-1 Hyperion in the overall classification accuracy. But the difference in spatial resolution and the use of pixel-based classifier made the comparison less meaningful. Another commonly used Chinese satellite is the high-resolution GaoFen-2 (GF-2) launched on 19 August 2014 (Yin and Yang 2017; Zhang et al. 2017). It has 4 m resolution multispectral bands and 1 m panchromatic band, with spectral resolution comparable to IKONOS.

3.2 *Social Sensing Data*

Recent research progress in urban remote sensing also includes the use of social sensing, especially social media data for characterizing human activities (Cai et al. 2017; Zhang et al. 2017). Social sensing refers to the collection of geospatial data from humans or mobile devices (Wang et al. 2015). While remote sensor data are well designed for earth surface observation, they oftentimes lack the capability of sensing the social dimension of the earth system. Many social media platforms collect "geotagged" information that provides researchers with complementary information to refine the image analysis procedures with an additional human layer. For example, Cai et al. (2017) incorporated the check-in records of Weibo (microblogs) along with nighttime imagery for urban structure detection. To achieve spatial stability, they have included a dataset of over 5 million check-in observations recorded over a year. Zhang et al. (2017) combined Baidu Point-Of-Interest (POI), Weibo posts with high spatial resolution imagery in an effort to identify fine grain land use information in Beijing. However, their methods failed to achieve acceptable accuracy for the important urban land use classes, such as business and institutions.

While the increasing availability of social media data is promising for providing information on human activities for urban area mapping, future research efforts are highly demanded. Processing the large amount of social media data usually requires knowledge and experiences of handling big data in order to extract the most relevant information. The various attributes, such as check-in records, account registration locations, geotagged posts, may be of interests for different tasks in the context of urban structure detection. Research efforts are needed to compare their usefulness for mapping urban structure and components across scales. Data integrity issues need to be further explored for the various social media platforms. For example, a user's account registration location may be recorded as the post location if the GPS is not turned on at the time of the post. Similar issues need to be explicitly reported and discussed before the usefulness of social sensing data can be fully recognized.

4 Quantified Dimensions

Land cover/use information derived from remote sensor data is often times used for further quantification of urbanization patterns across various scales. In the case of Beijing, researchers have attempted to quantify the spatiotemporal dynamics of urbanization from various aspects. We inventoried the articles that focused on the analysis of urbanization patterns and summarized the results in Table 2. The columns represent the different dimensions for defining urban forms and urbanization patterns (Galster et al. 2001; Jabareen 2006), as opposed to the actual quantification methods. We only included the dimensions or aspects that can be derived from remote sensor data. For example, Zheng et al. (2017) measured the vertical dimension of urban

Table 2 Summary of the dimensions of urbanization patterns quantified according to our literature survey

References	Dimensions of urbanization patterns						
	Extent/magnitude	Rate	Density	Contiguity	Direction	Nuclearity	Diversity
Cai et al. (2017)						•	
Huang et al. (2008)	•		•				
Liu et al. (2016)	•			•			•
Sun and Zhao (2018)	•	•		•			
Tian et al. (2014)	•	•					
Wu et al. (2015)	•	•		•	•		
Yu and Zhou (2017)	•	•		•			
Zhang et al. (2016)	•	•		•	•		
Zheng et al. (2017)	•		•		•		

growth based on building footprint and height data, which was not included as a dimension in Table 2.

Urban expansion extent or magnitude and growth rate (usually measured as annual average growth per unit area) are among the most commonly quantified aspect of urbanization. Although not included in the table, many technical articles have also reported the extent and rate of urban growth, which is readily available information from classified land use/cover maps using remote sensor data (e.g., Wu et al. 2006; Li et al. 2013b, 2015). These basic variables are also used as input to land change models for urbanization driving force analysis or land change prediction (e.g., Wu et al. 2006; Dong et al. 2008; Huang et al. 2008; Li et al. 2013b). In addition to the overall extent and magnitude of urban expansion, some researchers attempted to characterize the spatial variation of expansion along the urban-rural gradient (e.g., Wu et al. 2006, 2015; Zhang et al. 2016).

Density and contiguity are both useful indicators of urban land use efficiency (Galster et al. 2001). While density refers to the percentage of urban land within a unit area, contiguity concerns about the spatial patterns or connectivity of urban land. Density, however, is relatively less characterized which may have to do with the lack of well-defined unit areas for calculating density using remote sensor data. Huang et al. (2008) utilized a moving window approach that created a neighborhood for pixel-level density calculation. Zheng et al. (2017) incorporated ancillary data, i.e., neighborhood boundaries, as the unit areas for calculating building density. On the other hand, contiguity has been well examined through landscape metrics adopted from landscape ecology (Zhang et al. 2016; Yu and Zhou 2017) and geospatial indicators developed for urban studies (Wu et al. 2015; Liu et al. 2016). However, the calculation of these spatial metrics and indicators in the highly complex urban environments are largely dependent upon many factors, such as image resolution, classification scheme, and classification accuracy. For example, the different ways of handling the "mixed pixels" could create varying sizes of land "patches", the basic unit for calculating landscape metrics. Unfortunately, researchers made no attempt at addressing these issues, which made it less reliable to compare the results from these studies regarding landscape fragmentation and complexity.

Direction is another important aspect of urbanization, which characterizes the spatial variation and unevenness of urban growth. The direction-based analyses have all adopted the sector-based model that uses transects that cut across the city center to divide the city into sectors as representation of directions (Wu et al. 2015; Zhang et al. 2016, 2017). To compare the variation in different directions, the researchers then measured the extent and patterns for each sector separately. While the city is expanding in every direction, substantially more urban land growth was found in the southeast direction. This symmetrical definition of directional sectors can be easily implemented and enables the comparison among different cities. However, it may lose the power of characterizing the directional trends of cities with non-symmetrical shapes or special topographic patterns.

The nuclearity of Beijing was among one of the least quantified aspects, which involves the identification of nuclei. Huang et al. (2008) attempted to visually identify the multiple sub-expansion cores in Beijing based on classified maps. As a quantification effort, Cai et al. (2017) applied a set of spatial statistics, e.g., Moran's I and Geographically Weighted Regression, on nighttime imagery and social media data. Similar approach may be used for land cover map classified from remotely sensed data.

Diversity measures are often based on existing landscape metrics for measuring landscape diversity (Liu et al. 2016). However, information on the degree of mixed land use is of special interests as a more relevant measure of sustainable urban form (Jabareen 2006). Most classified maps derived solely from remote sensor data lack the detailed land use information, such as residential, commercial, industrial, institutional land, etc. In order to develop meaningful measurements of land use diversity in urban areas, research efforts need to be made in order to extract accuracy land use information from remote sensing and other readily available data sources and techniques.

Despite the efforts of quantifying urbanization patterns in Beijing, there are several dimensions of urbanization deserve further research (Bhatta et al. 2010). It is also necessary to develop standard methods and procedures across scales for comparing urbanization patterns at different geographic locations. The quantification of urbanization patterns from remote sensor data also largely depends on the conceptual and technical issues discussed in previous sections.

5 Conclusions

In this chapter, we surveyed research articles on monitoring urbanization patterns of Beijing through remote sensing. We discussed the research progress and major challenges on several issues related to the research design and implementation of urban growth monitoring. At the spatial dimension, mapping urbanization patterns can be affected by both spatial extent and grain (i.e., data resolution). Interactions between extent and grain could lead to inconsistent results when quantifying urbanization patterns. Future research needs to present scale factors in a more explicit manner and focus on developing a multiscale analysis framework that can help understand both the complexities within the city and its interactions and integrations with surrounding urban areas (i.e. the Jing-Jin-Ji Urban agglomeration).

At the temporal dimension, both duration and frequency can affect the monitored urbanization trend. While high frequency and long term information is desirable, the procedures can be difficult to implement. A proper design of temporal frequency needs to consider the dynamics of policies and other socioeconomic factors affecting Beijing's urban land use.

At the data dimension, the usefulness of active remote sensors is relatively less researched for mapping the urban forms in Beijing. With the increasing availability of high resolution satellite data and social sensing data, research efforts to explore the possibilities of integrating multiple data sources for detailed information extraction are highly demanded.

Lastly, scholars have attempted to quantify various dimensions of urban expansion from remote sensor data. However, standardized metrics and indicators still need to be developed for comprehensive measurements of urban forms. All the issues are interrelated and deserve more attention in future urban remote sensing research in Beijing.

We believe our longitudinal study focusing on major English literature examining the urbanization pattern in Beijing through remote sensing can not only facilitate better research design but also assist formulating effective strategies and polices to deal with the challenges towards ecological sustainability in large metropolises.

References

Akiyama CM (2017) Beijing metropolitan area. In: Murayama et al (eds) Urban development in Asia and Africa. The urban book series. Springer, Singapore, pp 65–83

An K, Zhang J, Xiao Y (2007) Object-oriented urban dynamic monitoring—a case study of Haidian District of Beijing. Chin Geogr Sci 17(3):236–242

Bhatta B, Saraswati S, Bandyopadhyay D (2010) Urban sprawl measurement from remote sensing data. Appl Geogr 30(4):731–740

Brenner AR, Roessing L (2008) Radar imaging of urban areas by means of very high-resolution SAR and interferometric SAR. IEEE Trans Geosci Remote Sens 46(10):2971–2982

Cai J, Huang B, Song Y (2017) Using multi-source geospatial big data to identify the structure of polycentric cities. Remote Sens Environ 202:210–221

Chen J, Gong P, He C, Pu R, Shi P (2003) Land-use/land-cover change detection using improved change-vector analysis. Photogramm Eng Remote Sens 369–379
Dong J, Zhuang D, Xu X, Ying L (2008) Integrated evaluation of urban development suitability based on remote sensing and GIS techniques—a case study in JingJinJi area, China. Sensors 8(9):5975–5986
Dong Y, Forster B, Ticehurst C (1997) Radar backscatter analysis for urban environments. Int J Remote Sens 18(6):1351–1364
Franke J, Roberts DA, Halligan K, Menz G (2009) Hierarchical multiple endmember spectral mixture analysis (MESMA) of hyperspectral imagery for urban environments. Remote Sens Environ 113(8):1712–1723
Galster G, Hanson R, Ratcliffe MR, Wolman H, Coleman S, Freihage J (2001) Wrestling sprawl to the ground: defining and measuring an elusive concept. Hous Policy Debate 12(4):681–717
Grimm NB, Faeth SH, Golubiewski NE, Redman CL, Wu J, Bai X, Briggs JM (2008) Global change and the ecology of cities. Science 319(5864):756–760
Guo X, Fu D, Wang J (2006) Mesoscale convective precipitation system modified by urbanization in Beijing City. Atmos Res 82(1):112–126
Hu D, Chen S, Qiao K, Cao S (2017) Integrating CART algorithm and multi-source remote sensing data to estimate sub-pixel impervious surface coverage: a case study from Beijing Municipality, China. Chin Geogr Sci 27(4):614–625
Huang W, Liu H, Luan Q, Jiang Q, Liu J, Liu H (2008) Detection and prediction of land use change in Beijing based on Remote Sensing and GIS. In: The International archives of the photogrammetry, remote sensing and spatial information sciences, vol XXXVII. Part B6b. Beijing 2008
Jabareen YR (2006) Sustainable urban forms: their typologies, models, and concepts. J Plan Educ Res 26(1):38–52
Jia K, Liang S, Zhang N, Wei X, Gu X, Zhao X, Yao Y, Xie X (2014) Land cover classification of finer resolution remote sensing data integrating temporal features from time series coarser resolution data. ISPRS J Photogramm Remote Sens 93:49–55
Li P, Jiang S, Wang X, Zhang J (2013a) Building extraction using LiDAR data and very high resolution image over complex urban area. In: 2013 IEEE international geoscience and remote sensing symposium—IGARSS, Melbourne, VIC, Australia, 21–26 July 2013
Li X, Zhou W, Ouyang Z (2013b) Forty years of urban expansion in Beijing: what is the relative importance of physical, socioeconomic, and neighborhood factors? Appl Geogr 38:1–10
Li X, Gong P, Liang L (2015) A 30-year (1984–2013) record of annual urban dynamics of Beijing City derived from Landsat data. Remote Sens Environ 166:78–90
Li X, Wu T, Liu K, Li Y, Zhang L (2016) Evaluation of the Chinese fine spatial resolution hyperspectral satellite TianGong-1 in urban land-cover classification. Remote Sens 8(5):438
Li X, Zhou W, Ouyang Z, Xu W, Zheng H (2012) Spatial pattern of greenspace affects land surface temperature: evidence from the heavily urbanized Beijing metropolitan area, China. Landsc Ecol 27(6):887–898
Liu Z, He C, Wu J (2016) General spatiotemporal patterns of urbanization: an examination of 16 world cities. Sustainability (Switzerland) 8(1):1–15
Liu H, Zhou Q (2004) Accuracy analysis of remote sensing change detection by rule-based rationality evaluation with post-classification comparison. Int J Remote Sens 25(5):1037–1050
Myint SW (2006) Urban vegetation mapping using sub-pixel analysis and expert system rules: a critical approach. Int J Remote Sens 27(13):2645–2665
Priestnall G, Jaafar J, Duncan A (2000) Extracting urban features from LiDAR digital surface models. Comput Environ Urban Syst 24(2):65–78
Powell RL (2011) Characterizing urban subpixel composition using spectral mixture analysis. In: Yang X (ed) Urban remote sensing: monitoring, synthesis and modelling in the urban environment. Wiley, Chichester, UK, pp 111–128
Rottensteiner F (2010) Automation of object extraction from LiDAR in urban areas. In: 2010 IEEE international geoscience and remote sensing symposium—IGARSS, Honolulu, HI, USA, 25–30 July 2010

Seto KC, Fragkias M, Güneralp B, Reilly MK (2011) A meta-analysis of global urban land expansion. PLoS ONE 6(8):e23777

Shao M, Tang X, Zhang Y, Li W (2006) City clusters in China: air and surface water pollution. Front Ecol Environ 4(7):353–361

Song W, Pijanowski BC, Tayyebi A (2015) Urban expansion and its consumption of high-quality farmland in Beijing, China. Ecol Ind 54:60–70

Song X, Zhang J, AghaKouchak A, Roy SS, Xuan Y, Wang G, He R, Wang X, Liu C (2014) Rapid urbanization and changes in spatiotemporal characteristics of precipitation in Beijing metropolitan area. J Geophys Res Atmos 119(19):11250–11271

Sun Y, Zhao S (2018) Spatiotemporal dynamics of urban expansion in 13 cities across the Jing-Jin-Ji urban agglomeration from 1978 to 2015. Ecol Ind 87:302–313

Tan M, Li X, Xie H, Lu C (2005) Urban land expansion and arable land loss in China—a case study of Beijing–Tianjin–Hebei region. Land Use Policy 22(3):187–196

The World Bank (2018) http://databank.worldbank.org/data/home.aspx. Accessed 17 Feb 2018

Tian Y, Yin K, Lu D, Hua L, Zhao Q, Wen M (2014) Examining land use and land cover spatiotemporal change and driving forces in Beijing from 1978 to 2010. Remote Sens 6(11):10593–10611

Turner MG, Gardner RH (2015) Landscape ecology in theory and practice, 2nd edn. Springer, New York

Turner MG, O'Neill RV, Gardner RH, Milne BT (1989) Effects of changing spatial scale on the analysis of landscape pattern. Landsc Ecol 3(3–4):153–162

Wang D, Abdelzaher T, Kaplan L (2015) Social sensing: building reliable systems on unreliable data. Morgan Kaufmann, San Francisco, CA

Wen H, Sun J, Zhang X (2014) Study on traffic congestion patterns of large city in China taking Beijing as an example. Procedia—Social Behav Sci 138:482–491

Wu Q, Li HQ, Wang RS, Paulussen J, He Y, Wang M, B-H Wang, Wang Z (2006) Monitoring and predicting land use change in Beijing using remote sensing and GIS. Landsc Urban Plan 78(4):322–333

Wu W, Zhao S, Zhu C, Jiang J (2015) A comparative study of urban expansion in Beijing, Tianjin and Shijiazhuang over the past three decades. Landsc Urban Plan 134:93–106

Xie Y, Fang C, Lin GCS, Gong H, Qiao B (2007) Tempo-spatial patterns of land use changes and urban development in globalizing China: a study of Beijing. Sensors (Basel, Switzerland) 7(11):2881–2906

Xu M, He C, Liu Z, Dou Y (2016) How did urban land expand in China between 1992 and 2015? a multi-scale landscape analysis. PLoS ONE 11(5):e0154839

Xue J, Zhong W (2003) Unemployment, poverty and income disparity in urban China. Asian Econ J 17(4):383–405

Yang Y, Ma M, Tan C, Li W (2017) Spatial recognition of the urban-rural fringe of Beijing using DMSP/OLS nighttime light data. Remote Sens 9(11):1141

Yin W, Yang J (2017) Sub-pixel vs. super-pixel-based greenspace mapping along the urban–rural gradient using high spatial resolution Gaofen-2 satellite imagery: a case study of Haidian District, Beijing, China. Int J Remote Sens 38(22):6386–6406

Yu W, Zhou W (2017) The spatiotemporal pattern of urban expansion in China: a comparison study of three urban megaregions. Remote Sens 9(1):45

Yu W, Zhou W, Qian Y, Yan J (2016) A new approach for land cover classification and change analysis: integrating backdating and an object-based method. Remote Sens Environ 177:37–47

Zhang Q, Wang J, Peng X, Gong P, Shi P (2002) Urban built-up land change detection with road density and spectral information from multi-temporal Landsat TM data. Int J Remote Sens 23(15):3057–3078

Zhang Y, Li Q, Huang H, Wu W, Du X, Wang H (2017) The combined use of remote sensing and social sensing data in fine-grained urban land use mapping: a case study in Beijing, China. Remote Sens 9(9):865

Zhang Z, Li N, Wang X, Liu F, Yang L (2016) A comparative study of urban expansion in Beijing, Tianjin and Tangshan from the 1970s to 2013. Remote Sens 8(6):496
Zheng Z, Zhou W, Wang J, Hu X, Qian Y (2017) Sixty-year changes in residential landscapes in Beijing: a perspective from both the horizontal (2d) and vertical (3d) dimensions. Remote Sens 9(10):992

Dr. Ting Liu is an Assistant Professor in the Department of Geography and Environmental Studies at Northeastern Illinois University, USA. Her research centers on the development and applications of geographic information science (GIScience) and related technologies to support geographic inquiries in urban and environmental domains.

Dr. Xiaojun Yang Editor of this volume, is a tenured Full Professor of Geography in the College of Social Sciences and Public Policy at Florida State University, USA. He has also had a few visiting academic positions with the Chinese Academy of Sciences and Jinan University, China. His research interests include the development of remote sensing and geospatial technologies with applications in the urban and environmental domains. Yang has authored or co-authored six English books and over 100 articles in these areas.

Diatoms as an Evaluation Tool for the Ecological and Environmental Conditions of Rivers and Streams in China: A Retrospective Study

Yuanda Lei, Yasu Wang, Richard William Jordan and Shijun Jiang

Abstract Diatoms play a crucial role as primary producers at the base of aquatic ecosystems, and their community dynamics and diversity patterns are sensitive indicators of ecological and environmental changes. Diatoms have been increasingly used in ecological assessments of rivers and streams in China from 1980 to 2017, especially after 2000. In this chapter, we review the development history, basic rationale and technical application of the typical approaches, attempting to provide an overall perspective of the applications of diatoms to environmental assessments in China. Benthic diatoms are preferentially used in river assessments with increasing incorporation of planktonic species. Diatom-based indicators including biomass, diversity, species dominance and autecology have been well used as traditional methods for decades in China, while the diatom biotic indices and multimetric indices have been tested across the country and widely accepted as more effective means. This indicates that diatoms are a useful tool in river biomonitoring programs for water resource management across the world, and its future applications are promising as global environmental change continues to escalate.

Keywords Diatom · Ecological assessment · Water quality · River · China

Y. Lei · Y. Wang · S. Jiang (✉)
Institute of Groundwater and Earth Sciences, Key Laboratory of Eutrophication and Red Tide Prevention of Guangdong Higher Education Institutes, Jinan University, Guangzhou, Guangdong 510632, China
e-mail: ssj0047@my.fsu.edu

R. W. Jordan
Department of Earth & Environmental Sciences, Faculty of Science, Yamagata University, 1-4-12 Kojirakawa-machi, Yamagata 990-8560, Japan

X. Yang and S. Jiang (eds.), *Challenges Towards Ecological Sustainability in China*, https://doi.org/10.1007/978-3-030-03484-9_5

1 Introduction: Diatoms as Ecological Indicators

China is facing the most severe water crises (namely, freshwater shortage and pollution) that have ever been documented (Han et al. 2016). This is largely a result of the relative shortage and uneven distribution of water resources, and a large population, which is exacerbated by widespread water pollution due to high-speed economic development, inadequate management, and outdated infrastructure (Liu and Yang 2012). As reported in the latest China Environmental State Bulletin (2016), nearly 30% of the water bodies in the seven major river systems, namely, the Yangtze River, Yellow River, Zhu (Pearl) River, Huai River, Hai River, Songhua River, and Liao River, have been ranked Category IV or worse in water quality, which is considered polluted and undrinkable (MEPPRC 2017). Consequently, river pollution issues have been listed among the top policy priorities in the national agenda of the Chinese government (Wang et al. 2016). The Chinese water administration eagerly calls for comprehensive monitoring methodologies to evaluate the actual environmental and ecological states of rivers, attempting to promote better river management including water resource conservation, pollution restoration and water supply.

The current river quality standards regulated by the Chinese legislation (GB3838-2002, 2002) are limited to the monitoring of chemical (e.g., chemical oxygen demand, total phosphorus, total nitrogen, heavy metals, etc.) and microbial (e.g., fecal coliforms) parameters. These parameters provide a good indication of the water quality at the time of sampling, but are unable to reveal the long-term response of organisms to river pollution (Pignata et al. 2013; Oeding and Taffs 2015). On the other hand, the biomonitoring methods based on aquatic biological communities (fish, macroinvertebrates and algae) are designed to assess the integrated environmental and biological conditions of a river ecosystem, and therefore provide a supplement and extension for the traditional chemical and microbial techniques (Oertel and Salánki 2003; Bere and Tundisi 2010). In principle, any change in diversity structure and distribution pattern of living organisms is indicative of environmental variation in a river. Various kinds of biomonitoring techniques have been developed and are receiving more and more attention in China.

As an important group of organisms in rivers, diatoms have been widely proven to be a useful bio-indicator in China (Ding and Zhi 2006; Liu and Zhang 2009; Li et al. 2012a), as well as across the world (Stevenson et al. 2010; Rimet 2012; Vilmi et al. 2015). Diatoms are essential and fundamental primary producers in river ecosystems, and the following characteristics and advantages make them excellent bio-indicators.

(1) Broad dispersal. Resting stages of many diatom taxa can survive extreme dry conditions and long-distance transportation, and this life strategy helps them to colonize almost every corner of the Earth (Kristiansen 1996). Their global distribution enables diatom assessment to be a cross-geographical method (Fan et al. 2007; Tang et al. 2013a).
(2) High species richness. Diatoms are among the most species-rich organisms in aquatic environments, and a conservative estimation of their global diversity is

200,000 species (Mann and Droop 1996). Such a large number of species offers the possibility of assessing water quality in a numerical manner (Lei 2016).

(3) Short life cycle. As microalgae, diatoms grow and reproduce more rapidly than large animals such as macroinvertebrates and fish, providing a potential for an early warning of environmental disturbances (Barbour et al. 1999).

(4) Sensitive response to ambient environment. Diatoms respond sensitively to the physical, chemical and biological variations in their ambient environments (Pan et al. 1996; Liu et al. 2013; Wang et al. 2014), as well as upper-level factors such as land use, geographic and climate changes (Potapova and Charles 2002; Soininen 2007; Li et al. 2015b). Their response is possibly more sensitive than macrophytes and other algal groups (Schneider et al. 2012). In addition, aquatic diatoms spend their whole life cycle either passively floating in the water (planktonic), attached to submerged surfaces (e.g., epiphytic, epizooic, epilithic) or as mobile populations on the surface sediment (e.g., epipelic, epipsammic), thus their response directly reflects the actual impacts of environmental stressors on aquatic organisms in rivers (Xu et al. 2011).

(5) Low cost and ease of use. The sampling (collecting with brush, net or bottle), preparation (fixation and oxidant cleaning) and identification (a microscope and a camera) procedures during the process of diatom assessment are easily operated, and cost and human resource effective (Round 1991; Barbour et al. 1999; Lei 2012).

(6) Relatively well-explored autecology. The autecology of many diatoms is well established by previous studies. Nearly 20 autecological diatom indices have been developed for the investigation of aquatic environments across the world (see Sect. 2.5 for details), and 948 diatom taxa have been assigned specific ranked indicator values for pH, salinity, nitrogen uptake metabolism, oxygen, saprobity, trophic state and moisture (van Dam et al. 1994). The well-established autecology enables diatoms to be a rapid and accurate evaluation tool in river environmental surveys.

Despite many ecological merits, diatoms have become a bio indicator in China only in recent decades. To the best of our knowledge, pioneering studies based solely on diatoms in China first appeared in the 1980s (Zhu and Xie 1982; Luo and Zeng 1985), and river surveys utilizing the Kolkwitz and Marsson's (1908) typed saprobic systems with ranked aquatic taxa including diatoms have been carried out since the 1990s (Shen 1990). These early diatom assessments were limited to qualitative descriptions of environmental status based on dominant species or biodiversity. Since 2000, the relationship between diatom assemblage and environmental stresses (pollutants) has been intensively explored and quantitatively confirmed in numerous multivariate statistical studies (Wang and Zhang 2004; Guo 2008; Wang et al. 2009). During the past decade (from 2009 to the present-day), multiple assessing techniques with advanced numerical diatom indices and diatom-based index of biotic integrity (D-IBI) have been introduced and developed rapidly to meet the scientific and socioeconomic needs (Zhao et al. 2009; Li et al. 2012b).

To reveal the development history of diatom applications to environmental assessment in China, we investigated the number of papers published during the years 1980–2017 from China's largest scientific database (China Integrated Knowledge Resources Database, http://www.cnki.net) for publications in Chinese (Fig. 1a), and from the Web of Science (Thomson Reuters) for publications in English (Fig. 1b). The results from both databases show a similar, continuously increasing trend in the number of papers published on diatom assessment in China during this period, and especially a dramatic rise after 2000 (Fig. 1). This suggests that Chinese ecologists have recognized the potential and robustness of diatom assessments, and attempted to modify the original techniques for their own scientific and practical purposes in China. This review focuses on diatom-based river assessments, attempting to (1) offer an overview of the development history of diatom techniques for assessing river water quality in China over the past 40 years (from 1980 to the present-day), and (2) summarize the typical diatom-based methods and their current status of practical use in China.

2 Diatom Assessments in China

In China, several diatom monitoring techniques have been commonly adopted in river assessment programs. The reliability of diatom assessment depends on the selection of an appropriate approach. Here we review the basic concepts, process designs and technical operations for the diatom methods including biomass, diversity indices, dominant species, autecological indices and multiple metrics.

2.1 Planktonic Versus Benthic Diatoms

Principally, diatoms in rivers or streams can be categorized as benthic and planktonic according to their habitats. Benthic diatoms mainly grow on submerged substrates, while planktonic ones generally live in open waters (Round et al. 1990). These two forms can be regarded as alternatives in environmental surveys, and which one is selected depends on the sampling situation (e.g., size of the river section, availability of sampling objects) and research goals (e.g., a point or areal pollution source detection, or a large-scale investigation within a river network) (Stevenson et al. 2010). Generally, it is better to choose planktonic diatoms as a bio-indicator when surveying large rivers, while for the assessment of streams or brooks it is advised to sample benthic diatoms. In addition, planktonic diatom assemblages are susceptible to seasonal variations and usually show waxing and waning related to other phytoplankton groups (Reynolds 2006), which may lead to biased assessment if using only planktonic diatoms rather than the entire phytoplankton community. Since benthic diatoms almost occupy the dominant position within the river benthic biofilm throughout the year, seasonality has little effect on them.

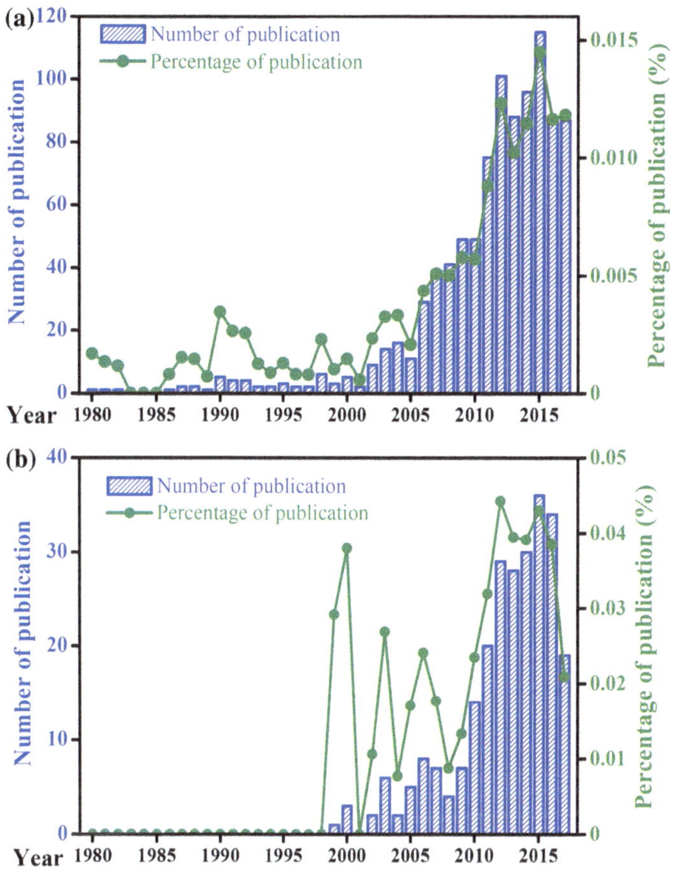

Fig. 1 Annual publication number on diatom assessment of China in the China Integrated Knowledge Resources Database (**a**) and the Web of Science (**b**), and their proportions to the total scientific articles in the databases

For the above reasons, most river assessments in China focus on benthic and attached diatoms, and seldom on planktonic diatoms (Lei 2012; Zhang et al. 2016). Surveying different diatom life forms shares the same principle. A new trend has started to emerge recently. Huang (2013) and Zhang et al. (2015) used the planktonic diatoms to build the trophic diatom index (TDI), a numerical ranking originally designed using benthic diatoms (Kelly and Whitton 1995; Wu et al. 2014), and found that the revised TDI was applicable in trophic status evaluation for the rivers in the Ganjiang River Basin, Southeast China. In fact, the functional groups (FGs) of planktonic diatoms are better established relative to benthic diatoms (Reynolds 2006). Zhang et al. (2016) revealed that planktonic diatom taxa in the Ganjiang rivers (Southeast China) could be assigned to 11 functional groups, and their seasonal succession was driven by several factors including temperature, water level, discharge regime and habitat stability. If budget allows, a combined assignment of benthic and planktonic samples is strongly recommended to achieve better completeness and accuracy in river assessment (Aloi 1990; Stevenson et al. 2010).

2.2 Diatom Biomass

Algal biomass is defined as the mass of all living algae in a water body at a given time. Traditionally, algal biomass is a common indicator of environmental changes in aquatic habitats, thus diatom biomass is frequently employed by Chinese researchers in ecological investigations (Table 1). The diatom biomass can be estimated by a number of metrics including dry mass (DM), ash-free dry mass (AFDM), chlorophyll α (Chl α), cell density (Dens.) and cell volume (Vol.), and each of them has a specific range of applications (Pei 2006; Stevenson et al. 2010). Chen (2012) suggested that Chl α was a reliable estimation for the periphytic diatom biomass in two subtropical streams, but warned that it should be used with caution under low-illumination or nutrient-deficient conditions. Both DM and AFDM are gravimetric measurements from dried bulk samples; however, DM is not an appropriate biomass proxy when the samples are rich in inorganic matters, and AFDM may be an overestimation when the samples contain a large amount of non-algal organic detritus (Pei 2006). It is worth noting that the biomass estimation with Chl α, DM and AFDM generally provides the total biomass, not the true diatom biomass, of a raw sample, and this inevitably biases the diatom biomass assessment (Stevenson et al. 2010). Cell density and volume require counting the diatom cell number, and are therefore specialized for estimating diatom biomass. Considering the relatively narrow range of diatom cell size (volume) and the time-consuming counting procedure (research cost), most Chinese studies prefer using density as the only proxy to estimate diatom biomass (Table 1) (Guo 2008; Huang 2009; Pei and Liu 2011).

As shown in Table 1, the values of diatom biomass are highly variable, whether within a specific research area (numerical range spanning over one or two orders of magnitude) or across different biogeographic regions (numerical range spanning over several orders of magnitude). As a matter of fact, algal biomass has been long criticized for its high variability over a large spatial scale and low sensitivity to environmental stresses, and consequently its reliability is cautioned when evaluating the water quality in rivers and streams (Leland 1995; Whitton and Kelly 1995; Stevenson et al. 2010). In a case study in Central China (i.e., the Ganhe River Watershed), the benthic diatom biomass, as expressed by Chl α and AFDM, showed no relation with water quality (Yang et al. 2015). On the other hand, the estimated biomass was found to be related to the environmental variations; for instance, nutrients stimulate the diatoms and increase their biomass (Sun 2015), but heavy metals and other toxicants can suppress the growth rates of diatoms and lead lower biomasses (Ding et al. 2012; Peng and Li 2016). Interestingly, You et al. (2013) found that the diatom biomass was positively related to chromium (Cr) and arsenic (As) concentrations, and interpreted this unusual response as a result of increased abundance of the metal-tolerant,

Table 1 Diatom biomass as an indicator in river assessments in China

References	Diatom biomass	Substrate type	Sampling location	Region
Li et al. (2002)	Dens.: 15–86.3 × 10^3 cell/cm^2	Rock	Laoyao River	North China
Guo (2008)	Dens.: 7.03–11.42 × 10^3 cell/cm^2	Rock	Wujiang River	Southwest China
Huang (2009)	Dens.: 0.3–66.79 × 10^3 cell/cm^2	Rock, Macrophyte and Mud	Zhujiang River	Southwest China
He (2011)	Dens.: 0.6–6.95 × 10^7 cell/cm^2	Artificial substrates	Dongjiang River	South China
Pei and Liu (2011)	Dens.: 1.1–15 × 10^5 cell/cm^2	Rock	Niyang River	West China
Chen (2012)	Chl α: 2–66 mg/L/cm^2	Artificial substrates	Zengjiang River	South China
Wu et al. (2012)	Dens.: 6.16–162 × 10^8 cell/cm^2; Chl α: 2.92–39.91 mg/m^2	Rock	Xiangxi River Watershed	Central China
Zhou et al. (2012)	Dens.: 0.10–8.65 × 10^6 cell/cm^2	Rock	Lijiang River	South China
You et al. (2013)	Dens.: 0.47–30.96 × 10^2 cell/cm^2	Rock	Chishui River	Southwest China
Ren et al. (2013)	Dens.: 5.92–7.93 × 10^5 cell/cm^2	Rock	Shanmu River	Southwest China
Sun (2015)	Dens.: 19.53–48.39 × 10^3 cell/cm^2	–	Wuyang River	Southwest China
Yang et al. (2015)	AFDM: 0.75–88.79 g/m^2 Chl α: <50 μg/cm^2	Rock	Ganhe River Watershed	Central China
Tang and Zhi (2016)	Dens.: 7.06–8.41 × 10^3 cell/cm^2	Rock	Xiaoche River	Southwest China

Abbreviations: *AFDM* ash-free dry mass, *Chl* α Chlorophyll α, and *Dens.* diatom cell density dominant species *Gomphonema pseudosphaerophorum* Kobayasi and *Synedra gaillonii* (Bory) Ehrenberg. Alternatively, heavy metals can act as essential micronutrients to promote algal growth when in low concentrations (Rai et al. 1981).

2.3 Diatom Diversity

Biodiversity refers to the variability among living organisms, from genes to species to ecosystems (Singh et al. 2017). Generally, diatom diversity changes with community structure in response to variations in environmental stress. Therefore, diversity offers a useful numerical tool to infer aquatic stressors (Stevenson et al. 2010; Lei 2012). In the early 1970s, Whittaker (1972) introduced 3 terms into the concept of species diversity for different research scales, with α-diversity for intra-community, β-diversity for two ecosystems, and γ-diversity for several ecosystems in a region. Since only α-diversity is involved in diatom assessment in China, here we restrict the definition of diversity index to the α-diversity type. An α-type diversity primarily consists of two components: richness and evenness, and accordingly 3 forms of index have been developed to explore the diversity characteristics of a diatom community: richness index for the number of taxa (e.g., Margalef's Index), evenness index for the abundance distribution of taxa (e.g., Pielou's Index), and composite index both for taxa richness and evenness (e.g., Shannon-Wiener Index).

In recent years, diatom-based diversity indices have been employed in many aquatic surveys in China (Li et al. 2002; Miao 2007; Ren et al. 2013; Yi et al. 2016). Table 2 lists the 10 diversity indices commonly used in the 30 typical Chinese aquatic studies, and one can find that Shannon-Wiener Index (H), Number of Species (R_S), and Pielou's Index (J) are the most used diversity metrics in diatom assessment. As a matter of fact, Shannon-Wiener Index is possibly the most popular biodiversity index in ecology, though species number is an easy-to-acquire indicator that forms the foundation of biodiversity measurement. The calculation formula of the evenness metric Pielou's Index is based on the scores of the former two indices and benefits from their widespread use. Though expressions and implications vary among different categories of diversity indices, a statistically significant correlation (correlation coefficient $r > 0.9$, and significance level $p < 0.01$ or 0.05) between many of them has been observed in several studies (Lei 2012; Yi 2014; Lei 2016). In other words, different indices usually tell similar stories, and therefore redundancy is introduced when using multiple diversity indices, highlighting the need for careful selection of diversity indices. In particular, the estimation of diatom richness (species number) is highly dependent on the sample size, which in diatom assessment refers to the diatom valve numbers acquired through counting under microscopic examination (Stevenson et al. 2010). The amount of effort in counting positively influences the evenness and representativeness of the sampling, and ultimately determines the accuracy for measuring the biodiversity in a sample. The low counting requirement (400–600 valves/individuals in most Chinese surveys) has been criticized for underestimating the diversity of a diatom community (Qiu et al. 2016). A better measurement of species richness can be achieved by developing a *Species Accumulation Curve* to establish the minimum number to reach the threshold counting requirement (Stevenson et al. 2010). Furthermore, *rarefaction curves* for comparing the richness among different samples under the condition of same sampling size (number of counts) are strongly recommended when the samples differ greatly in abundance. The rarefied species number (richness) helps build a better diversity index for ecological investigations (Chen et al. 2013; Lei 2016).

The typical hypothesis made for the diversity metrics to be applied in river quality assessments is that unstressed rivers favor higher biodiversity, while impaired rivers are featured with lower biodiversity (Allan and Castillo 2007; Li et al. 2010). In fact, the diversity indices do not always follow the above assumption, and their performance is somewhat unpredictable in river investigations. For example, Wang et al. (2015) confirmed that diatom diversity decreases along with the degradation gradient in rivers, while Yang et al. (2015) found increased diversity of diatom assemblages in response to higher nutrient levels, or there is no relationship between diatom diversity and water quality in running waters (Yi 2014). Three possible causes may be responsible for the uncertainty in diversity metrics. First, the type and concentration of pollutants in rivers impact diatom diversity in different ways (Patrick 1973; Chen et al. 2016), possibly resulting in multiple response models (linear, unimodal, bimodal and non-fitting) of diatom diversity to a complex pollution gradient. One typical example is the intermediate disturbance hypothesis, which suggests that moderate levels of stress stimulate the diatoms and lead to the highest biodiversity, and such response pattern has been observed in several diatom assessments in China (Yi 2014; Liu 2016). Second, different diatom species may respond differently to a certain stressor. In particular, the diversity metrics do not differentiate the sensitive or tolerant species in the diatom communities. In a given diatom community, an increase either in sensitive species or in tolerant species may result in a similar biodiversity value, but will lead to an opposite implication for water quality assessment (Lei 2012). Fang et al. (2007) argued that diversity indices do not take into account the tolerance of diatom taxa to environmental disturbance, thus offer a better-than-actual evaluation for the water quality. Moreover, biodiversity is related to not only water quality parameters, but also some other physical and habitat characteristics of rivers. The influence of some characteristics on diatom diversity has been discussed intensely, such as water velocity (Wang and Zhang 2004), geographical location (Pei and Liu 2011) and river order (Liu 2013, 2016). Due to these complications, it is strongly recommended to use diversity indices in combination with other more accurate metrics in river assessments (Li et al. 2010).

Table 2 Common diatom based diversity indices used in China, and their formulas, criteria for water quality and application frequency in the 30 typical case studies in Chinese freshwaters

Category	Diversity index	Formula	Range for water classification	Application frequency (%)	Reference number
Composite	Shannon-Wiener (H)	$H = -\sum_{i=1}^{S}\left(\frac{N_i}{N}\right)\log_2\left(\frac{N_i}{N}\right)$ or $H = -\sum_{i=1}^{S}\left(\frac{N_i}{N}\right)\ln\left(\frac{N_i}{N}\right)$	$H < 1$: heavily polluted; $1 \le H < 2$: moderately polluted; $2 \le H < 3$: slightly polluted; $H > 3$: clean	93	(1, 2, 3, 4, 5, 6, 7, 8, 9, 10, 11, 12, 13, 15, 16, 17, 18, 19, 20, 21, 22, 23, 24, 25, 26, 27, 29, 30)
	True Shannon-Wiener (H_t)	$H_t = 2^H$ or $H_t = e^H$	–	7	(14, 28)
	Simpson (D' & D)	Form 1: $D' = \sum_{i=1}^{S}\left(\frac{N_i}{N}\right)^2$ or $D' = \sum_{i=1}^{S}\frac{N_i(N_i-1)}{N(N-1)}$ Form 2: $D = 1 - D'$ Form 3: $D = 1/D'$	$D' > 0.8$: heavily polluted; $0.8 > D' \ge 0.6$: moderately polluted; $0.6 > D' \ge 0.4$: slightly polluted; $D' < 0.4$: clean	33	(3, 10, 12, 17, 18, 19, 26, 27, 30)
	McIntosh (M)	$\mathrm{M} = \frac{N-\sqrt{\sum_{i=1}^{S} N_i^2}}{N-\sqrt{N}}$	–	7	(18, 30)
Richness	The number of species (R_S)	$R_S = S$	–	43	(1, 2, 6, 7, 10, 11, 13, 14, 17, 20, 23, 25, 26)
	The number of genera (R_G)	$R_G = G$	–	10	(10, 17, 20)

(continued)

Table 2 (continued)

Category	Diversity index	Formula	Range for water classification	Application frequency (%)	Reference number
	Margalef (R_{mar})	$R_{mar} = \frac{S-1}{\ln N}$ or $R_{mar} = \frac{S-1}{\log_2 N}$	$R_{mar} < 1$: heavily polluted; $1 \leq R_{mar} < 2$: moderately polluted; $2 \leq R_{mar} < 3$: slightly polluted; $R_{mar} > 3$: clean **OR** $R_{mar} < 3$: heavily polluted; $3 \leq R_{mar} < 4$: moderately polluted; $4 \leq R_{mar} \leq 6$: slightly polluted; $R_{mar} > 6$: clean	23	(3, 10, 12, 17, 18, 23, 30)
	Menhinick (R_{men})	$R_{men} = \frac{\ln S}{\ln N}$ or $R_{men} = \frac{S}{\sqrt{N}}$	–	3	(10)
Evenness	Lloyd-Ghelardi (E)	$E = \frac{S}{S_t}$	$E \leq 0.3$: heavily polluted; $0.3 < E \leq 0.4$: moderately polluted; $0.4 < E \leq 0.5$: slightly polluted; $E > 0.5$: clean	27	(1, 3, 4, 8, 9, 15, 22, 24)
	Pielou (J)	$J = \frac{H}{\ln S}$ or $J = \frac{H}{\log_2 S}$	$J \leq 0.3$: heavily polluted; $0.3 < J \leq 0.5$: moderately polluted; $0.5 < J \leq 0.8$: slightly polluted; $J > 0.8$: clean	53	(2, 5, 6, 7, 11, 13, 17, 18, 19, 21, 22, 23, 25, 26, 29, 30)
Terms that used in the formulas: S = the number of species at a sampling site G = the number of genera at a sampling site S_t = the total number of species at all sampling sites N_i = the total number of specimens/individuals of a particular species N = the total number of specimens/individuals of all species					

(continued)

Table 2 (continued)

Category	Diversity index	Formula	Range for water classification	Application frequency (%)	Reference number
References					
(1) Li et al. (2002)		(11) Yin et al. (2012)	(21) Liu et al. (2015)		
(2) Wang and Zhang (2004)		(12) Zhou et al. (2012)	(22) Ren (2015)		
(3) Miao (2007)		(13) Chen et al. (2013)	(23) Sun et al. (2015)		
(4) He (2011)		(14) Liu (2013)	(24) Wang et al. (2015)		
(5) Pei and Liu (2011)		(15) Ren et al. (2013)	(25) Yang et al. (2015)		
(6) Chen et al. (2012)		(16) Tang et al. (2013b)	(26) Chen et al. (2016)		
(7) Lei (2012)		(17) Liu et al. (2014)	(27) Liu (2016)		
(8) Li (2012)		(18) Yi (2014)	(28) Qiu et al. (2016)		
(9) Li et al. (2012b)		(19) Zhou (2014)	(29) Xiang et al. (2016)		
(10) Wu et al. (2012)		(20) Dong et al. (2015)	(30) Yi et al. (2016)		

2.4 Dominant Species as Indicators

Like many other biological communities, diatom communities are generally defined by their dominant species. The dominant species are usually the most competitive ones under the existing environmental conditions. Once the relationship between one or more particular diatom species and water quality is determined and established, the dominant diatom species can be developed as a simple and accessible indicator of water quality. In fact, the indication analysis with dominant species has a long history dating back to the early 20th century. In 1908, Kolkwitz and Marsson established a hierarchical system, which grouped aquatic organisms into 4 zones to define the degree of saprobity and water quality (i.e., oligosaprobic—clean; β-mesosaprobic—slightly polluted; α-mesosaprobic—moderately polluted; polysaprobic—heavily polluted). Diatoms have been involved at all times in the original saprobic system (Kolkwitz and Marsson 1908) as well as in the later modified versions (Dresscher and van der Mark 1976), and have played a significant role in the biological classification of rivers and streams for over a century (Patrick 1973; Stevenson et al. 2010).

To the best of our knowledge, diatoms as eco-environmental indicators in China were first used in the early 1980s (Zhang et al. 1981; Zhu and Xie 1981, 1982), and the saprobic systems similar to that of Kolkwitz and Marsson (1908) with ranked diatom taxa were proposed in the 1990s (Shen 1990). These non-quantitative, descriptive evaluation methods have prevailed in aquatic research across China for decades (Zhang and Chen 1987; Bao et al. 1989; Qi et al. 1998). Table 3 lists a series of diatom species according to the 4 zones in the saprobic system in Chinese studies. Some indicative species have a cross-zone presence within the system (Table 3). For instance, *Cyclotella comta* (Ehrenberg) Kützing simultaneously occupies the oligosaprobity and β-mesosaprobity Zones; both α-mesosaprobity and β-mesosaprobity Zones contain *Nitzschia fonticola* Grunow and *N. palea* (Kützing) W. Smith; the eurysaprobic species *Melosira varians* Agardh shows a wide range from the oligosaprobity to α-mesosaprobity Zones. Obviously, the definition of each hierarchy and which diatom species belong to it rely heavily on the knowledge and experience of the investigators or the adopted references (Tolkamp and Gardeniers 1988). These subjective aspects seem to have weakened the advantages of the approach and incurred much criticism. Dresscher and van der Mark (1976) bluntly pointed out that the saprobic system lacks a fundamental basis despite its many applications to river assessments. Some studies were concerned about the inexplicit results from the saprobic system, due to its focus only on the optima of a species but ignorance of its tolerance range (Miao 2007). As a simple appraisal technique, using dominant diatoms as indicators in the saprobic system can hardly get rid of its descriptive label and qualitative nature. Consequently, most Chinese researchers have stopped using the method in river assessments since 2000, and have turned to numerical techniques such as autecological indices and multiple metrics as detailed in the following sections.

Table 3 Diatom saprobic system built from some Chinese studies

Pollution spectrum	Water quality	Indicating species (* for notes)	Reference number
Oligosaprobic	Unpolluted	*Achnanthes biasolettiana* Grunow*[1]	(5)
		A. lanceolata (Brébisson) Grunow*[2]	(5)
		Ceratoneis arcus (Ehrenberg) Kützing	(3)
		Cocconeis diminuta Pantocsek	(5)
		Cyclotella comta (Ehrenberg) Kützing	(7)
		C. kuetzingiana Thwaites	(4, 7)
		C. striata (Kützing) Grunow	(4)
		Cymbella naviculiformis Auerswald*[3]	(7)
		Diatoma hyemale (Roth) Heiberg	(6)
		D. vulgaris Bory	(3)
		Fragilaria construens Ehrenberg*[4]	(7)
		F. construens (Ehrenberg) Grunow var. *subsalina* Hustedt	(4)
		Gomphonema acuminatum Ehrenberg	(5)
		G. clavatum Ehrenberg	(5)
		Gyrosigma scalproides (Rabenhorst) Cleve	(7)
		Melosira varians Agardh	(6)
		Navicula heufleri var. *leptocephala* (Brébisson) Patrick	(4)
		Tabellaria fenestrata (Lyngbye) Kützing	(3, 7)
β-mesosaprobic	Slightly polluted	*Asterionella formosa* Hassall	(2)
		Cocconeis placentula Ehrenberg var. *placentula*	(5)
		Cyclotella comta (Ehrenberg) Kützing	(6)
		C. glomerata Bachmann	(6)
		C. meneghiniana Kützing	(3, 4, 6, 7)
		C. ocellata Pantocsek	(6, 7)
		Cymbella amphicephala Naegeli*[5]	(7)
		Fragilaria capucina Desmazieres	(7)

(continued)

Table 3 (continued)

Pollution spectrum	Water quality	Indicating species (* for notes)	Reference number
		F. virescens Ralfs*[6]	(2)
		Gomphonema olivaceum (Hornemann) Brébisson	(6)
		G. parvulum Kützing	(7)
		Melosira granulata (Ehrenberg) Ralfs*[7]	(3, 4, 7)
		M. varians Agardh	(3, 4)
		Navicula rhynchocephala Kützing	(7)
		Synedra acus Kützing*[8]	(6, 7)
α-mesosaprobic	Moderately polluted	*Achnanthes sublaevis* Hustedt	(7)
		Craticula cuspidata (Kützing) Mann	(2)
		Cymbella turgidula Grunow	(5)
		Diatoma hyemale (Roth) Heiberg	(7)
		Fragilaria brevistriata Grunow*[9]	(7)
		F. capucina Desmazieres	(2)
		Gomphonema parvulum Kützing	(4)
		Melosira granulata (Ehrenberg) Ralfs*[7]	(2)
		M. granulata (Ehrenberg) Ralfs var. *angustissima* O. Muller	(2)
		M. islandica subsp. *helvetica* O. Muller	(3)
		M. varians Agardh	(2)
		Navicula accomoda Hustedt*[10]	(7)
		Nitzschia angustata Grunow	(6, 7)
		N. apiculata (Gregory) Grunow	(6, 7)
		N. fonticola Grunow	(6)
		N. palea (Kützing) W. Smith	(5, 6)
		Synedra acus Kützing*[8]	(2)

(continued)

Table 3 (continued)

Pollution spectrum	Water quality	Indicating species (* for notes)	Reference number
Polysaprobic	Heavily polluted	*Gomphonema gracile* Ehrenberg	(7)
		Nitzschia clausii Hantzsch	(4)
		N. fonticola Grunow	(7)
		N. palea (Kützing) W. Smith	(1, 4, 7)
References			
(1) Zhang et al. (1981)		(5) Li et al. (2002)	
(2) Zhang and Chen (1987)		(6) Miao (2007)	
(3) Bao et al. (1989)		(7) Xu et al. (2011)	
(4) Qi et al. (1998)			

*Notes**
1 Now used more as *Achnanthidium biasolettiana* (Grunow) L. Bukhtiyarova
2 Now used more as *Planothidium lanceolatum* (Brébisson ex Kützing) Lange-Bertalot
3 Now used more as *Cymbopleura naviculiformis* (Auerswald) Krammer
4 Now used more as *Staurosira construens* Ehrenberg
5 Now used more as *Cymbopleura amphicephala* Krammer
6 Synonymous with *Fragilariforma virescens* (Ralfs) Williams & Round
7 Synonymous with *Aulacoseira granulata* (Ehrenberg) Simonsen
8 Synonymous with *Ulnaria acus* (Kützing) Aboal & *Fragilaria ulna* (Nitzsch.) Lange-Bertalot var. *acus* (Kützing) Lange-Bertalot
9 Now used more as *Pseudostaurosira brevistriata* (Grunow in Van Heurck) Williams & Round
10 Now used more as *Craticula accomoda* (Hustedt) Mann

2.5 Autecological Indices

A diatom autecological index applies the autecological traits, such as relative abundance of species and ecological preferences (optima and tolerance in response to an environmental gradient), to a numeric, statistical assessment of water quality. These quantitative metrics originated from Europe and USA in the 1970s, and relied on the relatively well-established diatom taxonomic lists of ecological preference in freshwater habitats (Slàdecek 1973; Lowe 1974; Lange-Bertalot 1979). At the early stage, the autecological data collections were handled by different algal ecologists using scattered field observations and laboratory experiments, and this inevitably introduced certain preference and bias into the statistical technology (Potapova and Charles 2007). Later in the 1990s, ecologists started to build the diatom indices on large-scale spatiotemporal datasets that could quantify the relationship between diatom species and specific water quality parameters. For instance, the popular Biological Diatom Index (IBD) was established by a large dataset based on a long-term (1977–1994) national monitoring program that involved 1332 sampling sections in rivers or streams in France (AFNOR 2000; Coste et al. 2009; Lei 2012).

During the past decades, many diatom-based autecological metrics have been developed across the world, especially in Europe. These metrics typically fall into two categories. One is the ecological ranking system, in which each diatom species is assigned to a relative group or ranked according to its response to a specific environmental stressor such as pH, trophic state and organic pollution (Lange-Bertalot 1979; Hofmann 1994; van Dam et al. 1994). The diatom ranking systems (Table 7) from Europe, such as the Lange-Bertalot system (Germany, 1979), van Dam system (Netherlands, 1994) and Hofmann system (Germany, 1994), have been tested and established as reliable autecological metrics in many other regions (Stevenson 2014). The other category of metrics is the biotic index, which is numerical in nature, and uses weighted average modeling with a typical formula (Zelinka and Marvan 1961):

$$Index\ value = \frac{\sum_{j=1}^{n} a_j s_j v_j}{\sum_{j=1}^{n} a_j v_j}$$

where a_j is the relative abundance of the species j, s_j is the sensitive (optimum) value of the species j, *and* v_j is the indicative (tolerance) value of the species j. Since the first diatom-based biotic index (DESCY, Descy Index) was established in 1979, more than a dozen similar numeric indices have been subsequently developed and used as a popular routine monitoring tool in river assessments of many European countries and other regions (Bere and Tundisi 2010; Besse-Lototskaya et al. 2011; Rimet 2012).

As shown in Table 4, all the globally popular biotic indices can be grouped into different categories according to the following different criteria. (1) Source region: Europe is a source land and hot spot for many diatom indices in use, thus there are European indices (e.g., IPS, Pollution Sensitivity Index originated in France) and Non-European indices (e.g., PDI, Pampean Diatom Index based on the data of the Pampean rivers and streams in Argentina); (2) Scientific purpose: there are eutrophication indicative indices (e.g., TDI, Trophic Diatom Index), saprobity indicative indices (e.g., DAIpo, Diatom Assemblage Index of organic pollution), and composite indices (e.g., IBD) that meet all the demands in river bioassessments; (3) Taxonomic resolution: most diatom indices are developed at the species level, and a few at the generic level (e.g., IDG, Generic Diatom Index; GI, Generic Index of Diatom Assemblages) are also used for a high-efficiency solution. The biggest differences among these indices are the diatom taxonomic list adopted in each index, and the autecological values (optimum and tolerance) assigned for each diatom taxon (Lavoie et al. 2009).

As mentioned above, several diatom autecological classification systems have been applied to define the environmental status in running waters, but only the van Dam system is widely used in Chinese studies (Zhao et al. 2009; Li et al. 2012c; Hong et al. 2014). The van Dam classification is a world-famous system with a comprehensive checklist, in which 948 diatom taxa are assigned with ecological indicator ranks (values) to reflect environmental traits including pH, salinity, nitrogen uptake metabolism, oxygen, saprobity, trophic state and moisture (van Dam et al. 1994). When the ranked system is used in an aquatic assessment, the top priority is to quantify the abundance distribution of relative ranks along an ecological spectrum

Table 4 Worldwide popular diatom biotic indices with their detail information

Code	Full name	Included in *Ominidia*	References	Source region
IBD	Biological Diatom Index	Yes	AFNOR (2000)	France
IPS	Pollution Sensitivity Index	Yes	CEMAGREF (1982)	France
IDG	Generic Diatom Index	Yes	Rumeau and Coste (1988)	France
DESCY	Descy Index	Yes	Descy (1979)	Belgium?
SLAD/SLA	Sládeček's Saprobic Index	Yes	Sládeček (1986)	Czech Republic
IDAP	Index Diatom for Artois Picardie	Yes	Prygiel et al. (1996)	France
EPI-D	Diatom-based Eutrophication/Pollution Index	Yes	Dell'Uomo (1996)	Italy
LOBO	Lobo's Trophic-Saprobic Index	Yes	Lobo et al. (2004)	Brazil
DI-CH	Swiss Diatom Index/Hürlimann Trophic Index	Yes	Hürlimann and Niederhauser (2002)	Switzerland
ROTT SI/SID	Rott's Saprobic Index	Yes	Rott et al. (1997)	Austria
ROTT TI/TID	Rott's Trophic Index	Yes	Rott et al. (1998)	Austria
CEE	Commission for Economical Community Metric	Yes	Descy and Coste (1991)	France and Belgium
WAT/DAIpo	Watanabe's Index/Diatom Assemblage Index of organic pollution	Yes	Watanabe et al. (1986)	Japan
TDI	Trophic Diatom Index	Yes	Kelly and Whitton (1995)	UK
PDI	Pampean Diatom Index	Yes	Gómez and Licursi (2001)	Argentina
SHE	Schiefele's Trophic Index	Yes	Steinberg and Schiefele (1988) Schiefele and Kohmann (1993)	Germany
PTI	Kentucky Diatom Pollution Tolerance Index	No	KYDEP (2002)	USA
GI	Generic Index of Diatom Assemblages	No	Wu (1999)	Taiwan

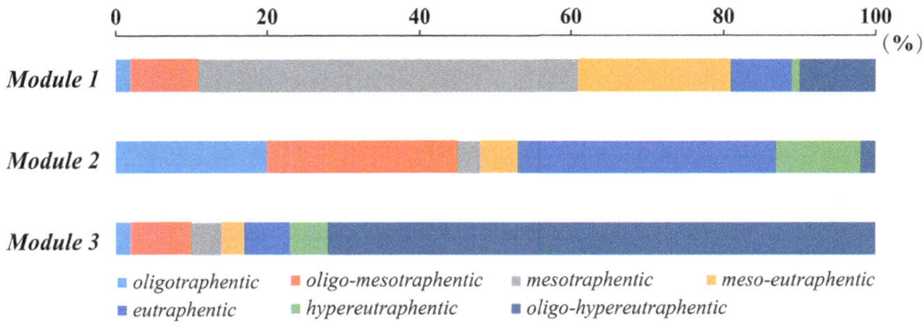

Fig. 2 Modules 1, 2 and 3 for diatom ranks to assess trophic degree in van Dam's system

(e.g., oxygen requirement, saprobic and trophic degree, etc.), and the proportion of each rank is commonly used to determine the water-quality level. Almost all the concerned investigations in China followed the rule of the van Dam metric system to assess water quality (Deng et al. 2012a; Tang et al. 2013b; Yi et al. 2016). To better explain the rule, we simulated 3 typical modules (Fig. 2) and assigned different diatom abundance percentages for the 7 ranks (from *oligotraphentic* to *hypereutraphentic*, and indifferent) along the ordinal scale of trophic conditions. *Module 1* presents a distributional pattern that is mainly characterized by *mesotraphentic* + *meso-eutraphentic* taxa (70%), and consequently the sampling site with an abundance pattern as *Module 1* should be defined by the median trophic status. In most cases, the abundant ranks are clear and help one to make a quick decision among water quality classes. However, when the key ranks are evenly distributed, the ranking metrics may offer a quite ambiguous implication for the literal judgment (Lei 2012; Li et al. 2015a). In *Module 2*, the nutrient sensitive taxa (*oligotraphentic* + *oligo-mesotraphentic*, 45%) and nutrient-tolerant taxa (*eutraphentic* + *hypereutraphentic*, 45%) occupy an identical share in the proportion scale, and there seems to be no explicit interpretation for trophic condition at a *Module 2*-like sampling site. There is actually a simple formula designed to quantify the ranking metrics (Zelinka and Marvan 1961):

$$Index\ value_{rank} = \frac{\sum_{i=1}^{n} p_i t_i}{\sum_{i=1}^{n} p_i}$$

where p_i is the relative abundance of all taxa in rank i, and t_i is the autecological rank value (t = 1, 2, … 7) for a specific stressor. However, Chinese investigators seldom apply this equation to their work when they are using the van Dam system; in addition, the van Dam system may not work under some extreme conditions. In *Module 3*, over 70% of the nutrient-indifferent taxa (*oligo-hypereutraphentic*) occur at the sampling site, which means that most of the constituent diatoms do not respond to nutrients, making the ranking method nearly useless in eutrophication evaluations due to an insufficient number of indicative species.

Qi et al. (1998) applied DAIpo to evaluate the water quality of Pearl River (Zhujiang), southern China, and revealed that this saprobity-indicative index had a negative correlation with BOD_5. This is believed to be the first time for a diatom biotic index to be used in China. From then onwards, biotic indices have been intensely tested across various geographic regions in China (Table 5), especially after 2009 when the characteristic software *OMNIDIA* was firstly adopted (Zhao et al. 2009). The *OMNIDIA* offers two main features: biological data management, and numerical statistical analyses of diatom indices and other autecological traits (Lecointe et al. 1993). The software was originally developed in France, and has subsequently spread to 50 other countries or regions including China, making itself a remarkable digital tool in diatom assessments (OMNIDIA 2018). The latest version of *OMNIDIA* (v 6.0) is capable of calculating 18 diatom indices and 33 ecological statistics from its large database, which, however, raises some scientific problems for Chinese ecologists. Which index is the best for river bioassessments in China? How can one select the best from a large pool of diatom indices? Why is the 'best' the best and the 'worst' the worst?

Much work has been done in China to answer the above questions. Deng et al. (2012c) used a series of statistical methods, including correlation coefficient analysis, factor analysis, cluster analysis, box plot analysis and discriminant analysis, in a stepwise test of seven diatom indices (IBD, TDI, SLA, IPS, IDG, DESCY and CEE) for river evaluation in Dongjiang Basin, South China and selected IBD and IDG as the most suitable indicators for their study region. Subsequently, several other studies used a similar methodology to examine more diatom indices in a search for better proxies to be applied to the monitoring projects of local rivers in China (Yi 2014; Yao et al. 2015; Li et al. 2017). After the statistical screening processes, three indices (IPS, IBD and CEE) were selected from ten candidates (checked in Table 5) for their better performance in water evaluation of the Liujiang River, South China (Li et al. 2015a). Li et al. (2017) tested 15 diatom indices (checked in Table 5) for water quality assessment in the Beijiang Watershed, southern China and confirmed that IBD is the best indicator. However, they also found that the IBD-derived water quality classes did not match well with the classification criteria of the Chinese national standard (GB 3838-2002) based on physicochemical parameters. Eventually, these multiple-index tests have reached consensus on the following issues (Deng et al. 2012c; Lei 2012; Liu et al. 2016; Xiang et al. 2016): (1) Some early established indices (e.g., DESCY 1979 and Sládecek 1986) need update modifications due to the recent advance in taxonomy and autecology of freshwater diatoms; (2) More tests of trophic indices (e.g., TDI) are required to further identify their applicability in China; (3) The well-trained indices from overseas (e.g., IBD and IPS are standard methods in French monitoring networks) needed to be tested to confirm their applicability in China; and (4) The genus-level indices (e.g., IDG) need to be tested to see if they perform as well as the species-level ones, since they are generally recommended for their simplicity by Chinese researchers. In addition, Tan et al. (2013, 2017) summarized 3 particular issues concerning the broader applicability of diatom biotic indices. The first is the similarity between the diatom taxa present in the study area and the species employed for developing each index. As found from their studies, IBD, IPS and WAT

Table 5 Diatom autecology indices used in Chinese river surveys

References	Scientific purpose	Adopted index	Study region	Notes
Qi et al. (1998)	Water quality assessment (especially for Saprobity)	DAIpo	South China	The first use of diatom autecology index in China
Li et al. (2002)	Water quality assessment	DAIpo	North China	
Tang et al. (2006)	Ecological condition evaluation	DAIpo, TDI, PTI	Central China	
Fan et al. (2007)	Water quality assessment	GI, TDI, DAIpo	South China	All indices based on genus level
Zhao et al. (2009)	Water quality assessment	IBD, IPS	South China	The first use of the software *OMNIDIA* in China
Deng et al. (2012a)	Assessing Habitats of *Ruppia maritime*	IBD, IPS	South China	
Deng et al. (2012b)	Water quality assessment	IBD, IPS	South China	The indices were related to physicochemical parameters, geographical and land use factors
Deng et al. (2012c)	Water quality assessment	IBD, TDI, SLA, IPS, IDG, DESCY, CEE	South China	The first study to select among multiple indices in China
Li et al. (2012c)	Water quality assessment	IBD, IPS	East China	
Zhou et al. (2012)	Trophic status assessment	TDI, PTI	South China	
Huang (2013)	Trophic status assessment	TDI	Central/South China	TDI was based on planktonic diatom assemblages
Tan et al. (2013)	Water quality assessment	CEE, DESCY, DI-CH EPI-D, IBD, IDAP, PDI, IPS, SHE, SID, TID, SLAD, TDI, WAT	Central China	
Hong et al. (2014)	Water quality assessment	IBD, IPS	East China	
(Yi 2014)	Water quality assessment	IDAP, EPI-D, IBD, SHE, SID, TID, WAT, IPS, SLA, DESCY, IDG, CEE, LOBO, PDI, DI-CH, TDI	South China	To select the best indices for water quality assessments

(continued)

Table 5 (continued)

References	Scientific purpose	Adopted index	Study region	Notes
Tan et al. (2014)	Water quality assessment	CEE, EPI-D, IBD, IPS, WAT	Central China	
Tang et al. (2014)	Water quality assessment	IPS, IBD, IDG	South China	
Li et al. (2015a)	Water quality assessment	IBD, IPS, IDG, SLA, SID, TID, EPI-D, CEE, SHE, TDI	South China	To select the best indices for water quality assessments
Liu et al. (2015)	River health assessment	IBD	North China	To compare the autecological indices with B-IBI
Zhang et al. (2015)	Trophic status assessment	TDI	Central/South China	TDI was based on planktonic diatom assemblages
Yao et al. (2015)	Water quality assessment	IPS, IBD, IDG, TDI	South China	
Yang et al. (2015)	Water quality assessment	IBD, IPS, TDI	Central China	To assess human impacts on streams across a rural to urban gradient
Tan et al. (2015)	River health assessment	CEE, DESCY, DI-CH, EPI-D, IBD, IDAP, PDI, IPS, SHE, SID, SLA, TDI, TID, WAT	Central China	To be candidate metrics for a D-IBI
Liu (2016)	Water quality assessment	TDI, IBD	North China	
Liu et al. (2016)	River health assessment	IBD, TDI, IPS, IDG, SLA, IDAP, EPI-D, LOBO, DI-CH, CEE, SHE, PDI	North China	To select the best indices for water quality assessments
Xiang et al. (2016)	River health assessment	IBD, IPS, IDG, CEE, TDI, DESCY	Northeast China	
Yi et al. (2016)	Water quality assessment	IBD, TDI	South China	
Li et al. (2017)	Water quality assessment	IBD, IPS, IDG, DESCY, SLA, IDAP, EPI-D, LOBO, DI-CH, TID, CEE, WAT, PDI, SHE, TDI	South China	To select the best indices for water quality assessments
Tian (2017)	Water quality assessment	IBD, IPS, IDG, DESCY, SLA, IDAP, CEE, PDI, TDI	Northeast China	To select the best indices for water quality assessments
Tan et al. (2017)	Water quality assessment	CEE, DESCY, DI-CH EPI-D, IBD, IDAP, PDI, IPS, SHE, SID, TID, SLAD, TDI, WAT	Central China and Southeast Australia	Two sampling streams: one in Australia and the other in China

performed well probably because their taxa list was large enough to cover most taxa that occurred in the study area. The second issue is that diatom indices may not be well adapted to broader geographic regions which possess completely different natural attributes from where they were originally created. The last issue is the main pollution type and stress gradient present in the sampling area, since a much lower ecological gradient (than that for which the diatom indices were originally created) may cause a weak response of diatom indices to water quality parameters in the study area.

2.6 Multimetric Methods

The emerging techniques such as multimetric indices (MMIs) have been increasingly used in river bioassessment worldwide (Stoddard et al. 2008; Herman and Nejadhashemi 2015; O'Brien et al. 2016), as well as in China (Tang et al. 2006; Wu et al. 2012; Tan et al. 2015). A multimetric approach can be regarded as a comprehensive assessment system that combines many of the metrics discussed in previous sections and other metrics (e.g., taxonomic similarity, functional guild and life history strategies). Specifically, an MMI is an integrated proxy measured from the standardized values of several individual metrics (Hawkins 2006). Stevenson et al. (2010) and Tang et al. (2006) highlighted that MMIs extract and condense information from the biological and ecological datasets, and therefore are more precise than a single metric, as evidenced by their sensitive, linear response to environmental changes along an anthropogenic disturbance gradient.

As early as 1977, Frey (1977) proposed a novel concept of biological integrity, which has later been widely accepted as the capability of a biosystem to maintain its structure, biodiversity, and functions comparable to that of the pristine habitat in the same region. Soon after, Karr (1981) proposed the Index of Biotic Integrity (IBI) to quantify biological integrity, and this index is regarded as the first MMI applied in aquatic environments. The IBI is devoted to performing a multiparameter assessment of a river's biotic integrity through comparing the ecological conditions between the disturbed and reference sites within a united geographical region (Karr et al. 1986; Karr 1991). After decades of modification and promotion, the IBI has become the best known and most popular MMI. Now it has been adopted as a standardized method by the US Environmental Protection Agency (USEPA) (Barbour et al. 1999; Paulsen et al. 2008), and as well listed as a popular approach in many countries within the European Union Water Framework Directive (EU WFD) (Ruaro and Gubiani 2013). Strictly speaking, an IBI is not a simple mixture of individual indices (variables). In contrast, the development of an IBI must follow certain technical rules involving distinction and determination of reference and impaired sites, examination and selection of candidate metrics, and standardization and assignment of metric scores. Therefore, literally there is no universal IBI. Instead, most IBIs are region-specific and elaborately built in line with virtual conditions to give accurate and effective results, making them a distinct assessment technique from those univariate indicators.

The IBI was originally developed for fish communities (Karr 1981). Several similar IBI systems were later established with other aquatic organisms such as macroinvertebrates, vascular plants and algae (Karr and Chu 1997; Griffith et al. 2005). After 2000, the strictly diatom-based IBIs emerged for biological integrity assessment in rivers (Fore and Grafe 2002; Wang et al. 2005), and the early practices were based primarily on the whole algal/periphyton assemblage (Hill et al. 2000; KYDEP 2002). It should be noted that not all the MMIs are literally IBIs. The multimetric indices, like the River Diatom Index (RDI) developed by Tang et al. (2006) and the Diatom Bioassessment Index (DBI) proposed by Liu et al. (2014), are in fact far from the typical concepts of IBIs, because they lack a discrimination decision for reference and disturbed sites, and use empirical preference, instead of statistical selection, of the imported metrics.

In China, most of the multimetric diatom methods directly focus on the IBI, and their procedures for constructing a diatom-based IBI (referred to as D-IBI hereafter) mainly follow the classic methods of Barbour et al. (1999), Wang et al. (2005), Stoddard et al. (2008) and Stevenson et al. (2010), which are modified when necessary. An appropriate methodology is crucial in creating a proper D-IBI, and the basic procedures are outlined below and concerned comments are provided to aid practical use. In addition, Table 6 summarizes the diatom-based multiple metrics used in various studies in China.

(1) Setting a goal for a D-IBI. As the name implies, the IBIs are generally produced to assess biological integrity, which covers the overall biological and environmental conditions in rivers (Stevenson et al. 2010). Most Chinese ecologists implement conventional applications of D-IBIs to river pollution assessments (Table 6), but some pioneers go further to explore the potentials of D-IBIs for more specific purposes. For instance, Dong et al. (2015) built the AMD-DIBI to assess and quantify the ecological damage of acid mine drainage (AMD) in a small river basin (Gaolan River, Central China), and Wu et al. (2012) established a D-IBI to detect the impacts of altered flow regimes caused by run-of-the-river dams in the Xiangxi Rivers (Central China). Both studies highlight that certain D-IBI can be designed for some specific goals. The study goal determines every step of metric processing during the development of a D-IBI. Taking Tang et al. (2006) and Wu et al. (2012) as examples, different diatom metrics were selected for different research purposes (ecological condition vs. flow regulation), though their study locations are in the same region (Xiangxi River catchment).

(2) Discrimination of reference and impaired sites. It is a challenge in most IBI studies to define a reference site and then measure the reference conditions of this site (Ruaro and Gubiani 2013). In China, several methods have been used in the establishment of reference conditions for D-IBIs , among which the commonly used ones include biodiversity metrics (Li 2012; Li et al. 2012b), physical and chemical water variables (Yin et al. 2012; Xiang et al. 2016), historical records of landuse type, urbanization degree, habitat characteristics

Table 6 Diatom-based multimetric indices (MMIs) in Chinese studies

References	Impairment type	Selected metrics	Study region	Notes
Tang et al. (2006)	Water pollution	pH-acidobiontic; Salinity-fresh; High oxygen requirement; Trophic-eutraphentic (v); Mobile groups percentage	Central China	5 selected metrics were used to build the River Diatom Index (RDI) for evaluating ecological conditions of the river
Li (2012)	Water pollution	Sensitive species percentage PDI, Pampean Diatom Index Shannon-Wiener Index	Northeast China	The indices of biological integrity established by diatom and fish, were simultaneously applied in water quality evaluation
Yin et al. (2012)	Water pollution	Mobile groups percentage *Nitzschia* percentage Sensitive species percentage Cell density Chl α	Northeast China	The IBI actually was established by periphyton assemblages, and only the diatom-related metrics were listed
Wu et al. (2012)	Flow regulation	*Gomphonema parvulum* individuals Numbers of species (Richness) Diatom cell density Chl α	Central China	A diatom-based index of biotic integrity (D-IBI) was developed to quantify the impacts of run-of-river dams in lotic environments
Li et al. (2012b)	Water pollution	Sensitive species percentage PDI, Pampean Diatom Index Shannon-Wiener Index	Northeast China	
Liu et al. (2014)	Water pollution	Sensitive species percentage Numbers of species (Richness) Shannon-Wiener Index PTI, Pollution Tolerance Index Similarity to reference sites	Northeast China	Diatom Bio-assessment Index (DBI) composed of 5 metrics, was made both at species and genus levels, to evaluate the river conditions

(continued)

Table 6 (continued)

References	Impairment type	Selected metrics	Study region	Notes
Liu et al. (2015)	Water pollution	Pielou index Numbers of genera *Nitzschia* percentage Mobile groups percentage *Achnanthidium minutissimum* individuals Diatom cell density	North China	To compare D-IBI with IBD (Biological Diatom Index) in river health assessments
Dong et al. (2015)	Acid mine drainage	PTI, Pollution Tolerance Index *Cocconeis* species numbers *Pinnularia* species numbers *Cymbella* species numbers Prostrate form *Achnanthes* percentage	Central China	To develop a diatom-based IBI in an assessment project for river ecosystems that were impacted by AMD (Acid mine drainage)
Tan et al. (2015)	Water pollution	Prostrate form *Amphora* percentage EPI-D, Diatom-based Eutrophication/Pollution index Saprobic-polysaprob (h)	Central China	

(Liu et al. 2015), and multivariate statistical analyses on diatom assemblages (Tan et al. 2015). Tan et al. (2015) introduced nonmetric multidimensional scaling (NMDS), a statistical ordination technique based on the similarity of species composition, to discriminate the reference and impacted sites during the D-IBI development, and the results revealed that a combination of multivariate and multimetric approaches benefits the development of a reliable D-IBI. On the other hand, it takes little effort to establish the reference conditions for some particular D-IBIs dependent on specific goal. The differentiating criteria of reference and impaired sites in AMD-DIBI is whether there exists an AMD input in the sampling sites (Dong et al. 2015). In Wu's D-IBI designed for indicating dam effects, upstream sites (50–100 m above dams) are regarded as free of dam impacts, defining the reference conditions; and the downstream sites located below dams or hydropower stations are legitimately considered as impacted (Wu et al. 2012).

(3) Determination and screening of candidate metrics. The selection of suitable metrics is the key step for the establishment of a reliable IBI (Karr and Chu 2000). Over the last two decades, a lot of investigations have helped collect and confirm those most-widely-used metrics as powerful candidates to develop the periphyton/diatom-based IBIs (Barbour et al. 1999; Hill et al. 2000; Wang

et al. 2005). Especially, Wang et al. (2005) initiatively compiled a large pool of biological and ecological attributes (potential metrics) tailored particularly for diatoms, which sets up a paradigm for many Chinese researchers to follow (Wu et al. 2012; Dong et al. 2015; Tan et al. 2015). Table 7 sums up 111 candidate metrics in D-IBIs from Chinese surveys, which can be grouped into 8 categories including biotic indices (e.g., IBD, TDI, PTI, etc.), ecological values (e.g., pH-acidobiontic and pH-alkaliphilic species in the van Dam system), diversity metrics (e.g., Shannon-Wiener, Simpson, Pielou Index, etc.), growth forms (e.g., prostrate and erect forms, etc.), similarity measures (e.g., species similarity to reference sites), biomass estimators (e.g., diatom cell density, AFDM, etc.), sensitive species (e.g., sensitive species percentage), and taxonomic composition (e.g., percentage of *Achnanthidium minutissimum*, ratio of *Cymbella* abundance to *Cymbella* and *Navicula* abundance, etc.). After the list of candidate attributes is determined, a series of statistical tests are performed to select the ultimate metrics to be used in the final D-IBI, and this screening procedure consists of: (i) a numerical range test to eliminate metrics with small ranges or a zero median for both reference and impaired sites; (ii) adjustment for natural conditions, as demonstrated in the random forest model adopted by Zhou (2014) to have metrics calibrated for natural gradients; (iii) discriminant judgment to be used to test the ability of a metric to distinguish the disturbed and undisturbed sites, including such statistical methods as non-parametric tests like the Kolmogorov-Smirnov test (Wu et al. 2012), Mann-Whitney *U* test (Tan et al. 2015) and Kruskal–Wallis test (Dong et al. 2015), and separation power defined by box plots and coefficient of variation (CV); (iv) redundancy check, which uses correlation analysis to exclude the highly correlated metrics. Generally, 3–6 metrics (as checked in Table 6) can eventually pass all these tests and will be used to build a practical D-IBI.

(4) Standardization and scoring in D-IBIs. Prior to scoring the selected metrics for a final D-IBI, it is necessary to convert them into a common scoring scale, typically for those differing greatly in units and absolute values (Karr and Chu 1997; Barbour et al. 1999). Three scoring methods can be found in Chinese studies, known as "Trisection" (Li 2012; Li et al. 2012b), "Quadrisection" (Yin et al. 2012; Liu et al. 2015) and "Percentage of Standard" (Zhou 2014; Tan et al. 2015). The former two are defined as scoring metrics with discrete, ranking values (e.g., 1, 3, 5), and the third one refers to scoring metrics along a continuous scale on the basis of a standardized value (commonly the 95th percentile value). After scoring the metrics, a final D-IBI for a site is generated by calculation of the sum of the scores from all the selected metrics.

Many Chinese studies have confirmed that the D-IBI is a suitable method in river assessments. However, the D-IBIs have their own drawbacks: (i) higher demand for taxonomic identification of diatoms compared to other aquatic organisms like fish and macroinvertebrates (Yin et al. 2012); (ii) technical limitations from the complex numerical calculations with a set of processes for discriminating, screening and scoring the candidate metrics (Xiang et al. 2016); (iii) possible loss of the biological

Table 7 Candidate diatom metrics for the D-IBIs in Chinese studies (Tang et al. 2006; Li 2012; Li et al. 2012b; Wu et al. 2012; Yin et al. 2012; Zhou 2014; Dong et al. 2015; Liu et al. 2015; Tan et al. 2015; Xiang et al. 2016)

Candidate metrics	Taxonomic level	Attributes	Predicted response	References
Biotic index				
IBD, Biological Diatom Index	Species	Weighted average method with tolerance and/or optima values for diatom species	↘	See Table 5
IPS, Pollution Sensitivity Index	Species	Same as above	↘	See Table 5
DESCY, Descy Index	Species	Same as above	↘	See Table 5
SLA, Sládeček's Saprobic Index	Species	Same as above	↘	See Table 5
IDAP, Index Diatom for Artois Picardie	Species	Same as above	↘	See Table 5
EPI-D, Diatom-based Eutrophication/Pollution Index	Species	Same as above	↘	See Table 5
DI-CH, Swiss Diatom Index	Species	Same as above	↘	See Table 5
SID, Rott's Saprobic Index	Species	Same as above	↘	See Table 5
TID, Rott's Trophic Index	Species	Same as above	↘	See Table 5
CEE, Commission for Economical Community Metric	Species	Method of a two-entries table	↘	See Table 5
WAT, Watanabe's Index	Species	Weighted average method with tolerance and/or optima values for diatom species	↘	See Table 5
TDI, Trophic Diatom Index	Species	Same as above	↘	See Table 5
PDI, Pampean Diatom Index	Species	Same as above	↘	See Table 5
SHE, Schiefele's Trophic Index	Species	Same as above	↘	See Table 5
PTI, Kentucky Diatom Pollution Tolerance Index	Species	Same as above	↘	See Table 5
GI, Generic Index of Diatom Assemblages	Genus	Ratio of *Achnanthes*, *Cocconeis* and *Cymbella*, to *Cyclotella*, *Melosira* and *Nitzschia*	↘	See Table 5

(continued)

Table 7 (continued)

Candidate metrics	Taxonomic level	Attributes	Predicted response	References
Ecological value (Ranking system)				
(1) van Dam's system				
pH-acidobiontic	Species	Relative abundance of acidobiontic species	↔	(van Dam et al. 1994)
pH-acidophilous	Species	Relative abundance of acidophilous species	↔	Same as above
pH-neutrophilous	Species	Relative abundance of neutrophilous species	↔	Same as above
pH-alcaliphilous	Species	Relative abundance of alcaliphilous species	↔	Same as above
pH-alcalibiontic	Species	Relative abundance of alcalibiontic species	↔	Same as above
Salinity-fresh	Species	Relative abundance of fresh species	↔	Same as above
Salinity-fresh brackish	Species	Relative abundance of fresh brackish species	↔	Same as above
Salinity-brackish	Species	Relative abundance of brackish species	↔	Same as above
N-autotrophic sensible	Species	Relative abundance of autotrophic sensitive species	↘	Same as above
N-autotrophic tolerant	Species	Relative abundance of autotrophic tolerant species	↔	Same as above
Facultatively N-heterotrophic	Species	Relative abundance of facultatively heterotrophic species	↔	Same as above
Obligately N-heterotrophic	Species	Relative abundance of obligately heterotrophic species	↔	Same as above
Continuously high oxygen	Species	Relative abundance of species preferring to live in 100% saturated oxygen conditions	↘	Same as above
Fairly high oxygen	Species	Relative abundance of species preferring to live fairly high oxygen conditions (>75% saturation)	↘	Same as above

(continued)

Table 7 (continued)

Candidate metrics	Taxonomic level	Attributes	Predicted response	References
Moderate oxygen	Species	Relative abundance of species preferring to live in moderate oxygen conditions (>50% saturation)	↔	Same as above
Low oxygen	Species	Relative abundance of species preferring to live in low oxygen conditions (>30% saturation)	↗	Same as above
Very low oxygen	Species	Relative abundance of species preferring to live in very low conditions (10% saturation)	↗	Same as above
Saprobic-oligosaprobous (v)	Species	Relative abundance of oligosaprobous species	↘	Same as above
Saprobic-mesosaprobous (v)	Species	Relative abundance of mesosaprobous species	↔	Same as above
Saprobic-α-mesosaprobous (v)	Species	Relative abundance of α-mesosaprobous species	↗	Same as above
Saprobic-α-meso-polysabrobous (v)	Species	Relative abundance of α-meso-polysabrobous species	↗	Same as above
Saprobic-polysaprobous (v)	Species	Relative abundance of polysaprobous species	↗	Same as above
Trophic-oligotraphentic (v)	Species	Relative abundance of oligotraphentic species	↘	Same as above
Trophic-oligomesotraphentic (v)	Species	Relative abundance of oligo-mesotraphentic species	↘	Same as above
Trophic-mesotraphentic (v)	Species	Relative abundance of mesotraphentic species	↘	Same as above
Trophic-meso-eutraphentic (v)	Species	Relative abundance of meso-eutraphentic species	↗	Same as above
Trophic-eutraphentic (v)	Species	Relative abundance of eutraphentic species	↗	Same as above

(continued)

Table 7 (continued)

Candidate metrics	Taxonomic level	Attributes	Predicted response	References
Trophic-hypereutraphentic (v)	Species	Relative abundance of hypereutraphentic species	↗	Same as above
Trophic-oligo- to eutraphentic (v)	Species	Relative abundance of oligo- to eutraphentic species	↔	Same as above
(2) Hofmann's system				
Trophic-Oligotraphent (h)	Species	Relative abundance of oligo- to eutraphentic species	↘	(Hofmann 1994)
Trophic-oligo-mesotraphent (h)	Species	Relative abundance of oligo-mesotraphent species	↘	Same as above
Trophic-oligo-α-mesotraphent (h)	Species	Relative abundance of oligo-α-mesotraphent species	↘	Same as above
Trophic-α-meso-eutraphent (h)	Species	Relative abundance of α meso-eutraphent species	↘	Same as above
Trophic-eutraphent (h)	Species	Relative abundance of eutraphent species	↗	Same as above
Trophic-tolerant (h)	Species	Relative abundance of tolerant species	↔	Same as above
Trophic-indifferent (h)	Species	Relative abundance of indifferent species	↔	Same as above
Trophic-saprotroph (h)	Species	Relative abundance of saprotroph species	↔	Same as above
Saprobic-oligosaprob (h)	Species	Relative abundance of oligosaprob species	↘	Same as above
Saprobic-mesosaprob (h)	Species	Relative abundance of mesosaprob species	↘	Same as above
Saprobic-meso-α-mesosaprob (h)	Species	Relative abundance of meso-α mesosaprob species	↘	Same as above
Saprobic-α-mesosaprob (h)	Species	Relative abundance of α mesosaprob species	↗	Same as above
Saprobic-α-meso polysaprob (h)	Species	Relative abundance of α–meso polysaprob species	↗	Same as above
Saprobic-polysaprob (h)	Species	Relative abundance of polysaprob species	↗	Same as above

(continued)

Table 7 (continued)

Candidate metrics	Taxonomic level	Attributes	Predicted response	References
(3) Watanabe's system				
Saprobic-Indifferent (w)	Species	Relative abundance of Indifferent species	↔	(Watanabe et al. 1986)
Saprobic-saprophile (w)	Species	Relative abundance of saprophile species	↗	Same as above
Saprobic-saproxene (w)	Species	Relative abundance of saproxene species	↗	Same as above
Diversity metric				
Shannon-Wiener Index	Species	See Table 2	↘	See Table 2
Simpson Index	Species	See Table 2	↔	See Table 2
Margalef Index	Species	See Table 2	↔	See Table 2
Menhinick Index	Species	See Table 2	↘	See Table 2
Pielou Index	Species	See Table 2	↘	See Table 2
Lloyd-Ghelardi Index	Species	See Table 2	↘	See Table 2
Numbers of species	Species	See Table 2	↔	See Table 2
Numbers of genera	Genus	See Table 2	↘	See Table 2
Growth form				
Prostrate form	Genus	Relative abundance of prostrate genera	↗	(Wang et al. 2005)
Erect form	Genus	Relative abundance of erect genera	↘	Same as above
Stalked form	Genus	Relative abundance of stalked genera	↔	Same as above
Unattached form	Genus	Relative abundance of unattached genera	↗	Same as above
Motile form	Genus	Relative abundance of motile genera	↗	Same as above
Similarity measure				
Similarity to reference sites	Species	Mean Bray–Curtis similarity in species composition to reference sites	↘	Same as above
Reference species percentage	Species	Percentage of species found in reference sites that occurred in impaired sites	↘	Same as above
Numbers of distinct reference species	Species	Numbers of species found primarily in reference sites not in impaired sites	↘	Same as above

(continued)

Table 7 (continued)

Candidate metrics	Taxonomic level	Attributes	Predicted response	References
Biomass estimator				
Diatom cell density	–	Total density of diatoms	↔	–
Chl α	–	Concentration of Chlorophyll α	↔	–
AFDM		Ash-free dry mass of benthic diatoms	↔	–
Sensitive species				
Sensitive species percentage	Species	Relative abundance of sensitive species (PTI value > 3)	↘	(Barbour et al. 1999)
Taxonomic composition				
Diatom class ratio	Class	Ratio of Centricae to Pennatae	↗	–
Mobile groups percentage	Genus	Sum of the relative abundances of *Navicula*, *Nitzschia* and *Surirella*	↗	(Tang et al. 2006)
Dominant species percentage	Species	Relative abundance of dominant species	↗	(Wang et al. 2005)
Achnanthidium minutissimum individuals	Species	Relative abundance of *Achnanthidium minutissimum*	↘	Same as above
Achnanthidium pyrenaicum individuals	Species	Relative abundance of *Achnanthidium pyrenaicum*	↘	Same as above
Achnanthidium subatomus individuals	Species	Relative abundance of *Achnanthidium subatomus*	↘	Same as above
Achnanthidium/(*Achnanthidium* + *Navicula*)	Genus	Ratio of *Achnanthidium* to *Achnanthidium* and *Navicula* combined	↘	Same as above
Cymbella/(*Cymbella* + *Navicula*)	Genus	Ratio of *Cymbella* to *Cymbella* and *Navicula* combined	↘	Same as above
Achnanthidium	Genus	Relative abundance of *Achnanthidium*	↘	Same as above
Amphora	Genus	Relative abundance of *Amphora*	↔	Same as above
Cocconeis	Genus	Relative abundance of *Cocconeis*	↘	Same as above

(continued)

Table 7 (continued)

Candidate metrics	Taxonomic level	Attributes	Predicted response	References
Cyclotella	Genus	Relative abundance of *Cyclotella*	↗	Same as above
Cymbella	Genus	Relative abundance of *Cymbella*	↘	Same as above
Fragilaria	Genus	Relative abundance of *Fragilaria*	↘	Same as above
Gomphonema	Genus	Relative abundance of *Gomphonema*	↘	Same as above
Navicula	Genus	Relative abundance of *Navicula*	↔	Same as above
Nitzschia	Genus	Relative abundance of *Nitzschia*	↔	Same as above
Surirella	Genus	Relative abundance of *Surirella*	↔	Same as above
Achnanthidium species numbers	Species	Percentage of *Achnanthidium* species numbers	↘	Same as above
Amphora species numbers	Species	Percentage of *Amphora* species numbers	↔	Same as above
Cocconeis species numbers	Species	Percentage of *Cocconeis* species numbers	↘	Same as above
Cyclotella species numbers	Species	Percentage of *Cyclotella* species numbers	↘	Same as above
Cymbella species numbers	Species	Percentage of *Cymbella* species numbers	↘	Same as above
Fragilaria species numbers	Species	Percentage of *Fragilaria* species numbers	↘	Same as above
Gomphonema species numbers	Species	Percentage of *Gomphonema* species numbers	↘	Same as above
Navicula species numbers	Species	Percentage of *Navicula* species numbers	↔	Same as above
Nitzschia species numbers	Species	Percentage of *Nitzschia* species numbers	↔	Same as above
Surirella species numbers	Species	Percentage of *Surirella* species numbers	↔	Same as above

The symbols for the predicted response to disturbance: ↗, to increase with impairment; ↘, to decrease with impairment; ↔, to respond variably and ecological information during the process reducing data capacity and combining many metrics into one (Tan et al. 2015); (iv) low comparability of D-IBIs among different regions because every D-IBI is developed on relatively localized conditions (Wu et al. 2012). Despite not being perfect, the diatom based MMI/IBI has served as a powerful tool to evaluate the biological integrity of lotic ecosystems in China for over a decade since the first attempt of Tang et al. (2006), yet are in strong need of more technical promotion and public attention.

3 Conclusions

In China, diatoms have been considered as bio-indicators for only a few decades, but their use in the assessment of environmental status and ecological integrity of rivers has been widely confirmed, and the related diatom techniques have gained a rapid development across the country. Benthic diatoms are usually the first choice in river assessments, but the potentials of planktonic diatoms are receiving more and more attention. The traditional diatom approaches involve biomass, diversity and dominant species, and their practical usability has helped maintain popularity for decades in China. The newly burgeoning methods of biotic indices and multimetric indices have undergone rigorous testing across the country and are widely accepted as more effective means in diatom assessments. A promising future is anticipated for diatom assessing techniques in China, and possibly across the globe.

Acknowledgements This study was supported by the Natural Science Foundation of China grants (No. 41672004 and 41672346).

References

AFNOR (2000) Détermination de l'Indice Biologique Diatomées (IBD) vol Norme française NFT 90-354. France

Allan JD, Castillo MM (2007) Stream ecology: structure and function of running waters. Springer, Netherlands. https://doi.org/10.1007/978-1-4020-5583-6

Aloi JE (1990) A critical review of recent freshwater periphyton field methods. Can J Fish Aquat Sci 47(3):656–670. https://doi.org/10.1139/f90-073

Bao W, Wang Q, Shi X (1989) Studies on the phytoplankton in the Gaoleng-Yilan Section of the Songhua River, an evaluation on the status of the quality of the polluted water in that section. Nat Sci J Harbin Normal Univ 5(1):75–93 (in Chinese)

Barbour MT, Gerritsen J, Snyder BD, Stribling JB (1999) Rapid bioassessment protocols for use in streams and wadeable rivers: periphyton, benthic macroinvertebrates, and fish, 2nd edn. Environmental Protection Agency, USA, Washington, DC

Bere T, Tundisi JG (2010) Biological monitoring of lotic ecosystems: the role of diatoms. Braz J Biol 70(3):493–502. https://doi.org/10.1590/S1519-69842010005000009

Besse-Lototskaya A, Verdonschot PFM, Coste M, Van de Vijver B (2011) Evaluation of European diatom trophic indices. Ecol Ind 11(2):456–467. https://doi.org/10.1016/j.ecolind.2010.06.017

CEMAGREF (1982) Etude des méthodes biologiques quantitative d'appréciation de la qualité des eaux

Chen X (2012) Seasonal variation of periphytic diatom communities in two small streams, Guangdong Province, southern China. Jinan University, Guangzhou, China (in Chinese)

Chen X, Liu J, He Q, Lin Q, Han B, Wei G (2012) Seasonal variations of periphytic diatom community in artificial substrata in Huizhou section of the Dongjiang River, Guangdong Province, Southern China. J Lake Sci 24(5):723–731 (in Chinese)

Chen Q, Xia W, Tang H, Jiang S (2013) Benthic diatom assemblages and their distribution pattern during a low water season in the main catchments of Hainan Island. Ecol Sci 32(1):1–7 (in Chinese)

Chen X, Zhou W, Pickett STA, Li W, Han L, Ren Y (2016) Diatoms are better indicators of urban stream conditions: a case study in Beijing, China. Ecol Ind 60:265–274. https://doi.org/10.1016/j.ecolind.2015.06.039

Coste M, Boutry S, Tison-Rosebery J, Delmas F (2009) Improvements of the Biological Diatom Index (BDI): description and efficiency of the new version (BDI-2006). Ecol Ind 9(4):621–650. https://doi.org/10.1016/j.ecolind.2008.06.003

Dell'Uomo A (1996) Assessment of water quality of an Apennine river as a pilot study for diatom-based monitoring of Italian water courses. In: Whitton BA, Rott E (eds) Use of algae for monitoring rivers II. Institut für Botanik Universität Innsbruck, Innsbruck, pp 64–72

Deng P, Lai G, Y-d Lei, Wang X, Liu W (2012a) Water quality integrated assessments of Ruppia maritime using epiphytic diatom. Ecol Sci 31(4):406–412 (in Chinese)

Deng P, Lei Y, Liu W, Huang S, Wang X, Zhao Y (2012b) Epilithic diatom assemblages distribution in Gui River basin, in relation to chemical and physiographical factors. Acta Ecologica Sinica 32(7):2196–2203 (in Chinese). https://doi.org/10.5846/stxb201107251096

Deng P, Lei Y, Liu W, Wang X (2012c) Exploration of benthic diatom indices to evaluate water quality in rivers in the Dongjiang basin. Acta Ecol Sin 32(16):5014–5024 (in Chinese)

Descy JP (1979) A new approach to water quality estimation using diatoms. Nova Hedwig Beih 64:305–323

Descy JP, Coste M (1991) A test of methods for assessing water quality based on diatoms. Verh Internat Verein Limnol 24:2112–2116

Ding L, Zhi C (2006) Environmental effects on diatom and its monitor of environment. J Guizhou Normal Univ (Nat Sci) 3:13–16 (in Chinese)

Ding T, Ni W, Zhang J (2012) Research advances in heavy metals pollution ecology of diatom. Chin J Appl Ecol 23(3):857–866 (in Chinese)

Dong X, Jia X, Jiang W, Wu N, Tan T, Cai Q (2015) Development and testing of a diatom-based index of biotic integrity for river ecosystems impacted by acid mine drainage in Gaolan River, China. Fresenius Environ Bull 24(11C):4114–4124

Dresscher TGN, van der Mark H (1976) A simplified method for the biological assessment of the quality of fresh and slightly brackish water. Hydrobiologia 48(3):199–201. https://doi.org/10.1007/bf00028691

Fan KW, Fok L, Lee JHW, Chen F (2007) Epilithic diatom as a biological indicator for the assessment of water quality in the East River. Sustain Safe Water Suppl 7(2):147–155. https://doi.org/10.2166/ws.2007.049

Fang Y, Tian C, Xiao Y (2007) Phytoplankton diversity and water quality evaluation in Suzi River. Hydraul Fish 27(1):57–58, 93 (in Chinese)

Fore LS, Grafe C (2002) Using diatoms to assess the biological condition of large rivers in Idaho (USA). Freshw Biol 47(10):2015–2037. https://doi.org/10.1046/j.1365-2427.2002.00948.x

Frey DG (1977) Biological integrity of water: an historical approach. In: Ballentine RK, Guarraia LJ (eds) The integrity of water. In: Proceedings of a symposium, 10–12 Mar 1975. US Environmental Protection Agency, Washington, DC, pp 127–140

Gómez N, Licursi M (2001) The Pampean Diatom Index (IDP) for assessment of rivers and streams in Argentina. Aquat Ecol 35(2):173–181. https://doi.org/10.1023/a:1011415209445

Griffith MB, Hill BH, McCormick FH, Kaufmann PR, Herlihy AT, Selle AR (2005) Comparative application of indices of biotic integrity based on periphyton, macroinvertebrates, and fish to southern Rocky Mountain streams. Ecol Ind 5(2):117–136. https://doi.org/10.1016/j.ecolind.2004.11.001
Guo Y (2008) Research quantitatively on relationship between benthic diatoms and water environmental factors in middle and upper of Wujiang River. Guizhou Normal University, Guiyang, China (in Chinese)
Han D, Currell MJ, Cao G (2016) Deep challenges for China's war on water pollution. Environ Pollut 218(Supplement C):1222–1233. https://doi.org/10.1016/j.envpol.2016.08.078
Hawkins CP (2006) Quantifying biological integrity by taxonomic completeness: its utility in regional and global assessments. Ecol Appl 16(4):1277–1294. https://doi.org/10.1890/1051-0761(2006)016%5b1277:QBIBTC%5d2.0.CO;2
He Q (2011) Species diversity and distribution of benthic diatom in Zeng River and in Huizhou Section of Dongjiang River. Jinan University, Guangzhou, China (in Chinese)
Herman MR, Nejadhashemi AP (2015) A review of macroinvertebrate- and fish-based stream health indices. Ecohydrol Hydrobiol 15(2):53–67. https://doi.org/10.1016/j.ecohyd.2015.04.001
Hill BH, Herlihy AT, Kaufmann PR, Stevenson RJ, McCormick FH, Johnson CB (2000) Use of periphyton assemblage data as an index of biotic integrity. J North Am Benthol Soc 19(1):50–67. https://doi.org/10.2307/1468281
Hofmann G (1994) Aufwuchs-Diatomeen in Seen und ihre Eignung als Indikatoren der Trophie. Bibliotheca Diatomologica 30:1–241
Hong J, Wang Z, Wang L, Yuan G (2014) Diatom-based biomonitoring method of water quality in the Jinhua River Basin. Resour Environ Yangtze Basin 23(9):1283–1288 (in Chinese)
Huang Q (2009) Study on distribution characteristics of benthic diatom communities and correlation between benthic diatom communities and environmental variables in rivers of Qiangui Zhujiang Watershed. Guizhou Normal University, Guiyang, China (in Chinese)
Huang D (2013) The study on river quality assessment of the mighty river in south China through the method of planktonic algae biological index. Nanchang University, Nanchang, China (in Chinese)
Hürlimann J, Niederhauser P (2002) Méthode d'étude et d'appréciation de l'état de santé des cours d'eau: Diatomées-niveau R (région). Office Fédéral de l'Environnment, des Forêts et des Paysages, Berne
Karr JR (1981) Assessment of biotic integrity using fish communities. Fisheries 6(6):21 27. https://doi.org/10.1577/1548-8446(1981)006%3c0021:AOBIUF%3e2.0.CO;2
Karr JR (1991) Biological integrity: a long-neglected aspect of water resource management. Ecol Appl 1(1):66–84. doi:citeulike-article-id:2789870
Karr JR, Chu EW (1997) Biological monitoring and assessment: using multimetric indexes effectively. EPA 235-R97-001. University of Washington, Seattle
Karr JR, Chu EW (2000) Introduction: sustaining living rivers. In: Jungwirth M, Muhar S, Schmutz S (eds) Assessing the ecological integrity of running waters: proceedings of the international conference, held in Vienna, Austria, 9–11 Nov 1998. Springer Netherlands, Dordrecht, pp 1–14. https://doi.org/10.1007/978-94-011-4164-2_1
Karr JR, Fausch KD, Angermeier PL, Yant PR, Schlosser IJ (1986) Assessing biological integrity in running waters. A method and its rationale Illinois Natural History Survey, Champaign, Special Publication 5
Kelly MG, Whitton BA (1995) The Trophic Diatom Index: a new index for monitoring eutrophication in rivers. J Appl Phycol 7(4):433–444. https://doi.org/10.1007/bf00003802
Kolkwitz R, Marsson M (1908) Ökologie der pflanzlichen Saprobien. Berichte der Deutschen Botanischen Gesellschaft 26(7):505–519. https://doi.org/10.1111/j.1438-8677.1908.tb06722.x
Kristiansen J (1996) Dispersal of freshwater algae—a review. Hydrobiologia 336(1–3):151–157. https://doi.org/10.1007/BF00010829
KYDEP (2002) Methods for assessing biological integrity of surface waters. Kentucky Department for Environmental Protection, Frankfort, Kentucky

Lange-Bertalot H (1979) Pollution tolerance of diatoms as a criterion for water quality estimation. Nova Hedwig Beih 64:285–304

Lavoie I, Hamilton PB, Wang Y-K, Dillon PJ, Campeau S (2009) A comparison of stream bioassessment in Quebec (Canada) using six European and North American diatom-based indices. Nova Hedwig Beih 135:37–56

Lecointe C, Coste M, Prygiel J (1993) "Omnidia": software for taxonomy, calculation of diatom indices and inventories management. Hydrobiologia 269(1):509–513. https://doi.org/10.1007/bf00028048

Lei Y (2012) Use of epilithic diatom to evaluate the water quality of Dongjiang River. South China Normal University, Guangzhou, China (in Chinese)

Lei Y (2016) Benthic diatom communities and their distributional pattern in the streams of southern China (Guangdong, Hainan, Fujian, and Taiwan Provinces). Jinan University, Guangzhou, China (in Chinese)

Leland HV (1995) Distribution of phytobenthos in the Yakima River basin, Washington, in relation to geology, land use, and other environmental factors. Can J Fish Aquat Sci 52(5):1108–1129. https://doi.org/10.1139/f95-108

Li G (2012) Study on index of biological integrity in Qing River Basin in Liaoning Province. Liaoning Universtiy, Shenyang, China (in Chinese)

Li W, Xu J, Xin X, Yan W (2002) Using diatoms to assess the water quality of Laoyao River. J Shanxi Coal-Min Admin Coll 2:75–76 (in Chinese)

Li L, Zheng B, Liu L (2010) Biomonitoring and bioindicators used for river ecosystems: definitions, approaches and trends. Procedia Environ Sci 2(Supplement C):1510–1524. https://doi.org/10.1016/j.proenv.2010.10.164

Li G, Liu L, Wang X, Li L (2012a) Applications of diatom in river health assessment: a review. The J Appl Ecol 23(9):2617–2624 (in Chinese)

Li G, Wang X, Liu L, Li F, Li L, Li X (2012b) Bioassessment of water quality based on diatom index of biotic integrity in the upstream of Liaohe River. Res Environ Sci 25(8):852–858 (in Chinese)

Li Z, Yuan G, Hao X, Liu W (2012c) Diatom-based biomonitoring method of water quality in Baisha River, a tributary of Jinhua River, Zhejiang Province. J Lake Sci 24(3):436–442 (in Chinese)

Li F, Cai D, Tang X, Yi R (2015a) Diatom index screening and multi-index system to assess water quality. J Yangtze River Sci Res Inst 32(8):26–33 (in Chinese)

Li L, Jin W, Wang B, Xiang Z, Yin X, Xu Z, Zhang Y (2015b) Relationship between land use types within riparian zones and community structure of diatom in Taizi River, China. Res Environ Sci 28(11):1662–1669 (in Chinese)

Li Z, Cao R, Mao J, Sheng X, Wang X, Deng P (2017) Study of proper diatom-based indices for bioassessment of water quality in Beijiang Watershed. Ecol Environ Sci 26(2):275–284 (in Chinese)

Liu J (2013) Diversity and metacommunity of benthic diatoms in the Dong River watershed. Jinan University, Guangzhou, China (in Chinese)

Liu H (2016) Distribution pattern of benthic diatom and its relationship to environmental factors in Yi-Shu-Si River Basin. Shanghai Normal University, Shanghai, China (in Chinese)

Liu J, Yang W (2012) Water Sustainability for China and Beyond. Science 337(6095):649–650. https://doi.org/10.1126/science.1219471

Liu J, Zhang C (2009) The application of diatom indices in aquatic environment biomonitoring and water quality assessment. Environ Monit Manage Technol 21(5):12–16 (in Chinese)

Liu C, Liu D, Chen S, Yao M (2013) Research progress of the periphytic diatoms. J Liaocheng Univ (Nat Sci) 26(3):47–51 (in Chinese)

Liu L, Tan B, Yin X, Zhang Y, Meng W (2014) Evaluation of relationship between environmental factors and diatom assemblages at genus and species levels in the Taizi River Basin, China. Acta Ecologica Sinica 34(22):6613–6621 (in Chinese)

Liu L, Song J, Wang B, Yin X, Xu Z, Wu W, Yan W (2015) Diatom community structure and aquatic system assessment of Wei River Basin. Res Environ Sci 28(10):1560–1569 (in Chinese)

Liu L, Xu Z, Yin X, Wu W, Min W (2016) Assessing water quality with diatom indices in the Wei River Basin. J Beijing Normal Univ (Nat Sci) 52(3):317–321 (in Chinese)
Lobo EA, Bes D, Tudesque L, Ector L (2004) Water quality assessment of the Pardinho River, RS, Brazil, using epilithic diatom assemblages and faecal coliforms as biological indicators. Vie et Milieu 54(2–3):115–125
Lowe RL (1974) Environmental requirements and pollution tolerance of freshwater diatoms. US Environmental Protection Agency, Cincinnati, Ohio
Luo S, Zeng S (1985) The application of the lognormal distribution of the diatom community in the assessment of the water quality of the Xiangjiang River. Environ Sci China 5(6):11–15 (in Chinese)
Mann DG, Droop SJM (1996) Biodiversity, biogeography and conservation of diatoms. In: Kristiansen J (ed) Biogeography of freshwater algae, vol 118. Developments in Hydrobiology. Springer Netherlands, pp 19–32. https://doi.org/10.1007/978-94-017-0908-8_2
MEPPRC (2017) 2016 China Environmental State Bulletin by Ministry of Environmental Protection of the People's Republic of China. MEPPRC (in Chinese)
Miao Z (2007) Research on the biodiversity of diatom and evaluation of the water quality in the Taiyuan Section of the Fenhe River. Shanxi University, Taiyuan, China (in Chinese)
O'Brien A, Townsend K, Hale R, Sharley D, Pettigrove V (2016) How is ecosystem health defined and measured? A critical review of freshwater and estuarine studies. Ecol Ind 69(Supplement C):722–729. https://doi.org/10.1016/j.ecolind.2016.05.004
Oeding S, Taffs KH (2015) Are diatoms a reliable and valuable bio-indicator to assess subtropical river ecosystem health? Hydrobiologia 758(1):151–169. https://doi.org/10.1007/s10750-015-2287-0
Oertel N, Salánki J (2003) Biomonitoring and bioindicators in aquatic ecosystems. In: Ambasht RS, Ambasht NK (eds) Modern Trends in Applied Aquatic Ecology. Springer US, Boston, MA, pp 219–246. https://doi.org/10.1007/978-1-4615-0221-0_10
OMNIDIA (2018) OMNIDIA. https://omnidia.fr/en/
Pan Y, Stevenson RJ, Hill BH, Herlihy AT, Collins GB (1996) Using diatoms as indicators of ecological conditions in lotic systems: a regional assessment. J North Am Benthol Soc 15(4):481–495. https://doi.org/10.2307/1467800
Patrick R (1973) Use of algae, especially diatoms, in the assessment of water quality. In: Cairns JJ, Dickson KL (eds) Biological methods for the assessment of water quality. American Society for Testing and Materials, Philadelphia, Pennsylvania, pp 76–95
Paulsen SG, Mayio A, Peck DV, Stoddard JL, Tarquinio E, Holdsworth SM, Sickle JV, Yuan LL, Hawkins CP, Herlihy AT, Kaufmann PR, Barbour MT, Larsen DP, Olsen AR (2008) Condition of stream ecosystems in the US: an overview of the first national assessment. J North Am Benthol Soc 27(4):812–821. https://doi.org/10.1899/08-098.1
Pei G (2006) Study on the ecology of benthic algae in freshwater lakes. Chinese Academy of Sciences, Wuhan, China (in Chinese)
Pei G, Liu G (2011) Distribution patterns of benthic diatoms during summer in the Niyang River, Tibet, China. Chin J Oceanol Limnol 29(6):1192–1198. https://doi.org/10.1007/s00343-011-0291-4
Peng T, Li J (2016) Advances in metal pollution assessment by diatom-based indicators. J Hydroecol 37(1):1–8 (in Chinese)
Pignata C, Morin S, Scharl A, Traversi D, Schilirò T, Degan R, Bartley P, Tu M, Liu H, Peres F, Coste M, Liu W, Gilli G (2013) Application of European biomonitoring techniques in China: are they a useful tool? Ecol Ind 29(Supplement C):489–500. https://doi.org/10.1016/j.ecolind.2013.01.024
Potapova MG, Charles DF (2002) Benthic diatoms in USA rivers: distributions along spatial and environmental gradients. J Biogeogr 29(2):167–187. https://doi.org/10.1046/j.1365-2699.2002.00668.x
Potapova M, Charles DF (2007) Diatom metrics for monitoring eutrophication in rivers of the United States. Ecol Ind 7(1):48–70. https://doi.org/10.1016/j.ecolind.2005.10.001

Prygiel J, Lévêque L, Iserentant R (1996) Un nouvel indice diatomique pratique pour l'évaluation de la qualité des eaux en réseau de surveillance. Revue Des Sciences De L'eau 1:97–113. https://doi.org/10.7202/705244ar

Qi Y, Huang W, Luo Y, Wu L (1998) Evaluation of water quality of the Pearl River by using Diatom Assemblage Index (DAIpo) and River Pollution Index (RPId). J Trop Subtrop Bot 6(4):329–335 (in Chinese)

Qiu L, Wei G, Li X, Shi L, Lin S, Han B (2016) Species diversity and temporal-spatial distribution of benthic diatoms in Jianjiang River, Guangdong Province. J Trop Subtrop Bot 24(2):197–207 (in Chinese)

Rai LC, Gaur JP, Kumar HD (1981) Phycology and heavy-metal pollution. Biol Rev 56(2):99–151. https://doi.org/10.1111/j.1469-185X.1981.tb00345.x

Ren H (2015) Distribution pattern of benthic diatom and correlation to its environmental factors in the Yinghe River, Huaihe Basin. Shanghai Normal University, Shanghai, China (in Chinese)

Ren Q, Chen C, Yan N, Liu Z (2013) Analysis on diatom biodiversity and water quality in drift section of Shanmu River in Shibing County of Guizhou. Guizhou Agric Sci 41(11):45–47 (in Chinese)

Reynolds CS (2006) The ecology of phytoplankton. Ecology, biodiversity, and conservation, vol 535. Cambridge University Press, Cambridge

Rimet F (2012) Recent views on river pollution and diatoms. Hydrobiologia 683(1):1–24. https://doi.org/10.1007/s10750-011-0949-0

Rott E, Hofmann G, Pall K, Pfister P, Pipp E (1997) Indikationslisten für Aufwuchsalgen Teil 1: Saprobielle indikation. Bundesministerium für Land- und Forstwirtschaft, Wien, Austria

Rott E, Pipp E, Pfister P, Van Dam H, Ortler K, Binder N, Pall K (1998) Indikationslisten für aufwuchsalgen in Österrichischen fliessgewässern. Teil2: Trophieindikation. Freshwater Biology. Arbeitsgruppe Hydrobotanik, Institut für Botanik, Universität Innsbruck, Austria

Round FE (1991) Diatoms in river water-monitoring studies. J Appl Phycol 3(2):129–145. https://doi.org/10.1007/bf00003695

Round FE, Crawford RM, Mann DG (1990) Diatoms: biology and morphology of the genera. Cambridge University Press, Cambridge

Ruaro R, Gubiani ÉA (2013) A scientometric assessment of 30 years of the Index of Biotic Integrity in aquatic ecosystems: applications and main flaws. Ecol Ind 29(Supplement C):105–110. https://doi.org/10.1016/j.ecolind.2012.12.016

Rumeau A, Coste M (1988) Initiation à la systematique des diatomées d'eau douce pour l'utilisation pratique d'un indice diatomique générique. Bulletin francais de la peche et de la pisciculture 309:1–69

Schiefele S, Kohmann F (1993) Bioindikation der Trophie in Fliessgewässern. Umweltforschungsplan des Bundesministers für Umwelt. Naturschutz und Reaktorsicherheit, Germany, pp 1–211

Schneider SC, Lawniczak AE, Picińska-Faltynowicz J, Szoszkiewicz K (2012) Do macrophytes, diatoms and non-diatom benthic algae give redundant information? Results from a case study in Poland. Limnologica—Ecol Manage Inland Waters 42(3):204–211. https://doi.org/10.1016/j.limno.2011.12.001

Shen F (1990) New monitoring techniques with microorganisms. China Architecture & Building Press, Beijing

Singh PK, Chauhan RS, Singh P (2017) Biodiversity and its conservation. In: Singh RL (ed) Principles and applications of environmental biotechnology for a sustainable future. Springer, pp 315–340

Sládeček V (1986) Diatoms as indicators of organic pollution. Acta Hydrochim Hydrobiol 14(5):555–566

Slàdecek V (1973) System of water quality from the biological point of view. Archiv für Hydrobiologie und Ergebnisse Limnologie 7:1–218

Soininen J (2007) Environmental and spatial control of freshwater diatoms—a review. Diatom Res 22(2):473–490. https://doi.org/10.1080/0269249X.2007.9705724

Steinberg C, Schiefele S (1988) Biological indication of trophy and pollution of running waters. Zeitschrift fur Wasser- und Abwasser-Forschung 21:(6)

Stevenson J (2014) Ecological assessments with algae: a review and synthesis. J Phycol 50(3):437–461. https://doi.org/10.1111/jpy.12189

Stevenson RJ, Pan Y, van Dam H (2010) Assessing environmental conditions in rivers and streams with diatoms. In: The diatoms: applications for the environmental and earth sciences, 2nd ed, vol 5785. Cambridge University Press, Cambridge

Stoddard JL, Herlihy AT, Peck DV, Hughes RM, Whittier TR, Tarquinio E (2008) A process for creating multimetric indices for large-scale aquatic surveys. J North Am Benthol Soc 27(4):878–891. https://doi.org/10.1899/08-053.1

Sun C (2015) Diatoms biological diversity and relationship among diatoms with heavy metals and TN & TP in Wuyang River. Guizhou Normal University, Guiyang, China (in Chinese)

Sun D, Zhi C, Li Y (2015) Analysis of planktonic diatom and water quality of ground cave in Bijie section of upper reaches of Wujiang River. Jiangsu Agric Sci 43(2):323–326 (in Chinese). https://doi.org/10.15889/j.issn.1002-1302.2015.02.105

Tan X, Sheldon F, Bunn SE, Zhang Q (2013) Using diatom indices for water quality assessment in a subtropical river, China. Environ Sci Pollut Res 20(6):4164–4175. https://doi.org/10.1007/s11356-012-1343-9

Tan X, Ma P, Xia X, Zhang Q (2014) Spatial pattern of benthic diatoms and water quality assessment using diatom indices in a subtropical river, China. Clean-Soil Air Water 42(1):20–28. https://doi.org/10.1002/clen.201200152

Tan X, Ma P, Bunn SE, Zhang Q (2015) Development of a benthic diatom index of biotic integrity (BD-IBI) for ecosystem health assessment of human dominant subtropical rivers, China. J Environ Manage 151:286–294. https://doi.org/10.1016/j.jenvman.2014.12.048

Tan X, Zhang Q, Burford MA, Sheldon F, Bunn SE (2017) Benthic diatom based indices for water quality assessment in two subtropical streams. Front Microbiol 8. https://doi.org/10.3389/fmicb.2017.00601

Tang S, Zhi C (2016) Benthic diatoms biodiversity of Xiaoche River. J Anhui Agric Sci 44(29):7–9, 54 (in Chinese)

Tang T, Cai QH, Liu JK (2006) Using epilithic diatom communities to assess ecological condition of Xiangxi River system. Environ Monit Assess 112(1–3):347–361. https://doi.org/10.1007/s10661-006-7666-6

Tang T, Wu N, Li F, Fu X, Cai Q (2013a) Disentangling the roles of spatial and environmental variables in shaping benthic algal assemblages in rivers of central and northern China. Aquat Ecol 47(4):453–466. https://doi.org/10.1007/s10452-013-9458-8

Tang X, Cai D, Li F, Yi R, Qin P (2013b) Ecological characteristics of diatom communities and water quality assessment in Beijiang River Basin. J China Three Gorges Univ (Nat Sci) 35(6):1–5 (in Chinese)

Tang X, Cai D, Yao W (2014) Grey correlation model based on diatom indices evaluating water quality of Longjiang River. In: Sun M, Zhang Y (eds) Renewable energy and environmental technology, Pts 1–6, vol 448–453. Appl Mech Mater 43–47. https://doi.org/10.4028/www.scientific.net/AMM.448-453.43

Tian Y (2017) Evaluation of water ecological health of Hunhe River by using diatom index. Water Resour Hydropower Northeast (5):44–47, 72 (in Chinese)

Tolkamp HH, Gardeniers JJ (1988) The development of biological water quality assessment in the Netherlands. Hydrobiol Bull 22(1):87–91

van Dam H, Mertens A, Sinkeldam J (1994) A coded checklist and ecological indicator values of freshwater diatoms from The Netherlands. Netherland J Aquat Ecol 28(1):117–133. https://doi.org/10.1007/bf02334251

Vilmi A, Karjalainen SM, Landeiro VL, Heino J (2015) Freshwater diatoms as environmental indicators: evaluating the effects of eutrophication using species morphology and biological indices. Environ Monit Assess 187(5). https://doi.org/10.1007/s10661-015-4485-7

Wang C, Zhang J (2004) Studies on DCCA of the attached diatom community in headwater rivers of Fenhe Reservoir. China Environ Sci 24(1):28–31 (in Chinese)
Wang Y-K, Stevenson RJ, Metzmeier L (2005) Development and evaluation of a diatom-based Index of Biotic Integrity for the Interior Plateau Ecoregion, USA. J North Am Benthol Soc 24(4):990–1008. https://doi.org/10.1899/03-028.1
Wang Q, Zhi C, Kang F (2009) Effect of major ions on the communities of benthic diatoms in Karst rivers from the Zhujiang Watershed. Acta Sci Circum 27(9):1517–1526 (in Chinese)
Wang X, Zheng B, Liu L, Li L (2014) Use of diatoms in river health assessment. Ann Res Rev Biol 4(24):4054–4074. https://doi.org/10.9734/ARRB/2014/11963
Wang B, Zhu M, Xiang Z, Li L, Yin X, Bai H (2015) Study on biodiversity of diatom community in Huihe River Watershed. Hebei Fish (1):13–16 (in Chinese). https://doi.org/10.3969/j.isn.1004-6755.2015.01.005
Wang Y, Mukherjee M, Wu D, Wu X (2016) Combating river pollution in China and India: policy measures and governance challenges. Water Policy 18(S1):122–137. https://doi.org/10.2166/wp.2016.008
Watanabe T, Asai K, Houki A (1986) Numerical estimation to organic pollution of flowing water by using the epilithic diatom assemblage-diatom assemblage index (DAIpo). Sci Total Environ 55(Supplement C):209–218. https://doi.org/10.1016/0048-9697(86)90180-4
Whittaker RH (1972) Evolution and measurement of species diversity. Taxon 21(2/3):213–251. https://doi.org/10.2307/1218190
Whitton BA, Kelly M (1995) Use of algae and other plants for monitoring rivers. Austral Ecol 20(1):45–56
Wu J-T (1999) A generic index of diatom assemblages as bioindicator of pollution in the Keelung River of Taiwan. Hydrobiologia 397:79–87. https://doi.org/10.1023/A:1003694414751
Wu N, Cai Q, Fohrer N (2012) Development and evaluation of a diatom-based index of biotic integrity (D-IBI) for rivers impacted by run-of-river dams. Ecol Ind 18(Supplement C):108–117. https://doi.org/10.1016/j.ecolind.2011.10.013
Wu N, Schmalz B, Fohrer N (2014) Study progress in riverine phytoplankton and its use as bio-indicator a review. Austin J Hydrol 1(1):1–9
Xiang Z, Chen H, Li C, Yin X, Xu Z, Zhang Y (2016) Application of diatom index in assessment of aquatic ecosystem health in Taizi River, China. J Dalian Ocean Univ 31(4):416–425 (in Chinese)
Xu X, Zhi C, Chen L, Ma J, Zhan J (2011) Application of diatoms in water quality monitoring and reconstruction of palaeoenvironment. J Anhui Agric Sci 39(9):5216–5217, 5386 (in Chinese)
Yang Y, Cao JX, Pei GF, Liu GX (2015) Using benthic diatom assemblages to assess human impacts on streams across a rural to urban gradient. Environ Sci Pollut Res 22(22):18093–18106. https://doi.org/10.1007/s11356-015-5026-1
Yao W, Cai D, Tang X, Wen H (2015) Distribution and evaluation of diatom community in Hejiang River, a branch of Xijiang River, Pearl River basin. J Lake Sci 27(1):86–93 (in Chinese)
Yi R (2014) Biological monitoring technology of water quality by using benthic diatoms in Longjiang River. Guangxi University, Nanning, China (in Chinese)
Yi R, Cai D, Zhang Y, Lin J, Zhong Y, Li H, Zhao F (2016) Evaluation of biomonitoring methods using benthic biatoms in Longjiang River, China. Chin J Environ Eng 10(6):3345–3353 (in Chinese)
Yin X, Qu X, Li Q, Liu Y, Zhang Y, Meng W (2012) Using periphyton assemblages to assess stream conditions of Taizi River Basin, China. Acta Ecologica Sinica 32(6):1677–1691 (in Chinese)
You Q, Zhi C, Wang L (2013) Relationship between benthic diatoms and heavy metals in Chishui River. Jiangsu Agric Sci 41(10):347–349 (in Chinese). https://doi.org/10.15889/j.issn.1002-1302.2013.10.004
Zelinka M, Marvan P (1961) Zur Präzisierung der biologischen Klassifikation des Reinheit fliessender Gewässer. Archiv für Hydrobiologie 57:389–407
Zhang B, Chen X (1987) Utilizing the fuzzy clustering analysis and phytoplankton to assess the water quality of the polluted water. J Nanjing Univ (Nat Sci Ed) 23(3):413–421 (in Chinese)

Zhang J, Li Y, Wang G (1981) Investigation on aquatic community structure in polluted waters of Wuli River in Jinxi County. J Jinzhou Med Coll Z1:9–15 (in Chinese)
Zhang M, Huang D, Liu Z, Li H (2015) Relationship between functional groups of planktonic diatom and environmental factors in Ganjiang River Basin. Jiangxi Sci 33(3):293–302 (in Chinese). https://doi.org/10.13990/j.issn1001-3679.2015.03.002
Zhang M, Deng D, Huang D, Zhou M, Wu J, Xu J, Liu Z (2016) Functional group diversity of planktonic diatoms and the driving pattern of water-period guild succession in Ganjiang River Basin. Res Environ Sci 29(5):680–691 (in Chinese)
Zhao X, Cai D, Liu W, Wang X (2009) Diatom biomonitoring method study on water quality in Lijiang River. J Guangxi Normal Univ (Nat Sci Ed) 27(2):142–147 (in Chinese)
Zhou J (2014) A study on the use of macroinvertebrates and benthic diatoms in water quality bioassessment. Nanjing Agricultural University, Nanjing, China (in Chinese)
Zhou J, Liu S, Cai D, Zhang Y, Wang B (2012) Water quality bioassessment with benthic diatom in Lijiang River, Guilin, China. J Guangxi Normal Univ (Nat Sci Ed) 30(4):123–129 (in Chinese)
Zhu Y, Xie S (1981) Investigation of water quality in Taiyuan Section of Fenhe Rivers using phytoplankton assemblages. Environ Sci 5:51–55 (in Chinese)
Zhu Y, Xie S (1982) Correlation between diatoms and organic pollution in rivers. J Shanxi Univ (Nat Sci Ed) 2:65–73 (in Chinese)

Dr. Yuanda Lei is a Postdoctoral Research Associate at the Institute of Groundwater and Earth Sciences, Jinan University, China. His research interests include taxonomy, ecology and biogeography of freshwater and marine diatoms, freshwater bio-assessment with river diatoms, and paleoenvironmental research with diatom subfossil in sediments.

Yasu Wang is currently a Ph.D. student at the Institute of Groundwater and Earth Sciences, Jinan University, China. Her research interests include systematic and molecular studies of freshwater diatoms and their applications to river biomonitoring, as well as paleolimnological research with sediments.

Dr. Richard Jordan is a Full Professor in the Faculty of Sciences, Yamagata University, Japan. His main interests include taxonomy, ecology and evolution of marine microalgae, including diatoms. He has also published papers on the diatoms in marine and freshwater environments. Jordan is currently President of the International Society for Diatom Research.

Dr. Shijun Jiang, Co-Editor of this volume, is a Full Professor of Paleoecology at the Institute of Groundwater and Earth Sciences, Jinan University, China. His research focuses on the use of fossils and geochemical measurements to examine the biotic responses to environmental changes and the dynamic processes of aquatic ecosystems over various time scales. Jiang has authored or co-authored more than 40 peer-reviewed journal articles.

Part II
Challenges Towards Ecological Sustainability in the Great Pearl River Delta

Part II (Chapters "Hyperspectral Remote Sensing of Vegetation Health at the Baiyun Mountain National Forest Park, China"–"Flooding Hazards and Risk Analysis in the Pearl River Delta, China") examines several issues challenging the transition towards ecological sustainability in the Great Pearl River Delta, the birthplace of China's economic reform initiated since 1978, and the home of the world's largest urban area in both size and population. Chapter "Hyperspectral Remote Sensing of Vegetation Health at the Baiyun Mountain National Forest Park, China" discusses a remote sensing approach to assess vegetation health in the Baiyun Mountain National Forest Park, which is perhaps the most popular recreational site in Guangzhou, the largest megacity in southern China with a population of more than 13 million. Chapter "Comparison of Urbanization and its Eco-environmental Effects in three Large Pearl River Delta Metropolises, China" examines the similarities and differences of three PRD megacities, i.e., Guangzhou, Shenzhen, and Hong Kong, in terms of their biophysical and socio-economic conditions. Chapter "Nutrient and Trace Metal Issues in the Pearl River Delta, China" discusses the impacts of urbanization and industrialization on the surface water and groundwater systems in the PRD region through several case studies targeting nutrients and trace metals pollution in different environmental settings. Chapter "Flooding Hazards and Risk Analysis in the Pearl River Delta, China" included in Part II discusses flooding hazards and risk in the PRD region and also describes some defense strategies and important infrastructures there.

Hyperspectral Remote Sensing of Vegetation Health at the Baiyun Mountain National Forest Park, China

Shuisen Chen, Weiqi Chen and Jia Liu

Abstract Forest pests, wilting disease, wood cutting and phenological changes can affect vegetation health. The traditional method for pigment extraction followed by spectrophotometric determination or high-performance liquid chromatography (HPLC) will have to destroy the measured leaves with high costs and a long processing time. Using hyperspectral EO-1 Hyperion remote sensor imagery and a spectral model of leaf pigment reflectance, we examined the two forest health related important parameters, anthocyanin and carotenoid, in the Baiyun Mountain national forest park, China. The remote sensing-derived outcome was validated through *in situ* sample analyses of canopy leaves. The result shows that the concentrations of anthocyanin and carotenoid indicating the vegetation stress can be quantified using the reflectance index derived from hyperspectral remote sensor imagery. The index has the potential to indicate the regional forest vegetation health. Furthermore, the forest phenological information can be retrieved when multi-temporal hyperspectral images are available.

Keywords Vegetation · Health · Anthocyanin · Carotenoid · EO-1 hyperion
Remote sensing

S. Chen (✉) · J. Liu
Guangdong Open Laboratory of Geospatial Information Technology and Applications, Guangdong Key Laboratory of Remote Sensing and GIS Technology Application, Guangdong Engineering Technology Center for Remote Sensing Big Data Applications, Guangzhou Institute of Geography, Guangzhou, Guangdong 510070, China
e-mail: css@gdas.ac.cn

W. Chen
Department of Geography and Anthropology, Louisiana State University, Baton Rouge, LA 70803, USA

X. Yang and S. Jiang (eds.), *Challenges Towards Ecological Sustainability in China*, https://doi.org/10.1007/978-3-030-03484-9_6

1 Introduction

Mapping vegetation health can help detect forest and wilting disease, and find the most suitable wood cutting area, and monitor changes of forest phenology (Zeng et al. 2002; Research Systems Inc. 2005; Ye and Win 2010; Lin et al. 2015; Dolling et al. 2017). Remote sensing has been recognized as an important tool for regional forest management. Low-stress forest is usually composed of healthy vegetations while high-stress forest is usually comprised of dry or death vegetations, overstocked or sparse vegetation with low light use efficiency (Gitelson et al. 2002). When incident light exceeds the need of vegetation photosynthesis, the carotenoid comprising of yellow pigment will disperse the excessive energy to avoid the vegetation photosynthesis being damaged. The degree of vegetation stress can be measured by leaf pigment-related concentrations of anthocyanin and carotenoid. For example, when vegetation is under stress or is in senescence phase, the chlorophyll content is adopted to lower faster than carotenoid content. Due to the importance of pigment to leaf function, the leaf pigment content can provide valuable insight to leaf physiological property (Sims and Gamon 2002; Dobrota et al. 2015).

The traditional method of pigment extraction followed by spectrophotometric determination and leaching or high-performance liquid chromatography (HPLC) will need to destroy the measured leaves with a high cost and a long processing time, and it may not be feasible to map the temporal and spatial changes of vegetation pigments for forest ecosystem assessment. In contrast, the spectral measurement by remote sensing is non-destructive and quick in measuring the leaf pigment contents while allowing large-scale mapping at different spatial scales (Gamon and Surfus 1999). For example, Gitelson et al. (2002) developed some reflectance spectral models of leaf pigment contents. However, their pigment models were seldom applied on satellite remote sensing of forest canopy and particularly rare in China. In the past, various research efforts on the forest community at the Baiyun Mountain national forest park were carried out concerning plant age, vegetation types, species richness index, and stand reform (Su et al. 1996, 2001; Peng and Fang 1995). For improving urban forest management, we mapped the spatial distribution of relative vegetation health at the Baiyun Mountain national forest park by examining spectral indices from hyperspectral remote sensing imagery in order to understand the relationship between forest community changes and urban development.

It may not be feasible to measure colorimetric time and space changes using traditional spectrophotometry and extraction or high performance liquid chromatographic methods because it needs to destroy the measuring blades. In addition, the conventional method takes much time making vegetation health assessment over a large area. In contrast, remote sensing-based spectroscopic measurements are not only non-destructive and rapid but also can be applied at different spatial scales (Gamon and Surfus 1999; Masek et al. 2015). In general, remote sensing spectral reflectance provides a large-scale, fast, non-destructive method for estimating leaf pigment content. By far, some leaf reflectivity spectral prediction models have been developed from pigment contents (Gitelson et al. 2002; Zhou et al. 2017). However, the application of these models to hyperspectral remote sensing of canopy is rare. In China, these applications are rarely conducted. In the past, research on the forest community at the Baiyun Mountain National Forest Park in Guangzhou was mainly focused on the flora composition, vegetation characteristics, species diversity index of forest community, and reforestation (e.g., Su et al. 1996, 2001; Peng and Fang 1995). In this research, we will use remote sensing-derived spectral indices to estimate the spatial distribution of relative vegetation health status at the Baiyun Mountain Scenic area for examining the phenological succession characteristics of the forest community and their relationship with urban development. Such information can help improve the management and protection of forest area and assist in afforestation decision-making.

2 Study Site and Data

2.1 Study Area

Located in the northeast of downtown Guangzhou (Fig. 1, Left), Baiyun Mountain covers a total area of 20.98 km^2, and is rated a national grade-5A scenic area and the only national scenic area in Guangzhou (Sun et al. 1997). It belongs to the Nankun mountain extension, Dayu mountain range of Southern China. It has been known as the most beautiful mountain in Guangzhou since ancient times (Cui et al. 2015). The geographical location is about N23°11′, E113°19′. Its highest peak, Moxing Ridge, stands at 382 m above sea level, where one can overlook the whole downtown Guangzhou and the Pearl River. With the development of Guangzhou's municipal construction, the ecological significance of Baiyun Mountain Scenic Spot has become increasingly prominent (Cui et al. 2015). It is not only a natural buffer in the northeastern part of Guangzhou, but also plays a positive environmental and ecological role in the area. It is also an excellent place for leisure activities such as recreation and vacations, and is considered as the "lungs" in Guangzhou. Vegetation is the most important natural landscape, and Baiyun Mountain provides rich ecological functions (Yu and Xie 2011). Today, Baiyun Mountain has become the National Scenic area with green forest and beautiful landscape and one of the country's "5A" attractions. However, some communities saw the development of Pinus massoniana lamb, and natural invaded broad-leaved species have predominated in the main strata, with traces of masson pine recession in the community (Su et al. 1996; Zeng et al. 2013). Plant growth there is also inhibited by vehicle exhaust and acid rain (Hu and Su 2000), and forest pests in pure forests also occurred from time to time (Li et al. 2003). Rapid urban and industrial development has brought potential risks to the conservation of vegetation, although an increasing awareness of the importance of vegetation has been seen among Chinese cities in recent years, including Guangzhou.

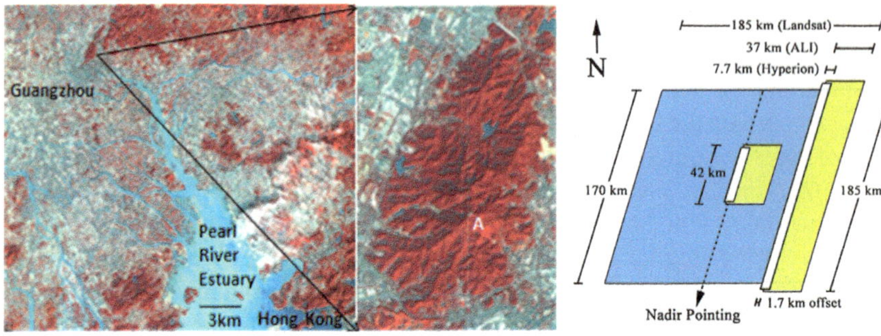

Fig. 1 The location map of study area. Left: Baiyun Mountain location in Pearl River delta (A: indicating the highest peak: Moxing ridge); Right: Footprint of Hyperion EO-1 data with respect to a Landsat row

2.2 *Data*

2.2.1 Field Data

Field leaf sampling was conducted on 22 December 2006. The leaf sample processing and analytical method was described by Tong and Zhou (1982) and Ma and Cheng (1984). The main chemical compositions in our analysis were carotenoid and anthocyanin for leaf samples at 11 sites across the study area. The arrangement of sample sites was based on the equal distribution method and the road accessibility in the study area. The sampling leaves are mainly from dominant species of plants in the study site (Su et al. 1996), such as acacia mangium, acacia, eucalyptus, pine and bamboo. Leaves were visually selected by their color differences, and only those being healthy and homogeneous in color without anthocyanin pigmentation or visible symptoms of damage were actually used in the experiment. The experimental results are presented in Table 1.

2.2.2 Hyperspectral Remote Sensor Data

A Hyperion image covering the Baiyun Mountain national forest park was acquired on 18 December 2005, almost one year after our field analytical data sampling (22 December 2006) but with same phenophase. The map in Fig. 1 (right) shows the

Table 1 Pigment content and composition in leaves studied (carotenoid: mg/100 cm^2; anthocyanin: nmol/cm^2)

Pigment	Range	Mean	Standard deviation
Anthocyanosides	3.22–11.64	7.21	2.74
Carotenoid	0.17–2.26	1.11	0.58

relevant footprint of a Hyperion (white) data, with respect to a Landsat WRS-2 row (dark grey) (http://eo1.usgs.gov/faq.php?id=2). Due to the orbit's limitation, the Hyperion image only covers about eighty percent of the Baiyun Mountain area, leaving twenty percent (mainly at the south-western part) not being covered. Besides, the study area includes the vegetation area only, and the reservoirs and the residential area were excluded.

Hyperion is the first Earth-orbiting imaging spectrometer acquiring 196 bands ranging from 426 to 2395 nm and with 10-nm bandwidth. The instrument is onboard the EO-1 satellite that was launched on 21 November 2000. The spatial resolution is 30 m, with the image swath of 7.7 km and a maximum of 185 km in length, which is appropriate for detecting complex spatial distribution of forest ecosystem. Its imagery data were recorded in 16-bit radiance values. Hyperion imagery includes several important bands for detecting forest pigments, such as 508.22, 548.92, 701.55, and 803.3 nm.

3 Research Methods

Carotenoid and anthocyanin are important parameters for forest monitoring with relatively high content in worse-growing vegetation. The function of carotenoid gives expression to the light absorption process of vegetation and avoids plants being damaged by strong sun light. Anthocyanin is a type of water-soluble pigment with relatively high content in new leaf and dead leaf of the vegetation. The pigment resulting from vegetation stress can be identified by using vegetation indices of hyperspectral remote sensing. Gitelson et al. (2002) found that the proportion change of carotenoid content with respect to chlorophyll content can make leaves' absorption characteristic change, which can be predicted by using the following two spectral reflectance indices of carotenoid (CRI):

$$\mathrm{CRI\,1} = (1/\rho 510) - (1/\rho 550) \tag{1}$$

$$\mathrm{CRI\,2} = (1/\rho 510) - (1/\rho 700) \tag{2}$$

Note that the Carotenoid Reflectance Index 1 (CRI 1) is a reflectance measurement that is sensitive to carotenoid pigments in plant foliage, and higher CRI 1 values indicate higher caratenoid concentration relative to chlorophyll; CRI 2 is a modification of CRI 1, which should provide better results in areas of high carotenoid concentration, and higher CRI 2 values indicate higher caratenoid concentrations relative to chlorophyll (Gitelson et al. 2002). According to the setting of Hyperion spectral bands, the above two indexes can be computed by using the following equations:

$$\mathrm{HCRI\,1} = (1/\rho 508.22) - (1/\rho 548.92) \tag{3}$$

$$\mathrm{HCRI\,2} = (1/\rho 508.22) - (1/\rho 701.55) \tag{4}$$

Table 2 Statistic of adjusted pigment spectral indexes and the measured values of anthocyanins and carotenoids

	HCRI 1	HCRI 2	HARI 1	Anthocyanins	Carotenoids
Max	33.5	99.7	31.5	11.6	1.84
Min	3.6	14	8.4	3.2	0.17
Mean	19.8	42.2	16.5	7.3	1.12
Stdev	12.0	24.9	8.7	2.8	0.58

The Anthocyanin Reflectance Index 1 (ARI 1) is a reflectance measurement that is sensitive to anthocynanin concentration in plant foliage. Increases in ARI 1 indicate canopy changes in foliage via new growth or death. The ARI 2 is a modification of the ARI 1 which detects higher concentrations of anthocynanins in vegetation. ARI 1 and ARI 2 are defined by the following equation:

$$\text{ARI}\,1 = (1/\rho550) - (1/\rho700) \quad (5)$$

$$\text{ARI}\,2 = \rho800(1/\rho510) - (1/\rho700) \quad (6)$$

Similarly, basing on the setting of Hyperion spectral bands, the two Anthocyanin Reflectance Index models can be adjusted as below:

$$\text{HARI}\,1 = (1/\rho548.92) - (1/\rho701.55) \quad (7)$$

$$\text{HARI}\,2 = \rho800(1/\rho508.22) - (1/\rho701.55) \quad (8)$$

Correspondingly, the statistic of adjusted pigment spectral indices and the measured values of anthocyanins and carotenoids can be estimated in Table 2.

In order to reduce the effect of non-vegetation factors, firstly the hyperspectral NDVI (HNDVI) was analyzed (Fig. 2). The study area was mapped with HNDVI range of 0.21–0.71. Note that HNDVI is defined as: (R793 − R681)/(R793 + R681). R793 and R681 are the reflectance of NIR (at 793 nm) and RED band (at 681 nm) (Pu and Gong 2004). Considering that the vegetation cover was greater than 95% (Chinese Jingwei Network 2009), the image area with pixel's HNDVI greater than 0.3 was windowed for the analysis of Carotenoid and Anthocyanin spectral indices HCRI and HARI.

4 Results and Discussions

4.1 Spectral Index Mapping of Vegetation Health

The spectral reflectance indices mappingof carotenoids (CRI 1, CRI 2) and anthocyanin (ARI 1, ARI 2) were produced by band math processing of pre-processed hyperspectral remote sensor imagery. The results are shown in Fig. 3 and Table 3.

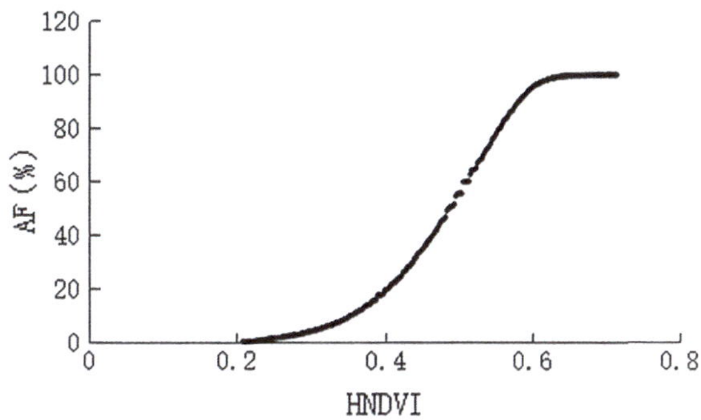

Fig. 2 Accumulated frequency distribution of hyperspectral vegetation index (HNDVI) in the Baiyun Mountain, Guangzhou

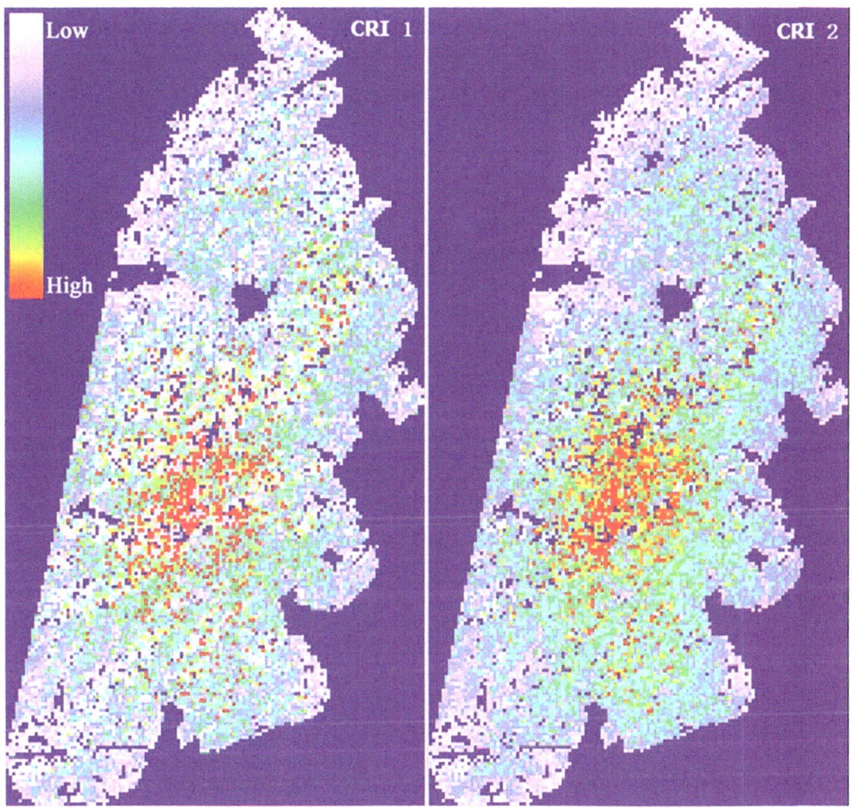

Fig. 3 The spatial distribution of the carotenoid reflectance index (CRI 1-left and CRI 2-right)

As shown in Fig. 3, the vegetation with poorer health status occurred in the Mount Peak (so called the Moxing ridge) and its surrounding area, especially at the southwest, while the vegetation with healthier status was distributed ring-zonally, followed by the mountain waist, with the best at the footslopes. Peak vegetation adapts

Table 3 Statistic of Hyperion mapping on relative content indices of carotenoids and anthocyanin

Hyperion index	Range	Average	Standard deviation
HCRI 1	0.05–96.1	23.4	22.9
HCRI 2	0.12–140.2	45.2	28.1
HARI 1	0.6–130.8	24.2	17.6

itself to environmental interference through its own anti-retrogradation reaction. For example, with the increase of the mountain altitude, the stress of moisture content in the peak is maximized. From the Moxing Ridge to the northeast ridge, a stress zone existed. Therefore, the ecological service function of vegetation will decline (such as cooling effect, purification, etc.). The high CRI values of CRI 1 and CRI 2 in Fig. 3 were mainly concentrated around the Moxing ridge at the core of the Baiyun Mountain Scenic site, which was likely caused by tourism-induced human activities or road engineering in the scenic area. The trend of increasing stress level with the mountain elevation was better observed through CRI 2, which may be related to the higher relative content of carotenoids in the preponderant broad-leaved tree species of Baiyun Mountain in this season. It also verified that CRI 2 can characterize the vegetation health better under higher carotenoid concentrations (Gitelson et al. 2002).

From the values in the CRI 1 and CRI 2 images (Fig. 3), the CRI 1 and CRI 2 values measured by Hyperion ranged from 0.05 to 96.1 (with the mean of 23.4 and the standard deviation of 22.9) and 0.12 to 140.2 (with the mean of 45.2 and the standard deviation of 28.1), respectively. As can be seen from the CRI 1 and CRI 2 maps in Fig. 3, the health status of most vegetation in the Baiyun Mountain was affected to varying extent except for the foothills, and the higher CRI values of the vegetation were distributed around the Moxing ridge and mountain ridge in the scenic area. In addition, the vegetation area with high CRI 2 values in the southern Baiyun Mountain is scattered in a patchy pattern as shown in Fig. 4, which may indicate that the south of the mountain is more affected by the polluted atmosphere from urban center or expressway around the city than the north due to closer to the center of the city.

The anthocyanin spectral index ARI 1 ranged from 0.6 to 130.8 (with the mean of 24.2 and the standard deviation of 17.6). In Fig. 4, the ARI value was high where the distribution of the ARI 1 peaked from the northeast to the northwest of Moxing Ridge in a ring-zonal distribution. Along this direction, vegetation index and soil moisture conditions should also demonstrate a declining tendency. Moreover, it is also obvious that the area with high ARI is mainly located in the southeastern Baiyun Mountain, such as around Guangzhou Forest of Steles and Moxing Ridge, indicating that the poor vegetation health around Baiyun Mountain are more or less caused by human activities in the surrounding areas. The low values of ARI are mainly concentrated in other areas of Baiyun Mountain where people and vehicles are less concentrated. The results of other works showed that plant photosynthesis system and cell membrane aside highway in the south of Baiyun Mountain (near Luhu lake) had obvious variation (Hu and Su 2000). The photochemical oxidant pollutants and acid

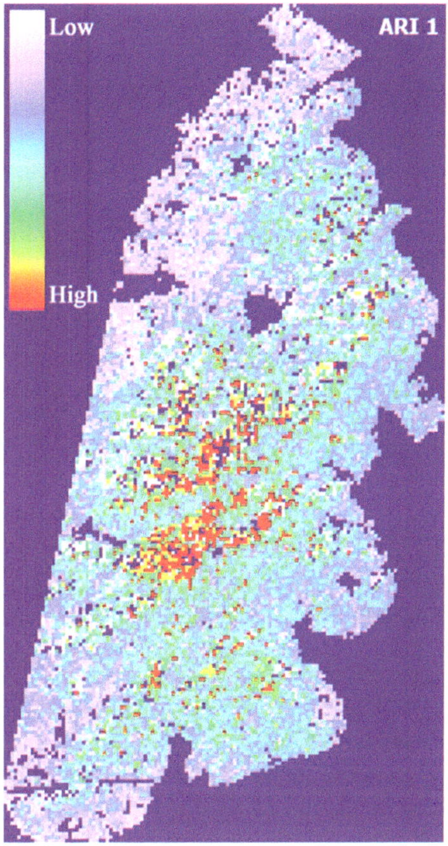

Fig. 4 The spatial distribution of Anthocyanin reflectance index (ARI 1)

rain derived from vehicle tail gas were the main influential factors. It was possible that the result of the mutation leads to the change of the chemical parameters in the vegetation. It shows that the air pollution in Guangzhou caused some damage to the forest vegetation in Baiyun Mountain. The damage effect is the long-term and comprehensive chronic effect of low dosage concentration that occurred during the period of time. In addition, due to the rolling of machinery and vehicles and human activities, the natural characteristics of road soil have undergone great changes from road construction and operation activities, adversely affecting the growth of vegetation. The environmental pollution caused by human activities has brought a greater stress to the vegetation.

4.2 Correlation Analysis Between Vegetation Health Indices and Measured Data

In the study area, 13 locations were selected as sampling points in the same phenophase period (22 December 2006, almost a year after the image acquisition), and the dominant species were sampled at sampling sites for a total of 12. Note that location 1 was discarded due to incorrect location coordinates and location 5 was also discarded because it was out of the experimental area. The types of vegetation were basically the types of main forest types in the main groups of Baiyun Mountain, including Acacia melanoleuca, Acacia mangium, Caribbean pine, Chinese tallow tree, bamboo leaves, Eucalyptus umbellatus, Cedar, broad leaf pine, Elaeocarpus apiculatus Masters, Podocarpus fleuryi (Nageia fleuryi), evodia, among others. The latitude and longitude of the GPS sampling points in the field were used to obtain the vegetation CRI and ARI spectral indices of the corresponding points from the pre-processed Hyperion images. The data of CRI 2 (based on the above analysis) and ARI 1 (due to the poor ARI 2 at the canopy level with many negative values) were selected to establish the correlation between the data of analysis (carotenoids, anthocyanins) and the vegetation health indices (CRI and ARI), and between CRI 2 and ARI 1.

The above vegetation indices and the measured values were statistically analyzed as follows: CRI 1 and measured carotenoids, CRI 2 and measured carotenoids; ARI 1 and measured anthocyanins (Table 4). The following results were obtained: the determination coefficients of CRI 1 and the measured carotenoids were 0.178, respectively; the determination coefficient of ARI 1 and the measured anthocyanin was 0.0013, indicating that the correlation of them was weak. As can be seen in Fig. 5, there is an exponential correlation between the carotenoid spectral reflectance index (CRI 2) and the measured carotenoid density ($R^2 = 0.5233$), indicating that the spectral index can be used to make an approximate estimate of the relative content of carotenoid in the vegetation canopy, and then the vegetation health status can be analyzed in the forest area. Meanwhile, the correlation between the measured carotenoid content and the carotenoid spectral reflectance index (CRI 1) was poor, indicating that the carotenoid concentrations in the study area may be high. It can be seen that there is little correlation between the pigments and spectral reflectance index CRI 1. Figure 6 shows that the carotenoid spectral reflectance index (CRI 2) is also naturally logarithmically correlated with the anthocyanin reflectance index (ARI 1) ($R^2 = 0.6778$), indicating that both indices CRI 2 and ARI 1 have a significant and consistent effect on characterizing the extent of vegetation wilting degree. However, they had different characteristic degrees of fractions of carotenoids and anthocyanins. From another viewpoint, Fig. 7 shows that the anthocyanin spectral index (ARI 1) has not good performance in predicting anthocyanin content in the study area, which may be related to the differences in spectral absorption characteristics caused by different anthocyanin and chlorophyll ratios between different tree species.

Table 4 The pigment spectral indices and measured values of anthocyanins and carotenoids

Sample No.	CRI 1	CRI 2	ARI 1	Anthocyanins	Carotenoids
1	–	–	–	9.185	1.3888
2	33.5	65	31.5	7.157	1.0214
3	33.5	99.7	26.2	5.428	1.8355
4	24.8	51.3	10	9.424	2.2614
5	3.6	11.6	8	6.024	1.2929
6	24.8	41.7	26.2	11.638	1.046
7	33.5	52.8	19.3	3.221	0.8186
8	15.5	25.7	10.2	10.02	1.3434
9	7.8	17.5	9.7	4.593	0.1732
10	12	22	10	10.796	0.7223
11	9.6	24.6	9	4.235	1.1823
12	5.7	14	8.4	6.322	0.3634
13	9.6	12.5	28	7.635	1.2008

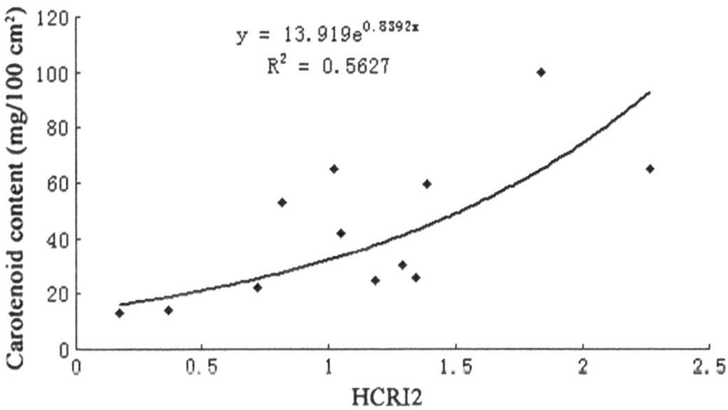

Fig. 5 Relationship between carotenoid reflectance index and ground measured carotenoid content

Correlation analysis between CRI 2 and ARI 1 shows that there is a significant logarithmic correlation between CRI 2 and ARI 1. However, there was no direct correlation between the measured carotenoids and anthocyanin because both CRI 2 and ARI 1 were the relative proportional relationship, not the absolute amount of content. Obviously, the correlation between vegetation health spectral reflectance index and measured pigment content values is affected by many factors, such as vegetation types, seasons, spatial scales, etc., which leads to the complicated relationship between the vegetation health spectral index and pigment content at the canopy level. Specifically, on one hand, the general spectral characteristics of green plants are mainly determined by their chemical and morphological characteristics, which are closely related to the vegetation growth and development stages, health conditions and phenological phenomena. There are differences in pigment content

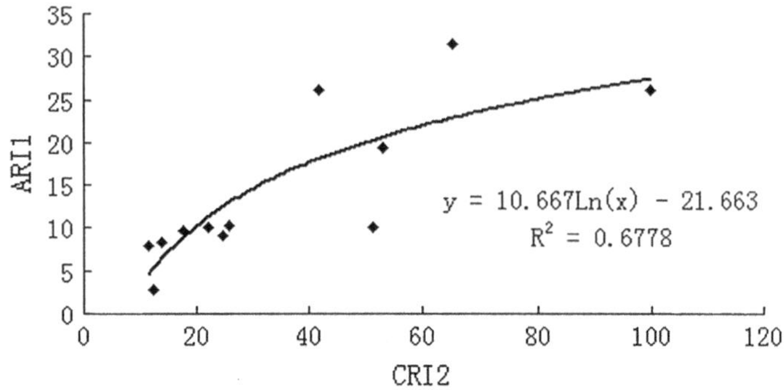

Fig. 6 Relationship between carotenoid reflectance index (CRI 2) and anthocyanin reflectance index (ARI 1)

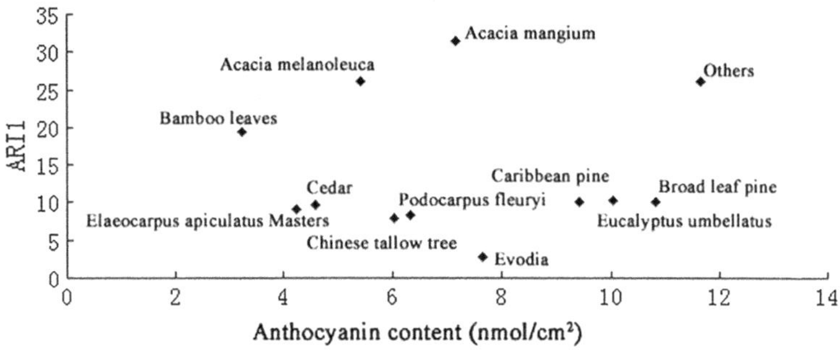

Fig. 7 Relationship between measured anthocyanin content and anthocyanin reflectance index (ARI 1)

and moisture content between different vegetation green leaves and green leaves in different parts of the same plant. Therefore, they have caused variation among their spectral curves (Su et al. 2001). Since during field sampling process, it is not necessarily the collection of the canopy leaves of the vegetation completely as a sample, nor is it easy to collect the leaves, and the canopy information of the vegetation was used while calculating the vegetation health index. This will to a certain extent affect the actual relationship between spectral index and pigment. On the other hand, the resolution of the Hyperion image is 30 m, which means that one pixel of the image represents the area of the surface of 30 m * 30 m. The 900 m^2 area actually covers many different plants and includes other factors. The information contained in a single pixel in the image may also include road or soil information, and information about different plants, etc. In fact, although we can set larger NDVI values for removal of noise to a certain degree, it represents the comprehensive information of the surface of the area. When comparing an area's combination of factors with single plant, there always exist some uncertain error factors among them. Also, the experiment used GPS for ground data acquisition, and the inaccuracy of GPS instruments will also cause the geographical position deviation of experimental data, which then lead to data deviation.

Another factor for inaccuracy is the difference between the acquirement time of the image data and the measured data. The date of image data acquisition was 18 December 2005, while the field measurement data collection time was 22 December 2006, and the time for obtaining the data was one year apart. Then during the year, the vegetation health status in the experimental area may change partially or the slight change of vegetation phenology due to the climate difference will affect the correlation between the spectral indices and the measured data.

In summary, although the above factors to some extent affect the result of our experimental data analysis, the sampling data are still representative of the typical vegetation in the study area. Therefore, the simulated results of the carotenoid and anthocyanin spectral reflectance index models in the application of canopy are of great significance, which is also consistent with the relevant geographical laws or research results (Sun et al. 1997; Zeng et al. 2002; Cui et al. 2015). Furthermore, the direct measurement and analysis of leaf spectrum may help obtain better results in the future.

5 Conclusions

Hyperspectral remote sensor data can greatly improve the accuracy of vegetation identification and classification. Existing spectroscopic techniques could be used to build a variety of spectroscopic exponential models of biophysical and chemical parameters from hyperspectral data. In order to achieve an efficient understanding of the regional forest phenology and health conditions, mapping canopy biochemical parameters with the field spectral model method was conducted by the parameter inversion and analysis technique. This can provide timely and accurate information for vegetation protection, maintenance management and afforestation decisions.

The carotenoid spectral reflectance index CRI 2 was applied to hyperspectral remote sensor image between Hyperion in the study. The remote sensing model between adjusted Hyperspectral canopy Spectral Reflectance Index (HCRI 2) and carotenoid content was established. In December, the health mapping of vegetation spectral reflectance index in Guangzhou's Baiyun Mountain National Forest Park, and the HCRI 1 mapping results of the carotenoid are basically in line with the field chemical analysis data showing that the vegetation carotenoid concentration in December was relatively higher.

The vegetation canopy differs from single vegetation in that it collects spectral information from a variety of different plants, and therefore is more complex than single vegetation. Hyperspectral data at different times can show the succession of phenology in vegetation communities. It is believed that with further development of spaceborne hyperspectral technology, remote sensing research on vegetation health status will have greater progress and higher practical applicability in the future.

Acknowledgements This research was partly funded by Guangdong Province's Science & Technology Plan Project (2016A020223011, 2018B030311059, 2015B070701020), and GDAS' Special Project of Science and Technology Development (2017GDASCX-0101), Production-Education-Research Collaborative Innovation Major Project of Guangzhou Municipality (201604020117) and Guangzhou Yuexiu District Science and Technology Project (2016-GX-059).

References

Chinese Jingwei Network (2009). http://www.huaxia.com/gdtb/zjny/rwjg/2009/09/1593568.html. Accessed 9 Jan 2018

Cui HS, Jiang Y, Lin MZ, Jiang T, Ji ST (2015) Forest naturalness evaluation in Baiyun Mountain of Guangzhou. J Subtrop Resources Environ 10(1):18–26

Dobrota C, Lazar L, Baciu C (2015) Assessment of physiological state of Betula pendula and Carpinus betulus through leaf reflectance measurements. Flora—Morphol Distrib Funct Ecol Plants 216:26–34

Dolling A, Nilsson H, Lundell Y (2017) Stress recovery in forest or handicraft environments—an intervention study. Urban For Urban Green 27:162–172

Gamon J, Surfus J (1999) Assessing leaf pigment content and activity with a reflectometer. New Phytol 143:105–117

Gitelson A, Zur Y, Chivkunova O, Merzlyak M (2002) Assessing carotenoid content in plant leaves with reflectance spectroscopy. Photochem Photobiol 75:272–281

Hu DQ, Su X (2000) Damage analysis of air pollution on the forest of Baiyun Mountain. Ecol Sci 19(3):67–72

Li YZ, Wei CH, Yi YH, Lu CC (2003) Investigation of insect species and control strategy of forest pests in Baiyunshan scenic spot, Guangzhou. J South China Agric Univ (Nat Sci Ed) 24(1):p34–p41

Lin MZ, Ji ST, Zhao JL, Fang BZ, Xie GW (2015) Comprehensive assessment of environmental quality of Baiyun Mountain scenic area. Ecol Sci 34(2):42–50

Ma ZB, Cheng Y (1984) Chemical determination of anthocyanin content on apple fruit surface. Chin Fruit Trees 4:49–51

Masek JG, Hayes DJ, Mj Hughes, Healey SP, Turner DP (2015) The role of remote sensing in process-scaling studies of managed forest ecosystems. For Ecol Manage 355:109–123

Peng SL, Fang W (1995) Dynamics of community composition structure of secondary evergreen broad leaved forest in Baiyunshan of Guangzhou. Chin Bull Botany 12(S2):p49–p54

Pu RL, Gong P (2004) Wavelet transform applied to EO-1 hyperspectral data for forest LAI and crown closure mapping. Remote Sens Environ 91:212–224

Research Systems Inc. (2005) Envi User's Guide

Sims D, Gamon J (2002) Relationships between leaf pigment content and spectral reflectance across a wide range of species, leaf structures and developmental stages. Remote Sens Environ 81(2–3):337–354

Su ZY, Chen BG, Gu YK, Xie ZS, Zeng SC (1996) Vegetation and main phytocommunity types of Baiyun Shan spot, Guangzhou. J South China Agr Univ 18(2):23–29

Su Z, Chen BG, Gu YK, Xie ZS, Zeng SC (2001) Species richness and diversity of forest communities in Baiyunshan, Guangzhou. J South China Agric Univ 22(3):5–8

Sun B, Mi J, Xie ZZ, Zhong F, Huang JP (1997) Spatial characteristic and future structure of urban forest in Guangzhou. Urban Environ Urban Ecol 2:50–54

Tong YA, Zhou HJ (1982) Fruit tree nutrition diagnosis. Agricultural Press, Beijing, pp p149–p150

Ye XL, Wen CY (2010) Study on the construction planning of biological protection forest belt in Baiyun Mountain Scenic Area. Prot Forest Sci Technol 5(98):101–104
Yu MM, Xie ZS (2011) Study on soil permeability capability of five forest types in Baiyunshan scenic spot of Guangzhou. Res Soil Water Conserv 18(1):153–156
Zeng F, Li XW, Chen HF (2013) Analysis on the characteristics of the planting design in typical scenic area of Guangzhou Baiyun Mountain. Landsc Plant Study Appl 1:49–54
Zeng SC, Xie ZS, Gu YK, Su ZY, Chen BG, Lin SH (2002) The biomass and water-holding capacities of some forest communities of Baiyunshan Scenic Spot, Guangzhou. J South China Agric Univ (Nat Sci Ed) 23(4):41–44
Zhou X, Huang WJ, Kong WP, Ye HC, Casa R (2017) Assessment of leaf carotenoids content with a new carotenoid index: development and validation on experimental and model data. Int J Appl Earth Obs Geoinf 57:24–35

Dr. Shuisen Chen is currently a Full Research Professor at the Guangzhou Institute of Geography, China. His research areas include remote sensing-based spatiotemporal analysis and modeling for the environment change precision agriculture, and agricultural information systems. Chen has co-authored over 100 articles in these areas.

Weiqi Chen is currently a doctoral student at the Louisiana State University, USA. Her research interests include remote sensing and GIS-based spatial analysis and modeling for urban and environmental applications. Chen has authored or co-authored one book chapter and over 15 articles.

Jia Liu is a postgraduate student at the University of Chinese Academy of Sciences. Her research areas include geographic information systems (GIS), environmental remote sensing, and Unmanned Aerial Vehicle (UAV) data processing and analysis.

Comparison of Urbanization and Its Eco-environmental Effects in Three Large Pearl River Delta Metropolises, China

Rongbo Xiao, Changguang Wu and Zhishan Li

Abstract Guangzhou, Shenzhen and Hong Kong are the three most representative large metropolises experiencing rapid urbanization in China. Through Landsat image interpretation, socio-economic and environmental monitoring data analysis, we constructed an index system for urbanization and eco-environmental quality assessment. We used this system to examine the urbanization process and evaluate the eco-environment quality changes in the three metropolises since 1980. Our goal was to provide a scientific understanding of green urban development in the Pearl River Delta and even in China. We found that the three metropolises are at different phases of urbanization. Hong Kong has the highest population density and the most intensive land use, followed by Shenzhen and Guangzhou. In terms of vegetation coverage, atmospheric environment quality, surface water environmental quality, and urban heat island effect, Guangzhou has the highest overall quality, followed by Hong Kong and Shenzhen. However, there are significant differences in ecological environment evolution for the metropolises, which are shaped by natural and geographical conditions, economic development pathways, and environmental protection policies.

Keywords Urbanization process · Eco-environmental effects
Pearl River Delta region · Evaluation index system

R. Xiao (✉) · Z. Li
Guangdong Provincial Academy of Environmental Science, 335 Dongfeng Middle Road, Guangzhou 510045, Guangdong, China
e-mail: ecoxiaorb@163.com

R. Xiao
School of Environmental Science and Engineering, Guangdong University of Technology, Guangzhou 510045, Guangdong, China

C. Wu
Department of Landscape Architecture, College of Horticulture and Forest Science, Huazhong Agricultural University, Wuhan 430070, Hubei, China

X. Yang and S. Jiang (eds.), *Challenges Towards Ecological Sustainability in China*, https://doi.org/10.1007/978-3-030-03484-9_7

1 Introduction

Since 1978 when the open-door policy initiated and economic reform began, China's urbanization has developed with a low starting point but a high speed driven by intensifying industrialization. From 1978 to 2013, the permanent urban population in China increased from 170 million to 730 million, and the urbanization rate increased from 17.9 to 53.7% with an average annual rate of 1.02% (Wu 2015). The number of cities also increased from 193 to 658 according to the 2014–2020 National New-Type Urbanization Plan. However, urbanization in China, problems such as loss of cultivated lands, decline of ecological quality and serious environmental pollution have emerged, which have restricted further development (Li et al. 2009). Therefore, how to maintain a good ecological quality in a rapid urbanization environment has become an important issue for China to promote the sustainable urban development.

There are obvious stage characteristics for urbanization and its eco-environmental influences, and the regional differences among cities may be due to the differences in the natural conditions, historical settings, location and policy orientations (Fang et al. 2016; Yang et al. 2017). To understand the eco-environmental effects of urbanization, it is necessary to conduct a comparative analysis of long-term changes in cities within a certain region. However, most of the existing studies largely focused on the overall impact (e.g., He and Zhuang 2006; Liu and Wang 2015; Fang and Ren 2017) or a single major city (Chang et al. 2012; Xie et al 2015), with few comparative studies considering different biophysical and socio-economic conditions. Such comparative studies can offer an important insight that can help urban planning and management.

Pearl River Delta has been the frontrunner in China's economic reform and the most representative region in terms of urban development in the country. Urbanization rate in this region has risen from 17.92% in 1978 to 82.72% in 2010. Meanwhile, the ecological environment has gradually degraded. In this study, we selected three most representative metropolises in the Pearl River Delta, namely, Guangzhou, Shenzhen and Hong Kong, as the study site to systematically analyze and evaluate the urbanization process and ecological environment quality since 1980, and discuss the influences of economic and social development models and urban management policies in order to provide a scientific understanding of the green urban development in the Pearl River Delta region and even in China.

2 Research Methods

2.1 Study Sites: Guangzhou, Shenzhen and Hong Kong

Guangzhou, the capital of Guangdong Province, is the largest city in southern China. Before the economic reform, the urban spatial structure of Guangzhou was relatively stable. It has experienced a rapid expansion in the 1980s and formed a huge agglomerate in the late 1990s (Xu 2004). With new "frog-leaping" development zones continued to emerge in the periphery, Guangzhou has become a typical urban space expansion model of China (Su et al. 2005). Therefore, for studying on urbanization in the Pearl River delta, Guangzhou should be a good case to include.

Shenzhen is located at the eastern coast of Pearl River estuary. With the advantage as one "special economic zone" and its geographical location being g adjacent to Hong Kong and Macao, Shenzhen has emerged from a small coastal fishing village to an international and modern metropolis since the city establishment in 1979. In 2004, Shenzhen completed the work changing all towns into sub-district offices and became the first and only megacity in China that does not include any rural area. Although Shenzhen is a megacity with a population of over 10 million, its physical size is rather small, and the land resources for construction are very scarce (Xie and Chen 2013). Shenzhen's "concentrated" and "fast-paced" urban development can be viewed as a typical case of urbanization in the Pearl River Delta.

Located at the eastern side of the Pearl River estuary and being adjacent to Shenzhen to the north, Hong Kong Special Administrative Region has evolved from a colonial port trade city into an international shipping center, trade center and financial center since 1841. Its development and urban construction are subject to the great influence of Western and Eastern international political, economic and cultural changes, which are different from the economic and cultural systems in Guangzhou and Shenzhen. In addition, Hong Kong was already a highly urbanized area in the 1970s, and therefore its urbanization process in the latter period has been stabilized, which is also different from the rapid urbanization pattern in Guangzhou and Shenzhen. Studying Hong Kong, a highly urbanized and land-intensive international city, can provide valuable historical experience for transformation and upgrading of other cities in the Pearl River Delta region.

As the three major metropolises in the Pearl River Delta region (Fig. 1), Guangzhou, Shenzhen and Hong Kong are very different in urbanization process and economic development models, and face different challenges in balancing urban development and ecological environment maintenance. Therefore, a comparative analysis of Guangzhou, Shenzhen and Hong Kong over the past 30 years can help understand and analyze the eco-environmental effects of urbanization in the Pearl River Delta Region and provide a reference for sustainable development of urban areas in the Pearl River Delta and even in other parts of China.

2.2 Data

The data for this study was derived from diverse sources. Socio-economic data, such as population, economic, industrial structure, were collected from official website published data of Guangzhou Statistical Bureau (http://www.gzstats.gov.cn/), Shenzhen Statistical Bureau (http://www.sztj.gov.cn/) and Hong Kong Census and Statistics Department (http://www.censtatd.gov.hk/). Water environment, atmosphere environment and related Environmental quality data came from Guangzhou Environmental Quality Report (Guangzhou Environmental Protection

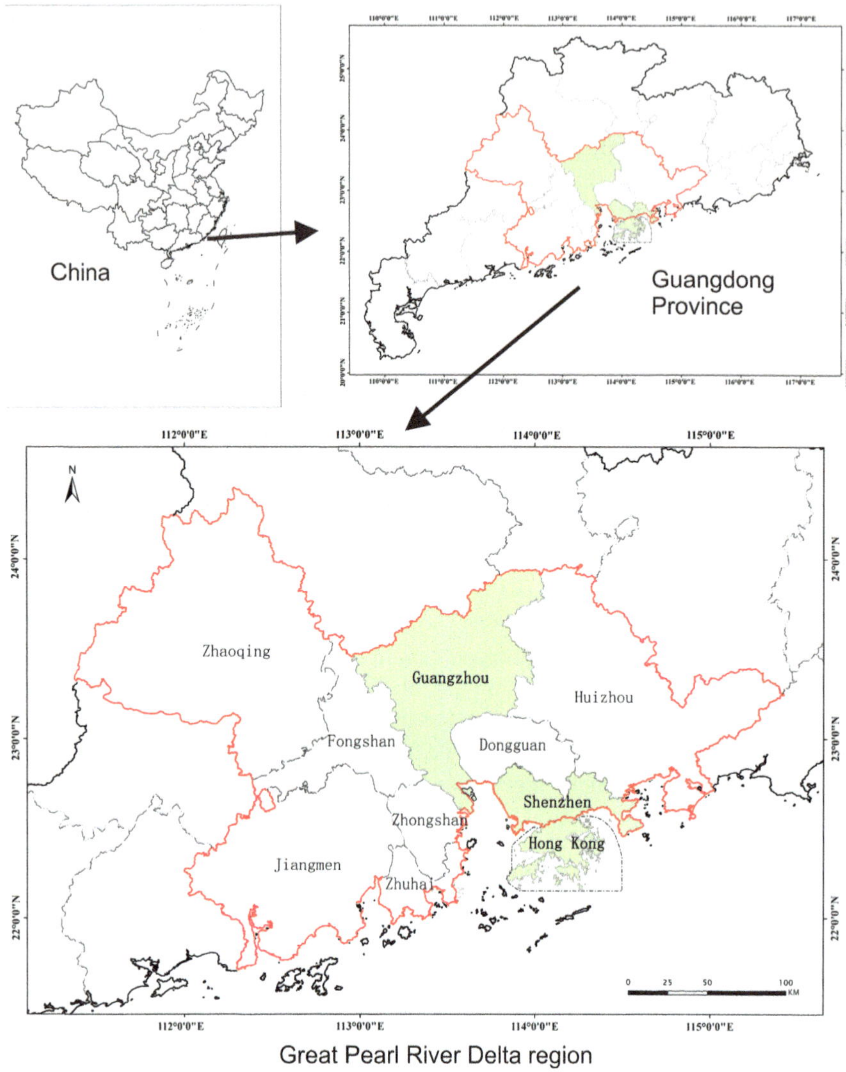

Fig. 1 Location of Guangzhou, Shenzhen, and Hong Kong

Bureau 2001–2011), Shenzhen Environmental Quality Report (Shenzhen Environmental Protection Bureau 2001–2011) and data published from the official websites of Hong Kong Environmental Protection Department (http://www.epd.gov.hk/).

Land-Use and Land-Cover Change (LUCC) data were derived from Landsat MSS (Multispectral Scanner)/TM (Thematic Mapper) images acquired in 1980, 1990, 2000, 2005, and 2010. Except Landsat MSS image data in 1980, all others were Landsat TM images. Object-oriented classification was conducted to derive land-use maps with several different land-use types including built-up area, forest, grassland, farmland, wetland, and others (Xiao et al. 2017).

In order to quantitatively evaluate the heat island intensity, we used Urban Heat Island Ratio Index (URI) (Xu and Chen 2003). Temperature data were obtained from China 1 km monthly night time surface temperature synthetic MODIS products (TERRA), which were downloaded from the website of the Computer Network Information Center, Chinese Academy of Sciences (http://www.gscloud.cn/). We composed the average annual surface temperature data by monthly data.

$$URI = \frac{1}{100m}\sum_{i=1}^{n} w_i p_i$$

where, m is the temperature normalized rank index, i is the ith temperature level above which the heat island is above the low temperature region, n is the temperature level number above the low temperature region, w is the weight value, and P_i is the percentage of the ith level. In this study, urban surface temperature rating is set at five levels, so m is 5; 5 and 4 levels are defined as the urban heat island range, so n is 2.

2.3 The Evaluation Index System

Considering the data availability, relevant research outcomes, and the actual situation in the Pearl River delta, we created an index system for urbanization and ecological environment quality evaluation in which the evaluation of urbanization levels includes three first-class indexes of land, population or economic urbanization, and the evaluation of ecological environment quality includes the proportion of natural vegetation coverage to urban area, biomass per unit area of forest, and the number of days with air quality above Grade II Standard, the proportion of rivers with case III water and above, and heat island intensity (Table 1).

Table 1 Urbanization and ecological environment evaluation index system

Items	Evaluation indexes
Urbanization indexes	Proportion of built-up area to urban area (%)
	Population density of built-up area (person/km^2)
	Proportion of output value of secondary and tertiary industries to GDP (%)
Eco-environment indexes	Proportion of natural vegetation coverage to urban area (%)
	Days with air quality above Grade II Standard (days)
	Proportion of rivers with case-III water and above (%)
	Heat island intensity

With equal weights, ecosystem quality index (EQI) uses the proportion of natural vegetation, biomass per unit area of forest, the proportion of the number of days with air quality reaching the standard, and the proportion of rivers with case-III water and above to establish the ecological quality index that is used to reflect ecological quality of urban agglomerations.

$$EQI_i = \sum_{j=1}^{n} w_j r_{ij}$$

where, EQI_i is the ecological quality index of the *i*th city, which is the relative weight of each index and the standardized value of each index of the *i*th city.

3 Results

3.1 Urbanization Indexes from 1980 to 2010

3.1.1 Land Urbanization

The change of built-up land can reflect the speed of urban expansion to a certain extent. The spatial expansion of urban built-up land in Guangzhou, Shenzhen and Hong Kong for the past 30 years mainly occurred between 1980 and 2000, with the most rapid expansion in 1990–2010 (Fig. 2). However, the patterns of urban spatial expansion are different for the three metropolises. Guangzhou mainly developed externally around the city center in circles, showing a trend of expansion from high density in the central area to medium and low density around circles of the outer area. The urban expansion in Shenzhen was concentrated on the built-up area in the Special Zone (including Nanshan, Futian, and Luohu districts), and gradually expanded and formed a multi-center axis-belt-cluster structure along mountains and water systems. Hong Kong changed little in the built-up land, which is mainly in a belt-like form along the coast, and discontinuous and shows an overall leap cluster expansion.

Based on the statistical data (see Table 2), the built-up land area in descending order was 1,492.08 km^2 in Guangzhou, 804.83 km^2 in Shenzhen and 229.88 km^2 in Hong Kong as of 2010. From 1980 to 2010, the annual growth of built-up land in Guangzhou, Shenzhen and Hong Kong was 26.76 km^2, 18.64 km^2, and 2.36 km^2, respectively. From the perspective of land urbanization, Guangzhou and Shenzhen were in the stage of rapid urbanization, especially during 1990–2000 when the expansion of built-up land was the fastest with the annual growth rate of 45.09 km^2 and 32.11 km^2, respectively. However, the land urbanization was basically stable in Hong Kong. In terms of land urbanization intensity, Shenzhen was the highest among the three metropolises, which was 42.61%, while the intensity in Guangzhou and Hong Kong was more or less identical, which was around 20%.

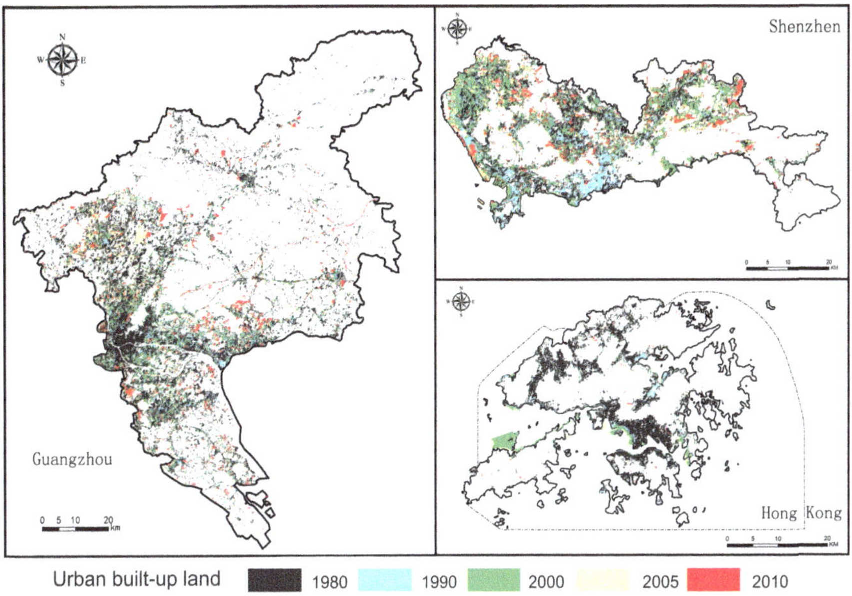

Fig. 2 The spatial distribution of urban built-up land in Guangzhou, Shenzhen and Hong Kong from 1980 to 2010

3.1.2 Population Urbanization

From 1980 to 2010, the population of Guangzhou, Shenzhen and Hong Kong shows an increasing trend. However, the average annual growth of 330,000 people in Shenzhen is equivalent to the population of a typical medium-sized Chinese city, which is much larger than the rate of 100,000 in Guangzhou and 68,000 in Hong Kong each year (Fig. 3). As of 2010, the population size of permanent residents in the descending order was 10,372,000 in Shenzhen, 8,042,000 in Guangzhou, and 7,052,000 in Hong Kong. However, from the perspective of population urbanization intensity, Hong Kong was still the highest among the three cities, and its the population density of the built-up area remained at 30,000 persons/km^2 in the past 30 years. As of 2010, the population density of the built-up area in Hong Kong was 30,677 persons/km^2, which is 2.4 times and 3.6 times that in Shenzhen (12,887 persons/km^2) and in Guangzhou (5390 persons/km^2), respectively.

3.1.3 Economic Urbanization

From 1980 to 2010, the GDP of the three cities shows an increasing trend, with an average annual growth of RMB35.636 billion in Guangzhou, RMB31.929 billion in Shenzhen, and RMB50.148 billion in Hong Kong (Table 3). As of 2010, the GDP in descending order was RMB1.548111 trillion in Hong Kong, RMB1.074828 trillion

Table 2 Summary of the urban built-up land growth in Guangzhou, Shenzhen and Hong Kong from 1980 to 2010

Cities	Indicators	Year				
		1980	1990	2000	2005	2010
Guangzhou	Built-up land area (km^2)	689.27	761.51	1212.37	1331.49	1492.08
	Average annual increase (km^2)	NA	7.22	45.09	23.82	32.12
	Percentage to urban area (%)	9.61	10.61	16.91	18.55	20.79
Shenzhen	Built-up land area (km^2)	205.56	333.20	654.26	712.15	804.83
	Average annual increase (km^2)	NA	12.76	32.11	11.58	18.54
	Percentage to urban area (%)	10.88	17.64	34.64	37.70	42.61
Hong Kong	Built-up land area (km^2)	159.12	175.07	214.60	227.88	229.88
	Average annual increase (km^2)	NA	1.60	3.95	2.66	0.40
	Percentage to urban area (%)	14.40	15.84	19.42	20.62	20.80

in Guangzhou, and RMB 0.958151 trillion in Shenzhen. From the perspective of industrial structure, the economic urbanization intensity of the three cities was more or less the same, but there are some differences in the industrial structure proportion. The proportion of tertiary industry in Guangzhou exceeded that of the secondary industry in 1990, and the industrial structure showed a transition from "industrial economy" to "service economy", and tertiary industry accounted for 61% of the total GDP in 2010. Shenzhen relied on its geographical advantages to actively undertake the development opportunity from relocation of Hong Kong's manufacturing industry to Mainland. The secondary and tertiary industries in Shenzhen were balanced developing. In 2010, a new industrial system was formed, which was adaptive to the modern city central functions based mainly on high-tech and advanced manufactur-

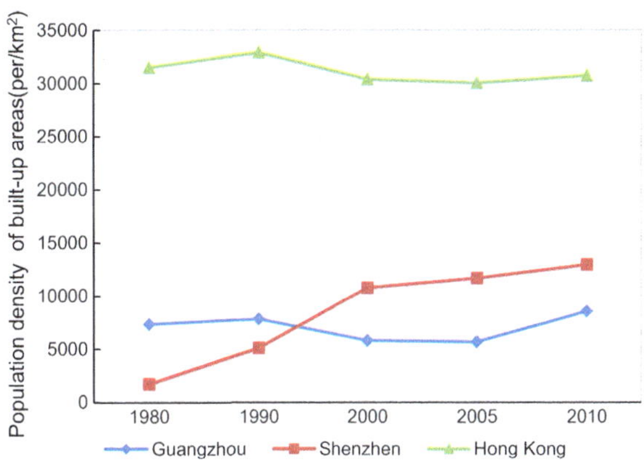

Fig. 3 Changes in the population density of the built-up area in Guangzhou, Shenzhen and Hong Kong from 1980 to 2010

ing industries supported by modern service industry. In the past 30 years, Hong Kong had been actively promoting industrial upgrading and rapidly developing modern service industries such as trade and logistics, financial services, professional services and tourism as its pillar industries, which accounted for 93% of GDP by 2010. Hong Kong is now a major service center in Asia and in the world.

3.1.4 Urbanization Stages

According to the three-stage theory of urban development (Fang et al. 2008), Guangzhou, Shenzhen and Hong Kong were at different stages of urbanization. Hong Kong was at the later stage of urbanization, while Guangzhou and Shenzhen had undergone the initial and mid-urbanization stages or the initial and accelerated development stage of urbanization since the 1980 s and maintained a high rate of built-up area expansion (Table 4). In particular, the built-up area of Shenzhen expanded 3.9 times from the early 1980s to 2010, with the permanent population and GDP increasing by 31.4 times and 3548.5 times, respectively. This remarkable, rapid population and economic growth has been labeled as the "Shenzhen Speed". However, in terms of land utilization, population density and GDP, Hong Kong had the highest level of land intensive use, followed by Shenzhen and Guangzhou. This may be attributed to the high proportion of hard-to-use land such as mountainous areas in Hong Kong and the efficient use of built-up land due to strict land management system in Hong Kong. Guangzhou and Shenzhen were at a period of rapid economic development. Although the population was highly concentrated, a large amount of land had been extensively used due to the imperfect land management system, and the population density was relatively small.

Table 3 Summary of the output value proportions for secondary and tertiary industries in Guangzhou, Shenzhen and Hong Kong from 1980 to 2010

Cities	Indicators	Year				
		1980	1990	2000	2005	2010
Guangzhou	GDP (RMB100 million)	57.55	319.60	2375.91	5154.23	10748.28
	Share of secondary industry to GDP (%)	54.53	42.65	43.44	39.68	37.24
	Share of tertiary industry to GDP (%)	34.63	49.30	52.59	57.79	61.01
Shenzhen	GDP (RMB100 million)	2.70	171.67	2187.45	4950.91	9581.51
	Share of secondary industry to GDP (%)	26.05	44.81	49.65	53.37	47.21
	Share of tertiary industry to GDP (%)	45.07	51.09	49.64	46.43	52.72
Hong Kong	GDP (RMB100 million)	436.75	3678.83	14201.59	14869.68	15481.11
	Share of secondary industry to GDP (%)	26.10	24.10	13.13	8.67	6.99
	Share of tertiary industry to GDP (%)	72.89	75.62	86.77	91.26	

Table 4 Comparison of the multiples of the expansion and population and economic growth for urban built-up areas in Guangzhou, Shenzhen, and Hong Kong from 1980 to 2010

Cities	Multiple of built-up land expansion	Multiple of resident population growth	Multiple of GDP growth
Guangzhou	2.2	1.6	188.3
Shenzhen	3.9	31.4	3548.5
Hong Kong	1.4	1.4	35.5

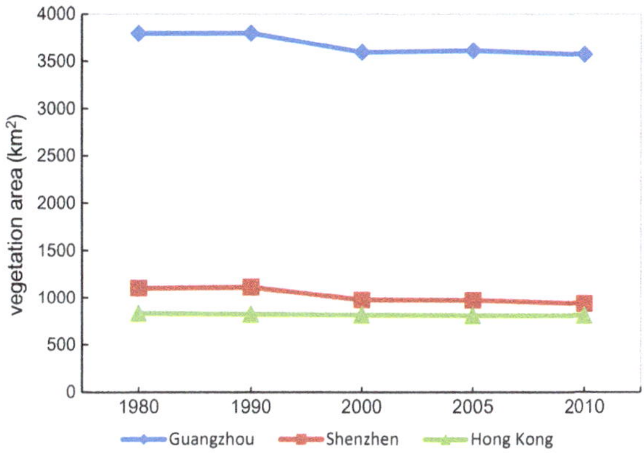

Fig. 4 Vegetation area of Guangzhou, Shenzhen and Hong Kong from 1980 to 2010

3.2 Changes of Ecological Environment Quality from 2000 to 2010

3.2.1 Vegetation Coverage

From the comparative analysis of vegetation coverage (Fig. 4), the natural vegetation area occupied by urban construction in descending order was 225 km^2 in Guangzhou, 169 km^2 in Shenzhen, and 20 km^2 in Hong Kong. Given a city's total land area, the reduced area of natural vegetation accounted for only 2–3% in Guangzhou and Hong Kong and 9% in Shenzhen. Comparatively analyzing forest land landscape patterns (Table 5) shows that landscape heterogeneity and fragmentation in Guangzhou and Shenzhen were high and landscape diversity was obviously higher than that of Hong Kong. The spatial correlation of land use types in Hong Kong was higher than that of Guangzhou and Shenzhen, and the spatial connectivity of all kinds of land was higher, and land use landscape stability was better.

3.2.2 Air Quality

The comparative analysis of sulfur dioxide and respirable particulate matter concentration in the air (Figs. 5 and 6) show that Guangzhou had the worst atmospheric quality, followed by Shenzhen and Hong Kong, but in general the atmospheric quality of the three cities from 2000 to 2010 tended to improve.

3.2.3 Surface Water Quality

Our comparative analysis (Fig. 7) shows that the water quality of Hong Kong streams had always been at the optimal level with the minimum and maximum proportions of waters above case-III being 67 and 83% from 2000 to 2010. Guangzhou ranked

Table 5 Forest landscape characteristic indices of Guangzhou, Shenzhen and Hong Kong from 1980 to 2010

Cities	Years	Patch number PN (pcs)	Mean patch size MPS (hm^2/pcs)	Patch edge density PED (m/hm^2)	Mean shape index MSI
Guangzhou	1980	10385	24.22	32.96	1.57
	1990	10087	24.93	32.74	1.57
	2000	7738	29.66	28.42	1.60
	2005	7389	31.19	27.98	1.61
	2010	7262	30.98	27.14	1.60
Shenzhen	1980	3402	22.38	36.36	1.51
	1990	3118	24.71	36.52	1.54
	2000	2630	23.83	30.52	1.60
	2005	2375	26.14	29.29	1.61
	2010	2384	24.46	27.90	1.60
Hong Kong	1980	651	126.81	23.47	1.47
	1990	772	105.69	24.39	1.53
	2000	862	93.61	24.66	1.53
	2005	854	93.88	25.01	1.53
	2010	877	91.31	25.40	1.53

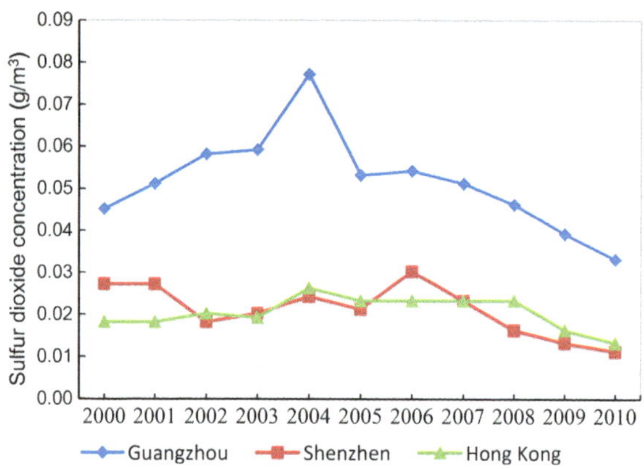

Fig. 5 Sulfur dioxide concentration in the atmosphere of Guangzhou, Shenzhen and Hong Kong from 2000 to 2010

the second with 50% of waters above case-III for 6 years and 35.7% of the minimum proportion in 2004. The water quality of rivers in Shenzhen was the worst with less than 8% of waters above case-III, especially in 2008, and the water quality of rivers for the whole city didn't meet the case-III standard.

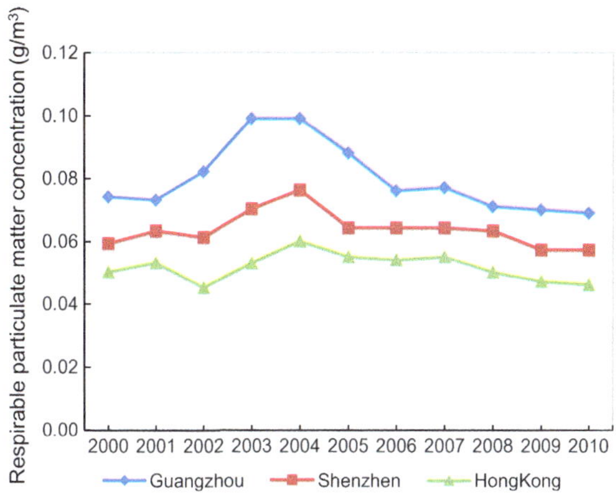

Fig. 6 Respirable particulate matter concentration in the atmosphere of Guangzhou, Shenzhen and Hong Kong from 2000 to 2010

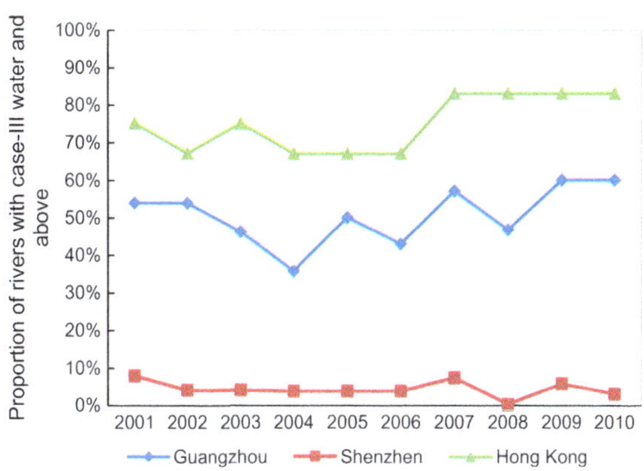

Fig. 7 Proportion of rivers with case-III water and above in Guangzhou, Shenzhen and Hong Kong from 2000 to 2010

3.2.4 Urban Heat Island

Our comparatively analysis shows that urban heat island ratio indexes in the three cities fluctuated slightly from 2000 to 2010 without significant change (Fig. 8). The nighttime urban heat island intensity in Shenzhen was basically the same to that in Guangzhou, and the lowest was in Hong Kong.

3.2.5 Ecological Quality

Our comparative analysis of the ecosystem quality index changes during 2000 to 2010 (Fig. 9) shows that Guangzhou ranked the first in the overall environmental quality, followed by Hong Kong and Shenzhen. During the period of 2000 to 2010, the overall quality of ecological environment in Guangzhou became better mainly due to the improved atmospheric and water environment quality. Although the atmospheric

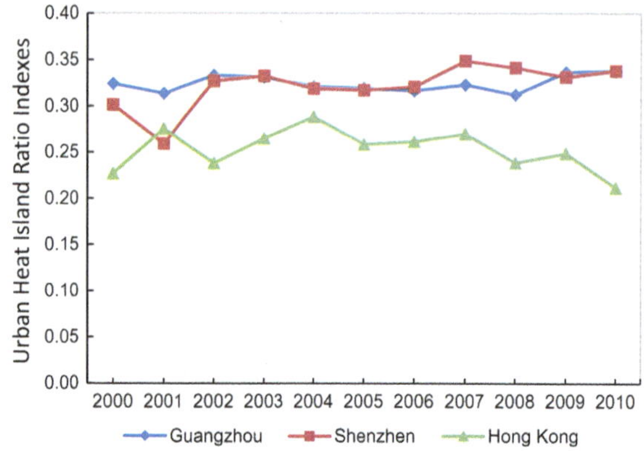

Fig. 8 Changes of Urban Heat Island Ratio Indexes in Guangzhou, Shenzhen and Hong Kong from 2000 to 2010

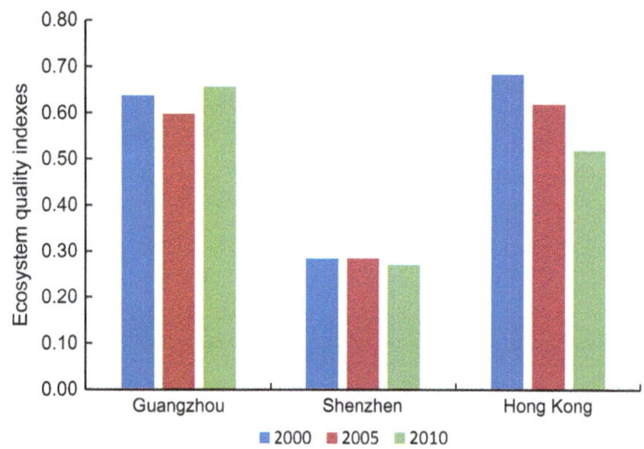

Fig. 9 Ecosystem quality indexes in Guangzhou, Shenzhen and Hong Kong from 2000 to 2010

environment improved in Shenzhen, the water environment problem was serious and the proportion of natural vegetation invaded was higher, which brought about less change to the comprehensive index of ecological environment. The atmosphere and water environment in Hong Kong were stable but the comprehensive index of ecological environment dropped significantly due to the vegetation damage and the increasingly serious heat island problem.

4 Discussion and Conclusions

4.1 Discussion

The urban expansion and eco-environment consequences are mainly controlled by factors such as nature, society and economy. In this chapter, we analyzed the driving forces for urban expansion and its eco-environmental effects in Guangzhou,

Shenzhen, and Hong Kong from the aspects of natural and geographical conditions, economic growth models, and environmental protection policies.

4.1.1 Natural and Geographical Conditions

As the fundamental condition for urban land expansion, geographical settings largely determine the overall trend of urban expansion, affect the direction, speed, model and spatial form of urban growth, and sometimes even become the limit of urban form evolution. The urban built-up land in Guangzhou and Shenzhen tended to expand into the landform types conducive to land development, such as plains and hills, which are mainly characterized by high-speed, extensive and epitaxial land use patterns. The inefficient and poorly coordinated use of urban land was a major issue. On the other hand, Hong Kong has enforced a strict land management system. The high-density and high-intensity comprehensive utilization of land makes the construction area to be mainly concentrated on about one quarter of the land. Most of the suburban environment is preserved in order to protect nature and biodiversity. It becomes a typical example of an intensive and livable city in the world.

4.1.2 Economic Growth Model

Economic development is a major force driving and shaping the dynamic process of urbanization, and the most important factor affecting the quality of regional ecological environment. Both Guangzhou and Shenzhen adopted the economic growth model that "makes money from land" at the early stage of urbanization and at the expense of losing a large amount of agricultural production resources. This model of development was also a natural choice given the economic fundamentals and environmental conditions. At the accelerating urban development period, the economic development models began to change in the two cities. Guangzhou continued to follow the path of traditional industrialization development and promoted urbanization through "investment". With limited land resources, Shenzhen tried hard to change its model of economic growth, taking the low-carbon development road and shifting to a low-carbon economy with low energy consumption, emission and pollution (Liu and Wang 2010). Hong Kong took the post-industrialization path in the 1980s and paid a great attention to improve land use efficiency in urban development. In a knowledge-intensive growth model with high technical content and highly added value, Hong Kong reduced consuming natural resources by in urbanization.

4.1.3 Environmental Protection Policies

The environmental problems presented in the rapid urbanization in Guangzhou and Shenzhen are characterized by accumulation, structure and compactness. Although the urban environmental quality has been improved by formulating a series of environmental protection policies, further improvement of the environmental quality has been under greater pressure as the rapid urbanization development stage, the large and continued increasing population of permanent residents and the continued increasing supply demand for land, energy and water resources will make the two cities to maintain large increases in pollutants. Hong Kong gradually established a set of sound environmental protection legal systems at 1950s so that the environmental governance in Hong Kong has achieved good results.

5 Conclusions

Guangzhou, Shenzhen and Hong Kong were at different stages of urbanization. Hong Kong was in the later stage of urbanization, while Guangzhou and Shenzhen had experienced the initial development stage and the accelerating development stage of urbanization since 1980s. In terms of the land use level, population density, and GDP, Hong Kong had the highest land intensive use level, followed by Shenzhen and Guangzhou. As for comprehensive vegetation coverage, atmospheric environment quality, surface water environmental quality and urban heat island effect, Guangzhou had the highest overall environmental quality, followed by Hong Kong and Shenzhen. From 2000 to 2010, Guangzhou generally became better in terms of its environmental quality due to the improved atmosphere and water quality. Although the atmospheric environment in Shenzhen improved, the comprehensive index didn't change much due to serious water environment problems. The atmospheric and water environment was stable in Hong Kong. However, the problems of vegetation deterioration and urban heat island became more and more serious, and thus comprehensive index dropped significantly. The evolution of urbanization and ecological environment quality in Guangzhou, Shenzhen, and Hong Kong has been controlled by natural and geographical conditions, economic growth models, and environmental protection policies.

Acknowledgements This work was supported by National Natural Science Foundation of China (#31470703).

References

Chang Q, LI S, Wang Y, Qiu Y (2012) Green landscape evolution and its diving factors in Shenzhen. Acta Geogr Sin 67(12):1611–1622

Fang CL, Liu XL, Lin XQ (2008) Stages correction and regularity analysis of urbanization of China. Arid Land Geogr 31(4):512–523

Fang CL, Ren YF (2017) Analysis of emergy-based metabolic efficiency and environmental pressure on the local coupling and telecoupling between urbanization and the eco-environment in the Beijing-Tianjin-Hebei urban agglomeration. Sci China Earth Sci 60:1083–1097

Fang CL, Wang Y, Fang JW et al (2016) A comprehensive assessment of urban vulnerability and its spatial differentiation in China. Acta Geogr Sin 26(2):153–170

Guangzhou Environmental Protection Bureau (2001) Guangzhou Environmental Quality Report—1996–2000, Guangzhou Municipality, Guangdong Province

Guangzhou Environmental Protection Bureau (2006) Guangzhou Environmental Quality Report—2001–2005, Guangzhou Municipality, Guangdong Province

Guangzhou Environmental Protection Bureau (2011) Guangzhou Environmental Quality Report—2006–2010, Guangzhou Municipality, Guangdong Province

He JF, Zhuang DF (2006) Analysis of the relationship between urban dynamic change pattern of the Yangtze River Delta and the regional eco-environment. Geogr Res 25(3):388–396

Li SC, Zhao ZQ, Wang YL (2009) Urbanization process and effects of natural resource and environment in China: research trends and future directions. Prog Geogr 28(1):63–70

Liu WL, Wang C (2010) Practice and patterns of low carbon city development. China Popul Resour Environ 20(4):17–22

Liu YY, Wang SJ (2015) Coupling coordinative degree and interactive coercing relationship between urbanization and eco-environment in Pearl River Delta. Hum Geogr 3:64–71

New national urbanization planning (2014–2020) from http://ghs.ndrc.gov.cn/zttp/xxczhjs/ghzc/201605/t20160505_800839.html.2016.05.05

Shenzhen Environmental Protection Bureau (2001) Shenzhen Environmental Quality Report—1996–2000, Shenzhen Municipality, Guangdong Province

Shenzhen Environmental Protection Bureau (2006) Shenzhen Environmental Quality Report—2001–2005, Guangzhou Municipality, Guangdong Province

Shenzhen Environmental Protection Bureau (2011) Shenzhen Environmental Quality Report—2006–2010, Shenzhen Municipality, Guangdong Province

Su JZ, Wen QQ, Guo HL (2005) The mechanism and adjustment of urban sprawl of Guangzhou. J Geog Sci 60(4):626–636

Wu J (2015) A research on the development path of new urbanization based on low-carbon perspective. Can Soc Sci 11(3):309–313

Xiao RB, Li ZS, Wu ZF et al (2017) Urbanization in the Pearl River Delta and its eco-environmental effects. Science Press, Beijing

Xie DX, Cheng HQ (2013) Land problems and reflecting in highly urbanized areas: a case study of Shenzhen. Green Econ (Chin) 4:48–51

Xie GD, Zhang B, Lu CX et al (2015) Rapid expansion of the metropolitan areas and impacts of resources and the environment. Resour Sci 37(6):1108–1114

Xu HQ, Chen BQ (2003) An image processing technique for the study of urban heat island changes using different seasonal remote sensing data. Remote Sens Technol Appl 18(3):129–133

Xu RS (2004) The history, current status and future of Guangzhou City. Constr Dyn Guangzhou 8:8–17

Yang KC, Bi RC, Sun RH et al (2017) The spatio-temporal changes of urbanization in Beijing-Tianjin-Hebei region in Northern China. Acta Ecol Sin 37(12):3998–4007

Dr. Rongbo Xiao is the Chief Engineer of the Guangdong Provincial Academy of Environmental Science, China, and an Adjunct Professor at the Guangdong University of Technology. Xiao's research interests include urban and regional ecology and ecological planning and restoration. He has authored or co-authored 3 books and over 40 articles in these areas.

Dr. Changguang Wu is an Associate Professor at the Huazhong Agricultural University. His research interests include urban ecology, ecological health, green space efficiency, and climate adaptability. Wu has authored or co-authored multiple peer-reviewed articles in these areas.

Zhishan Li is an Engineer at the Guangdong Provincial Academy of Environmental Science, China. His major research area includes urban and regional ecology.

Nutrient and Trace Metal Issues in the Pearl River Delta, China

Lichun Xie, Lei Gao and Jianyao Chen

Abstract The Pearl River Delta (PRD) has undergone rapid urban growth and industrial development during the past four decades, and human activities have become one of the most important factors affecting the environment in the region. Therefore, there is an urgent need to examine the influence of urbanization on surface water and groundwater systems. Based on data collection and analyses of industry, agriculture, environmental protection efforts, natural conditions, and population statistics in 2000, 2005, and 2010, we assessed the nitrogen (N) and phosphorus (P) budgets and their regional differences in the PRD between the three time periods. The N and P input and output varied greatly during 2000–2010, whereas the N and P surplus increased continuously. In addition, intense human activities induced severe trace metal pollution in the Shima River close to an important water supply source. Zinc and copper concentrations markedly exceeded the national water quality standards (Class I) in the dry season. Meanwhile, various pollution sources significantly contributed to metal accumulation in riverine sediments, leading to a slight enrichment in lead, manganese, and iron, and moderate-to-heavy enrichment of chromium, nickel, copper, zinc, and cadmium. Hierarchical cluster analysis indicates that sediment pollution caused by trace metals was mainly associated with industrial and agricultural activities.

Keywords Pearl River Delta · Environmental mediums · Nutrients budget
Trace metals pollution · Water safety

L. Xie
School of Geography and Tourism, Guangdong University
of Finance & Economics, Guangzhou 510320, Guangdong, China

L. Gao · J. Chen (✉)
School of Geography and Planning, Sun Yat-sen
University, Guangzhou 510275, Guangdong, China
e-mail: chenjyao@mail.sysu.edu.cn

X. Yang and S. Jiang (eds.), *Challenges Towards Ecological Sustainability in China*, https://doi.org/10.1007/978-3-030-03484-9_8

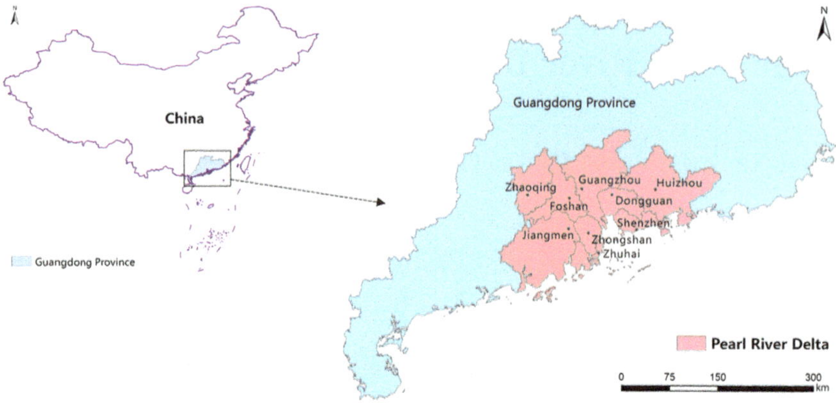

Fig. 1 Location of the Pearl River Delta

1 Introduction

The Pearl River Delta (PRD) is one of the highly developed areas in China. Geographically, it includes nine cities, i.e., Guangzhou, Shenzhen, Zhuhai, Foshan, Huizhou, Zhaoqing, Jiangmen, Zhongshan, and Dongguan, with a total area of 41,700 km^2 (Fig. 1). Since the beginning of China's economic reform in 1978, this region has undergone rapid industrialization and urbanization, making it one of the largest manufacturing centers in China. During 1978–2007, the gross domestic product (GDP) in the PRD region increased at an average annual rate of 21.2%, accounting for approximately 10.2% of the total national GDP in 2006 (Huang et al. 2012). However, rapid socioeconomic development comes at the cost of the environment. According to Chen et al. (2006a, b), more than 3×10^9 tons of industrial and domestic wastewater were annually discharged into surface water systems, which resulted in increasing levels of trace metals and nutrients, ultimately leading to severe environmental issues such as pollution and eutrophication of surface water bodies.

Nitrogen (N) and phosphorus (P) play crucial roles in many biogeochemical processes and functions as important controlling factors in terrestrial and aquatic ecosystems. The N and P cycles are the two major material cycles of the earth's bio-, atmo-, hydro-, and geo-spheres. In recent centuries, the development of human civilization, accompanied by rapid urbanization, explosive population growth, and industrial and agricultural revolutions, tremendously increased nutrient input into the environment, altering N and P cycles. While contributing to the growth of agricultural and industrial production, excessive N and P accumulate on land or are discharged into water or atmosphere, exacerbating environmental problems including the greenhouse effect, eutrophication, and acid rain, with visible effects on land and sea, particularly in coastal areas. Since about 60% of the human populations live within 100 km from coasts, eutrophication in coastal waters can cast an immediate impact on humans, making it a global concern (Valiela et al. 1992; Jonge et al. 2002).

Trace metal pollution has received increasing interest due to its unique physicochemical properties (e.g., persistence, non-degradability, toxicity, and bioaccumulation) (Li et al. 2013), can have adverse impacts on aquatic organisms through bioaccumulation and biomagnification, thus causing health problems after entering the food chain (Varol and Şen 2012). Furthermore, water pollution caused by heavy metals directly threatens human health via water consumption. Thus, heavy metals pollution is arousing growing public concerns worldwide.

In general, the PRD area suffered from complex sources of pollution (e.g. nutrients and heavy metals, organic compounds). Understanding the distribution, magnitude, potential sources and environmental behaviors of major pollutants in the different mediums is important for implementing pollution prevention and control measures. However, little research regarding the nutrients and metals pollution has been done in the highly urbanized watersheds of the PRD region. Consequently, the aims of our study were: (1) to estimate nutrients budgets in the whole PRD area, (2) to select Tangjiawan town as one of the study areas in Zhuhai city to investigate spatial and temporal characteristics of major nutrients (nitrogen and phosphorous) in different water bodies, (3) to assess trace metals pollution in the river and sediments from the Shima River, a highly urbanized watershed close to an important water supply source in Dongguan city, and (4) to reveal the impacts of human activities on the environmental behaviors of nutrients and trace metals in water bodies, and to preliminarily identify their potential sources.

2 Case Study: Nitrogen and Phosphorus Budgets in the PRD Area

The N and P budgets within a given region will need to be quantitatively evaluated before they can be effectively understood and managed. In the past 30 years, the PRD has experienced economic growth, urbanization, and human activities on an unprecedented scale, and the resulting changes and regional variations in N and P budgets and nutrient output in water deserve intensive studies. This section examines relevant data for the years of 2000, 2005, and 2010, and explores the N and P budgets of the PRD including their histories, regional variations, and potential impacts. All data related to N and P budget calculation were collected from the *Statistical Yearbook of Guangdong Province* for each year published by the provincial Bureau of Statistics, including land cover area, population, agriculture, industries, and environmental protection information for the nine cities of the PRD.

2.1 Nitrogen Budget Estimates and Regional Variation Analysis

Input, output, and surplus are three components that are necessary to estimate a regional N budget. The parameters and computation methods below are partially based on several existing studies (Galloway et al. 1996; Zhu 1997; Xing and Yan 1999; Xing and Zhu 2000, 2002; Galloway 2005; Nancy and Carly 2005; Liu et al. 2006; Deng et al. 2007; Russell et al. 2008; Chen and Jia 2009).

2.1.1 Nitrogen Input: General Features and Regional Variation

The N input of the nine PRD cities was calculated for 2000, 2005, and 2010 based on their statistics (Table 1). The total N input of the PRD did not vary much during the three periods, ranging from 84.71×10^4 to 91.16×10^4 t, and it was slightly lower in 2010 than in 2000. Each N source generally had a similar contribution rate in the three periods and across the regions, showing the descending order of fertilizer, human and livestock excreta, wet deposition, wastewater NH_3-N, farmland symbiotic fixation, crop residue, and farmland non-symbiotic fixation. This is consistent with the estimates by other studies on the Beijiang River Basin (Chen and Jia 2009), the Yangtze River Delta (Deng et al. 2007), the three major drainage basins of China (i.e., the Yangtze, Yellow, and Pearl Rivers) (Xing and Zhu 2002), and the North Atlantic Region (Howarth et al. 1996), where fertilizer, excreta, and wet deposition ranked as the top three N input factors. The results can be closely related to the current agricultural activities and population distributions in China, where economic prosperity and robust industrial activities in the three major catchments have led to the massive use of chemical fertilizers, making them the largest N input sources. As the economic boom attracts more people into the PRD, their demands for meats, eggs, and dairy products have contributed to N input through excreta. Wet deposition is another major N input, which can be likely attributed to NO_X compounds discharged into the atmosphere by fossil fuel combustion during vehicle transportation; the compounds are deposited back onto land and into water bodies via the region's ample precipitation.

The relative contributions of nine cities to the N input in the three periods were similar (Table 1), following the descending order of Zhaoqing, Guangzhou, Jiangmen, Huizhou, Foshan, Dongguan, Shenzhen, Zhongshan, and Zhuhai. Figure 2 compares the N loads in each city, providing an overview of the N input intensity, where the dotted line represents the PRD average, and the N load is the total N input per unit land area. The N load ranking of each city differs somewhat from its N input ranking. The N load values range from 114.96 to –303.73 kg/hm^2 a. Shenzhen always had the highest N load, which increased over the decade-long study period. Rapid population and industrial growth in Shenzhen were reflected by the fastest-increasing N sources, excreta and wastewater, which rose from 2.62×10^4 and 0.99×10^4 t in 2000 to 3.61×10^4 and 1.54×10^4 t in 2010, respectively. Conversely, the N load of Zhuhai fell from 246.37 kg/hm^2 a in 2000 to 131.67 kg/hm^2 a in 2010. The largest

Table 1 Nitrogen inputs in the Pearl River Delta by city and time period (in 10^4 t)

City (Region)	Year	Fertilizer	Farmland symbiotic fixation	Farmland Non-S fixation	Human and livestock excreta	Wet deposition	Waste water NH_3-N	Crop residue	Total input
Guangzhou	2000	6.18	0.26	0.51	4.44	2.60	2.20	0.30	16.50
	2005	5.34	0.11	0.55	4.07	2.87	1.78	0.15	14.86
	2010	5.78	0.09	0.53	5.55	3.10	1.84	0.11	16.99
Shenzhen	2000	0.22	0.00	0.02	2.62	0.98	0.99	0.00	4.82
	2005	0.19	0.00	0.01	2.92	0.83	1.13	0.00	5.08
	2010	0.14	0.00	0.01	3.61	0.63	1.54	0.00	5.93
Zhuhai	2000	2.31	0.01	0.11	0.57	0.82	0.23	0.03	4.07
	2005	0.46	0.00	0.06	0.63	0.72	0.20	0.02	2.10
	2010	0.39	0.00	0.08	0.80	0.59	0.30	0.01	2.18
Foshan	2000	3.12	0.05	0.22	2.61	1.12	0.60	0.14	7.86
	2005	2.67	0.03	0.19	2.76	1.28	0.68	0.04	7.64
	2010	2.39	0.02	0.20	3.34	1.53	1.20	0.02	8.71
Huizhou	2000	5.49	0.39	0.30	2.84	5.20	0.22	0.33	14.76
	2005	5.22	0.38	0.37	2.94	3.84	0.21	0.20	13.18
	2010	5.34	0.26	0.34	3.01	4.03	0.46	0.16	13.58
Dongguan	2000	1.32	0.01	0.09	2.95	0.99	0.77	0.07	6.21
	2005	0.60	0.00	0.15	2.87	0.90	1.07	0.00	5.60
	2010	0.54	0.00	0.12	2.94	0.87	1.12	0.00	5.59

(continued)

Table 1 (continued)

City (Region)	Year	Fertilizer	Farmland symbiotic fixation	Farmland Non-S fixation	Human and livestock excreta	Wet deposition	Waste water NH_3-N	Crop residue	Total input
Zhongshan	2000	1.48	0.01	0.11	0.96	0.72	0.20	0.08	3.55
	2005	1.17	0.00	0.08	1.03	0.64	0.27	0.04	3.23
	2010	1.00	0.00	0.10	1.28	0.68	0.52	0.01	3.60
Jiangmen	2000	7.11	0.29	0.32	2.80	3.43	0.46	0.44	14.84
	2005	6.03	0.24	0.53	2.85	3.36	0.37	0.28	13.66
	2010	6.21	0.16	0.42	2.92	4.33	0.46	0.26	14.77
Zhaoqing	2000	7.35	0.30	0.27	4.26	4.19	0.23	0.45	17.04
	2005	7.61	0.33	0.53	4.89	3.96	0.22	0.37	17.91
	2010	7.44	0.30	0.40	4.17	4.84	0.34	0.33	17.82
In total	2000	34.58	1.32	1.95	24.04	21.54	5.90	1.82	91.16
	2005	29.30	1.09	2.47	24.96	19.85	5.94	1.08	84.71
	2010	29.22	0.84	2.21	27.62	20.49	7.78	0.90	89.06

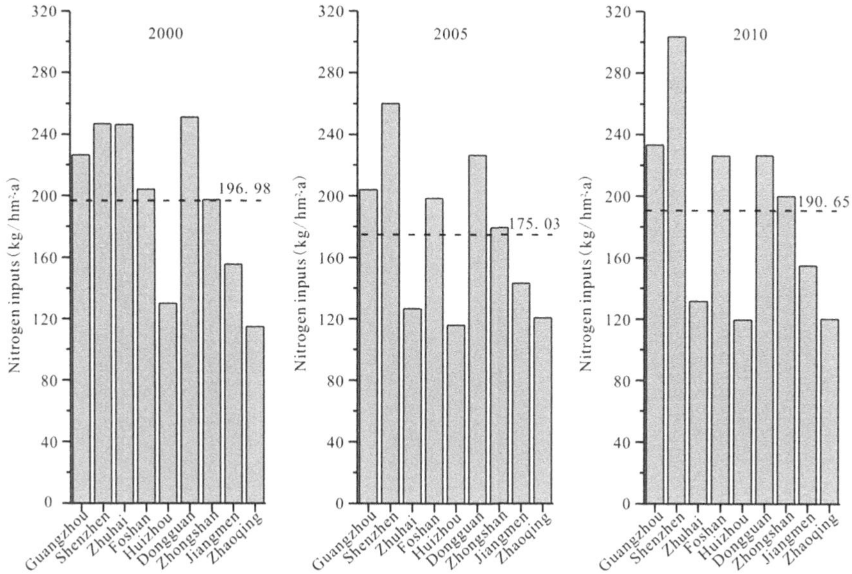

Fig. 2 Comparison of nitrogen (N) inputs per unit land area in the Pearl River Delta (PRD) by city and period

change occurred in fertilizer N, which decreased from 2.31×10^4 t in 2000 to 0.39×10^4 t in 2010. This can be explained by the increase in urbanization and reduction of farmland area from 12.68×10^4 hm^2 in 2000 to 1.74×10^4 hm^2 in 2010. The N loads of the other cities did not vary much. Overall, the PRD had a higher average N load than the national mean (~64 kg/hm^2 a) and the Pearl River drainage basin (~104.44 kg/hm^2 a) (Xing and Zhu 2002), but was lower than that of the Yangtze River Delta (~291 kg/hm^2 a) (Deng et al. 2007).

2.1.2 Terrestrial Nitrogen Flux

The terrestrial N flux (TNF) is an important indicator of the environmental impact of N in a region (Howarth et al. 1996), and is defined as the sum of fertilizer N and excreta N per unit land area within a year. The TNF values in the PRD were computed and compared (Fig. 3).

TNF values tend to exhibit high regional variation throughout the world due to differences in population density, human activities, and industrial technologies. Around the North Atlantic, the mean TNFs have been reported as 15–13 kg/hm^2 a in the North Sea and northwestern coasts of Europe, 11 kg/hm^2 a along the northeastern coast of the United States, and 8 kg/hm^2 a in the northern riverine regions of Canada (Howarth et al. 1996). In China, the mean TNF (Deng et al. 2007) increased from 1950s about 6 kg/hm^2 a to nearly 45 kg/hm^2 a in 1999. The TNF values for 2000,

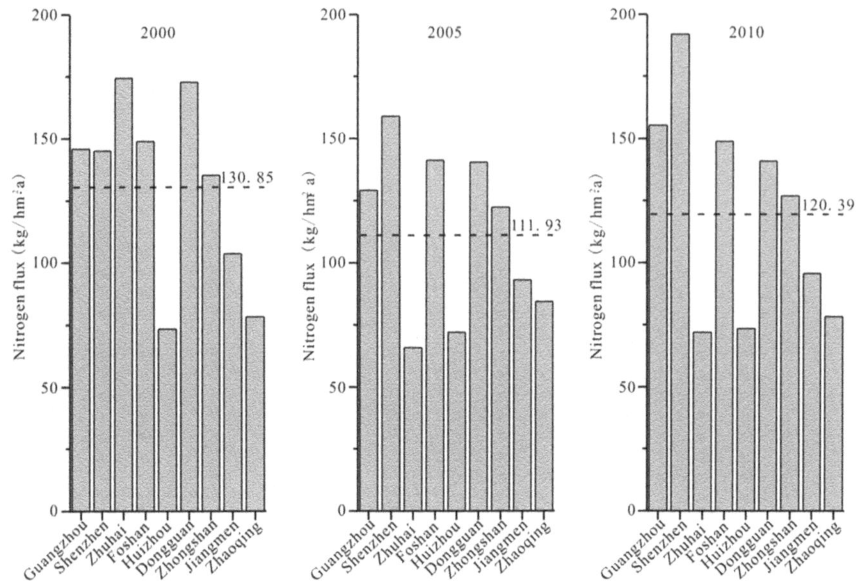

Fig. 3 Comparison of terrestrial N fluxes in the PRD by city and period

2005, and 2010 were 130.85, 111.93, and 120.39 kg/hm^2 a, respectively, exceeding the previous values but were still below that of the Yangtze River Delta (224 kg/hm^2 a). The city-specific TNF values were between 73.38 and 192.06 kg/hm^2 a, and Shenzhen, Guangzhou, Foshan, Dongguan, and Zhongshan had values higher than the average in all three periods, possibly due to their stronger economic growth and higher population densities. The TNF of Zhuhai decreased from 174.45 kg/hm^2 a in 2000 to 71.91 kg/hm^2 a in 2010. The other cities, Huizhou, Jiangmen, and Zhaoqing, had values significantly lower than the average. Based on these results, the five higher-than-average cities are at a greater risk due to N-related environmental issues.

2.1.3 Nitrogen Output: General Features and Regional Variation

Table 2 presents the estimated N outputs of the nine PRD cities in 2000, 2005, and 2010. The total N output in the PRD ranged from 56.82 × 10^4 to 67.38 × 10^4 t for the three periods. The major contributions were from water bodies and denitrification, followed by crop harvesting and vaporization. The N output via crop harvesting significantly declined over the study period (Table 2), from 14.50 × 10^4 t in 2000 to 7.18 × 10^4 t in 2010, with a decrease of over 50%. This provides additional evidence of the effects of the decrease in farmland area and increase in urbanization in the PRD. According to the Statistical Yearbook of Guangdong Province (2000 and 2010), the total cropland in the PRD region was 189.47 × 10^4 hm^2 in 2000 and 131.21 × 10^4 hm^2 in 2010, showing a decrease by about one-third over a period of 10 years. In

Table 2 Nitrogen outputs in the Pearl River Delta by city and time period (in 10^4 t)

City (Region)	Year	Vaporization	Harvested crops	Export by water bodies	Denitrification	Total output
Guangzhou	2000	1.55	2.38	4.82	3.30	12.06
	2005	1.39	1.17	4.41	2.97	9.95
	2010	1.73	0.88	5.48	3.40	11.49
Shenzhen	2000	0.55	0.00	2.04	0.96	3.56
	2005	0.61	0.00	2.38	1.02	4.01
	2010	0.74	0.00	2.96	1.19	4.88
Zhuhai	2000	0.35	0.22	1.08	0.81	2.47
	2005	0.19	0.12	0.55	0.42	1.27
	2010	0.20	0.08	0.58	0.44	1.30
Foshan	2000	1.12	1.08	2.53	1.57	6.30
	2005	0.97	0.28	2.49	1.53	5.28
	2010	1.09	0.18	2.80	1.74	5.82
Huizhou	2000	1.71	2.62	2.77	2.95	10.06
	2005	1.29	1.65	2.85	2.64	8.43
	2010	1.54	1.27	3.07	2.72	8.60
Dongguan	2000	0.88	0.53	2.24	1.24	4.89
	2005	0.63	0.03	2.07	1.12	3.85
	2010	0.68	0.02	2.51	1.12	4.32
Zhongshan	2000	0.46	0.61	1.14	0.71	2.93
	2005	0.42	0.27	1.06	0.65	2.40
	2010	0.44	0.10	1.21	0.72	2.46
Jiangmen	2000	2.16	3.48	3.42	2.97	12.03
	2005	1.28	2.21	3.19	2.73	9.41
	2010	1.68	2.04	3.29	2.95	9.96
Zhaoqing	2000	2.50	3.57	3.32	3.41	12.80
	2005	1.85	2.93	3.57	3.58	11.93
	2010	2.10	2.62	3.54	3.56	11.83
In Total	2000	11.29	14.50	23.37	18.23	67.38
	2005	8.64	8.67	22.57	16.94	56.82
	2010	10.21	7.18	25.44	17.81	60.65

addition, there was an increasing trend in the export of N via water bodies, which was mainly attributed to artificially activated N and N in waste directly discharged into water, another sign of the increasing human impacts on the N cycle.

The contribution of each city to the N output was similar over the study period (Table 2), and the ranking showed the same decreasing trend as the N input: Zhaoqing, Guangzhou, Jiangmen, Huizhou, Foshan, Dongguan, Shenzhen, Zhongshan, and Zhuhai. In terms of the N output fluxes (total N output per unit land area) shown in Fig. 4, where the dotted lines represent the PRD average, Shenzhen had the highest

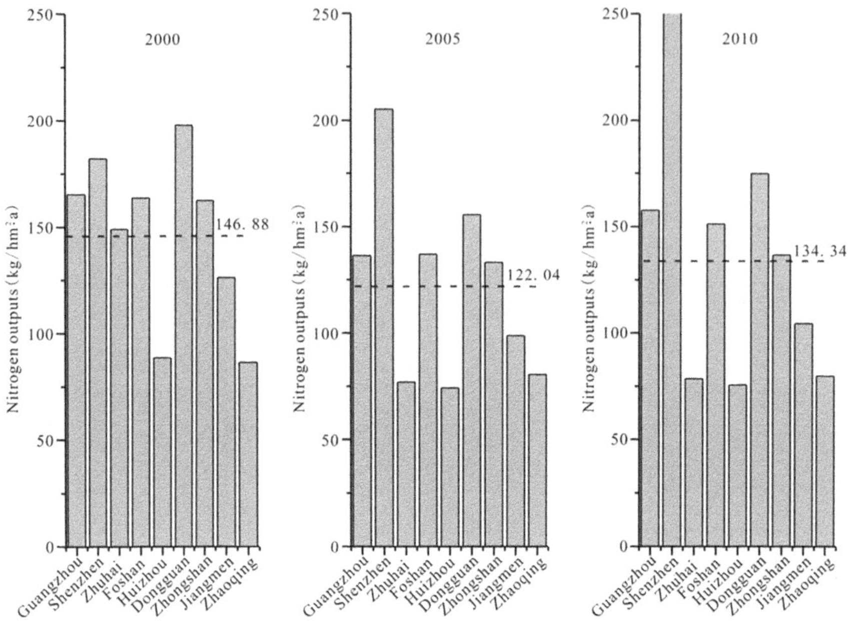

Fig. 4 Comparison of N outputs per unit land area in the Pearl PRD by city and period

values, followed by Dongguan, Guangzhou, Foshan, Zhongshan, Jiangmen, Zhuhai, Zhaoqing, and Huizhou. In particular, the N output flux of Shenzhen exhibited a rapidly increasing trend between 2000 and 2010.

2.1.4 Nitrogen Surplus

The N surplus (total N input minus total N output) in the PRD increased (Table 3) from 23.77×10^4 t in 2000 to 28.42×10^4 t in 2010, which was significantly higher than that of the neighboring Beijiang River Basin (~9.67×10^4 t/a) (Chen and Jia 2009), but substantially lower than that of the Yangtze River Delta ($99 \sim 128 \times 10^4$ t/a) (Deng et al. 2007). The N surplus per unit land area increased from 43.43 kg/hm^2 a in 2000 to 50.95 kg/hm^2 a in 2005 and 51.92 kg/hm^2 a in 2010. Zhaoqing had the highest total N surplus, followed by Guangzhou, Huizhou, Jiangmen, Foshan, Dongguan, Zhongshan, Shenzhen, and Zhuhai. However, Guangzhou had the highest surplus per unit land area, followed by Zhuhai, Shenzhen, Dongguan, Foshan, Zhongshan, Jiangmen, Huizhou, and Zhaoqing. Surplus N per unit land area, mostly held in soils, plants, and water bodies, has an important influence on any potential N-related pollution; therefore, Guangzhou, Zhuhai, Shenzhen, Dongguan, and Foshan are at greater risk of the effects of N-related pollution.

Table 3 Nitrogen budgets in the Pearl River Delta in the three time periods

Year	Input (10^4 t)	Output (10^4 t)	Surplus (10^4 t)	Surplus per unit land area (kg/hm^2 a)
2000	91.16	67.38	23.77	43.43
2005	84.71	56.82	27.89	50.95
2010	89.06	60.65	28.42	51.92

Table 4 Phosphorus budgets in the Pearl River Delta in the three time periods

Year	Input (t)	Output (t)	Surplus (t)	Surplus per unit land area (kg/hm^2 a)
2000	144506.38	52635.80	91870.58	16.79
2005	138771.14	45584.81	93186.34	17.03
2010	147317.40	39484.53	107832.88	19.70

2.2 *Phosphorus Budget Estimates*

Similar to N, input, output, and surplus are the three components of a regional P budget. The parameters and computation methods below are partially based on several existing studies (e.g., Xing and Yan 1999; Xing and Zhu 2002; Yang et al. 2006; Russell et al. 2008; Wang et al. 2009; Fan et al. 2010; Liu et al. 2011a, b).

2.2.1 Phosphorus Budget: General Features

The P input of the nine PRD cities was calculated for 2000, 2005, and 2010 based on the statistical data in Table 4. The total input was slightly lower in 2005 than in 2000 but slightly higher in 2010 than in 2000, with an increase of about 6% from 2005 to 2000. However, the total P output significantly declined over the study decade, decreasing by about 25% from 2000 to 2010. These changes considerably increased the P surplus of the PRD. Such an unbalance between P input and output is likely to lead to eutrophication of water bodies. Because both the total amount and per unit area amount of the P surplus increased, the PRD is under great threat of P pollution, and management of the P budget is an important task for local authorities.

2.2.2 Phosphorus Input and Its Regional Variation

The largest source of P input in all three periods was human and livestock excreta, accounting for more than 50% of the contribution in each year (Fig. 5). This was followed by fertilizer, with an average contribution of around 43%. The effects of atmospheric deposition and crop residue were small, contributing less than 5% of the total P input. These results are similar to those reported in other regions of

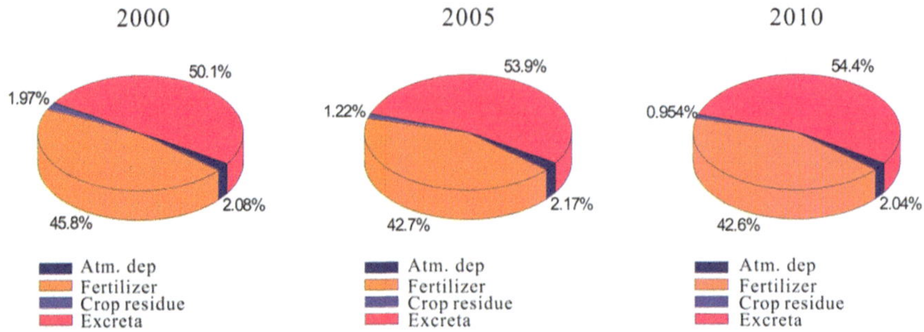

Fig. 5 Phosphorus (P) input sources in the PRD in the three studied periods

China (Wang et al. 2009; Liu et al. 2011a, b), where excreta and fertilizers are the most important P input sources. This can be related to the large population and large numbers of livestock required to meet the demand for meats, eggs, and dairy products. Fertilizer was another major source of P due to the high rate of fertilizer use in the region, which loses P via soil erosion. Another feature was the increase in excreta P, with 72,434.87 t, 74,834.97 t, and 80,074.78 t in 2000, 2005 and 2010, respectively. In contrast, crop residue P showed a declining trend of 2,850.72 t, 1,693.14 t, and 1,405.76 t in the three respective periods, accounting for a 50% decrease over the decade. The former was related to the booming population, particularly migrant workers from other regions, and the latter to a reduction in farmland area due to urbanization.

The P input differed significantly across the cities, as shown in Fig. 6, where the dotted lines indicate the PRD average. Zhaoqing had the highest P inputs, followed by Guangzhou, Jiangmen, Huizhou, Foshan, Dongguan, Shenzhen, Zhongshan, and Zhuhai. This was due to the larger area of farmland in Zhaoqing, Jiangmen, and Huizhou, which required more fertilizer, the most important source of their P inputs. Foshan, Dongguan, Shenzhen, Zhongshan, and Zhuhai had smaller areas of farmland, and therefore, excreta became the most important P source. Guangzhou had both a large population and a large farmland area, which explains why the two sources were relatively comparable.

In terms of the P input (Fig. 6), Shenzhen, Dongguan, and Zhuhai showed a significant variation among years, which was not the case for the other cities. The P input per unit area in Shenzhen increased from 37.68 kg/hm^2 a in 2000 to 48.86 kg/hm^2 a in 2010. Excreta had the highest contribution rate, which also increased over the study period. Specifically, the total P input in Shenzhen was 6,837.24, 7,417.0, and 9,130.89 t in 2000, 2005, and 2010, showing an increase of approximately 33% over the decade. The P input per unit area in Dongguan and Zhuhai decreased in 2010 compared to those in 2000, predominantly driven by the decrease in fertilizer P use. Notably, fertilizer P in Zhuhai decreased from 4,430.55 t in 2000 to 843.08 t in 2010,

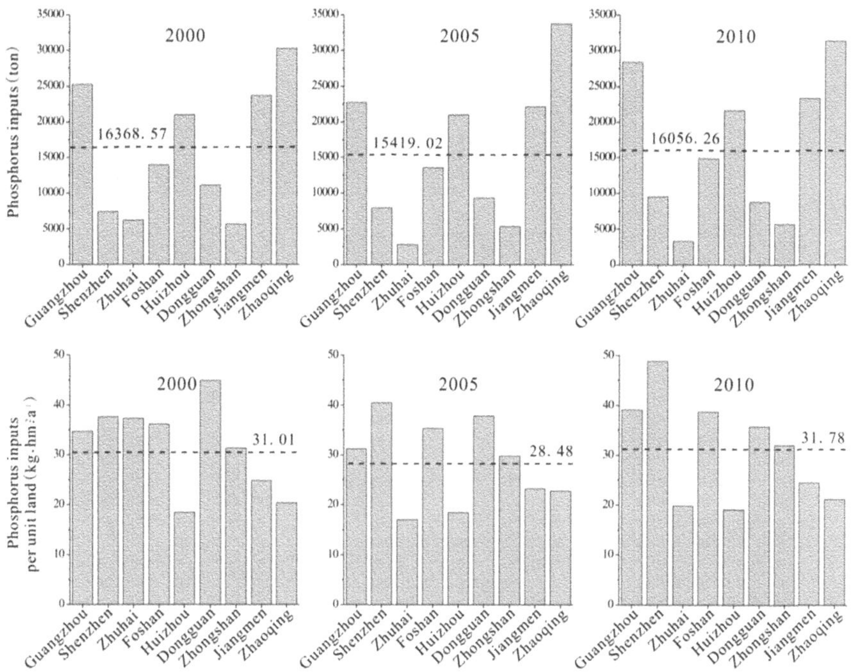

Fig. 6 Comparison of total P inputs and inputs per unit land area in the PRD

an approximately 80% decrease over the 10-year study period, in accordance with its urbanization and reduction in farmland area.

2.2.3 Phosphorus Output and Its Regional Variation

The most frequent destination for P output in the PRD was crop harvests and P stored in livestock, contributing to over 60% and 23% of the total P output respectively (Fig. 7). Soil erosion had a smaller effect (~11%), and export via runoff had the lowest contribution (~2%), which are similar to previous reports s (e.g., Wang et al. 2009; Liu et al. 2011a, b). Since PRD is a flood plain with good vegetation coverage and effective soil retention measures in place, soil erosion should not be a major problem.

Figure 8 shows the P output in the three periods, where the dotted lines represent the average values. The average P output decreased from 5,848.42 t in 2000 to 4,387.16 t in 2010. Crop harvests made the greatest contribution, decreasing from 36,150.91 t in 2000 to 24,753.81 t in 2010, indicating a reduction in farmland for crop production. The next greatest contribution was P stored in livestock, which

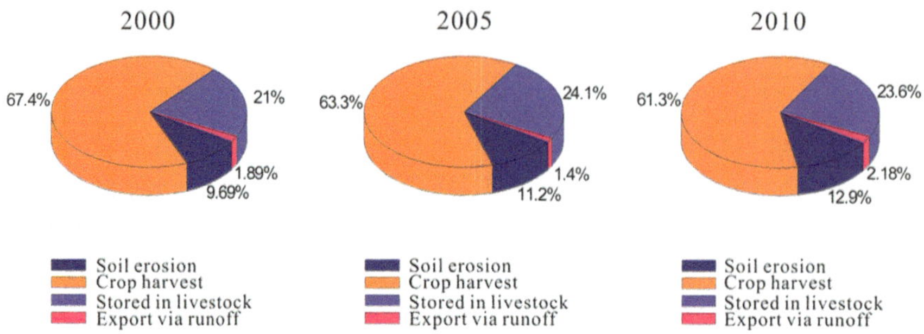

Fig. 7 Phosphorus output destinations in the PRD in the three studied periods

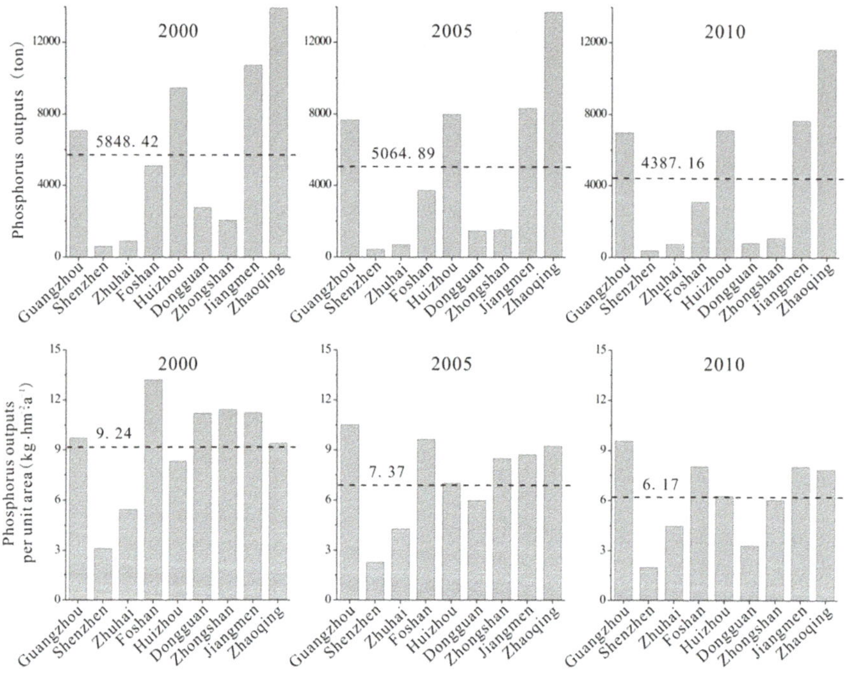

Fig. 8 Comparisons of total P outputs and outputs per unit land area in the PRD

decreased by about 1,800 t from 2000 to 2010. The region with the largest P output was Zhaoqing, followed by Jiangmen, Huizhou, Guangzhou, Foshan, Zhongshan, Dongguan, Zhuhai, and Shenzhen, with farmland area, livestock, and soil retention as the most important determining factors. In particular, Zhaoqing, Jiangmen, and Huizhou displayed considerable P outputs via crop harvests. Although the P output per unit area decreased significantly in the PRD, the values were still relatively high in Foshan, Guangzhou, Jiangmen, Zhaoqing, and Zhongshan.

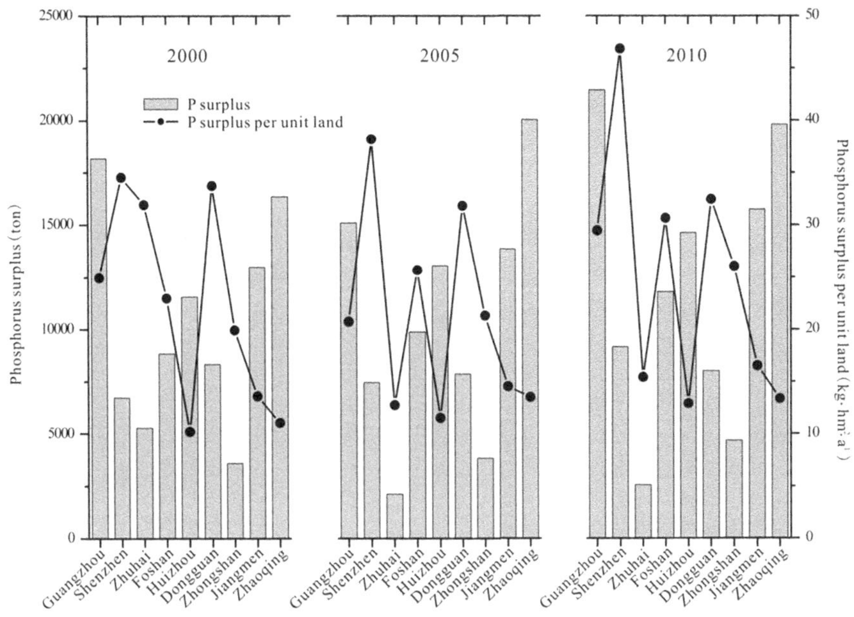

Fig. 9 Comparison of the total P surplus and surplus per unit land area in the PRD

2.2.4 Phosphorus Surplus

The P surplus results (Table 4 and Fig. 9) showed a gradual increase in the total P surplus, from 91,870.58 t in 2000 to 107,832.88 t in 2010. The P surplus per unit area also increased, and was 16.79, 17.03, and 19.70 kg/hm^2 a in 2000, 2005, and 2010, respectively. The P surplus per unit area was relatively low, which was even lower than the national average (Cao et al. 2009). A significant regional variation was observed among cities, with Zhaoqing showing the highest total surplus, followed by Guangzhou, Jiangmen, Huizhou, Foshan, Dongguan, Shenzhen, Zhongshan, and Zhuhai. In terms of the average surplus per unit area over the three periods, Shenzhen had the highest surplus, followed by Dongguan, Foshan, Guangzhou, Zhongshan, Zhuhai, Jiangmen, Zhaoqing, and Huizhou. In particular, Shenzhen, Dongguan, Foshan, Guangzhou, and Zhongshan had a higher surplus per unit area compared to the PRD average, indicating a higher risk of P-related pollution.

3 Case Study: Nitrogen and Phosphorus Pollution in Groundwater of the Tangjiawan Town, Zhuhai

3.1 Study Area and Water Sampling

Zhuhai is a prefecture-level city located at the west bank of the Pearl River in southern Guangdong Province. It is separated from Shenzhen and Hong Kong to the east by the sea, and is connected to Macau to the south by land. Located at 21°43′–22°29′N and 113°03′–114°24′E, Zhuhai has a subtropical monsoon climate, with an average annual temperature of 22.4 °C, average annual precipitation of 2011 mm, and average annual evaporation of around 1,469 mm. It has two very different seasons, the rainy season (April–September) and the dry season (October–March). Surrounded by mountains and ocean, the region has diverse geographical and ecological environments, including 731 km of coastlines and good vegetation coverage with predominantly subtropical plant species.

Located in northern Zhuhai, Tangjiawan is a town with 130 km^2 in size and a total population of about 100,000. It is governed by 16 neighborhood committees. It includes five university campuses housing 15,000 students, including the Zhuhai Campus of Sun Yat-sen University (ZCSYSU) and that of Beijing Normal University. It is also the home to high-tech industrial parks, such as the Tsinghua (Zhuhai) Technology Park and National Software Base. In this study, water samples were collected in two areas, ZCSYSU and the Tangjiawan residential area (Fig. 10), which are located 1 km from each other and have similar hydrogeological settings but differ in terms of the intensity of human activities. ZCSYSU is a small catchment (3.4 km^2) facing the sea. It is enclosed by hills on the other three sides and experiences a relatively lower human impact. Here, we set up flow weirs A and B in the upper and lower ends of the fault gully in the upper segment of the catchment, and 14 observation wells along the stream outside the gully (Table 5), including 3 in the recharge area (R1–R3), 5 in the middle area (M1–M5), and 6 in the discharge area (D1–D6). The Tangjiawan residential area is located at the eastern side of Eling Mountain to the north of ZCSYSU, and also faces the sea. Other than several public facilities, the area largely contains residential houses and protected historical buildings. Public wells are found throughout the mountain slope alongside residential buildings, and the local groundwater has experienced varying degrees of human-induced pollution.

During the period from July 2009 to June 2010, water samples were collected once per month from flow weirs A and B (surface water), 14 observation wells in the campus (groundwater), and 7 public wells in the Tangjiawan residential area (groundwater), with some parameters measured in situ, including temperature, pH, electric conductivity, dissolved oxygen, and oxidation-reduction potential. In total, 24 surface water samples and 262 groundwater samples were obtained. From 2006 to 2007, 41 rainwater samples were collected at the campus as well. The samples were analyzed for cations and anions by the IRIS Advantage (HR) spectrometer (Thermo Jarrell Ash) and the DX-600 ion chromatographer (Dionex) respectively at

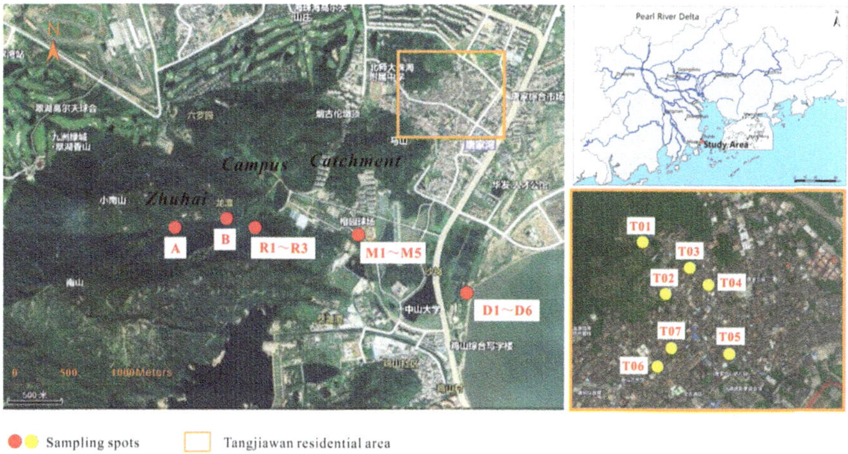

Fig. 10 Study areas and sampling sites (based on images from Google Earth)

Table 5 Observation wells in the Zhuhai Campus catchment

Area	Observation well	Depth/m	Type
Recharge	R1	4.2	Unconfined water
	R2	10	Base rock fissure water
	R3	13.2	Base rock fissure water
Middle	M1	5	Unconfined water
	M2	8.5	Confined water
	M3	10	Confined water
	M4	11.3	Confined water
	M5	17.8	Base rock fissure water
Discharge	D1	5	Unconfined water
	D2	8	Unconfined water
	D3	10	Unconfined water
	D4	15	Confined water
	D5	20	Confined water
	D6	34	Base rock fissure water

the Laboratory Center of Sun Yat-sen University, and bicarbonate by 0.01 NH_2SO_4 titration method.

3.2 Characteristics of Nitrogen and Phosphorus Pollution in Groundwater Under Minor Human Disturbance

The ZCSYSU, established in 2000 with around total 9000 students, had a relatively minor impact on the groundwater. N and P content in the groundwater (Table 6) indicated that NO_3^- was low, with a maximum concentration of only 4.11 mg/L. The NO_3^- concentration was higher in the wells of shallow unconfined aquifers, following the descending order of recharge area, discharge area, and middle area. NH_4^+ was similarly concentrated in shallow unconfined aquifer wells, following the descending order of middle area, discharge area, and recharge area, and with a maximum concentration of 4.74 mg/L in well M1. The NO_2^- and PO_4^{3-} concentrations were both relatively low, and did not differ greatly among the sites.

The dissolved inorganic N ($DIN = NO_3^- + NO_2^- + NH_4^+$) values were further analyzed using depth diagrams (Fig. 11), and the following results were obtained.

(1) The DIN concentration varied by groundwater types and showed an ascending order of unconfined water, confined water, and base rock fissure water. DIN pollution was largely concentrated in the shallow wells of unconfined waters (R1, M1, D1, and D2).
(2) For the shallow unconfined aquifer wells, the DIN contents were higher in the dry season than those in the rainy season, and showed a large variation. Low DIN was detected in the other wells with little seasonal variation. The lower concentrations in the wet season could be explained by a dilution process due to rainwater permeation and enhanced N consumption by microorganisms under higher temperatures (Jin et al. 2004).
(3) The highest DIN concentration was found in the middle area and the lowest in the discharge area. The correlation analysis between DIN and its components, NO_3^- and NH_4^+ (Fig. 12) revealed a significant positive correlation between DIN and NO_3^- in the recharge area ($R = 0.99$), indicating that DIN principally existed as NO_3^-, particularly in the unconfined water of R1, where NO_3^- accounted for 94% of DIN. In the middle area, DIN was significantly correlated with NH_4^+ ($R = 0.98$), which accounted for 87.5% of DIN, exclusively in the unconfined water of M1 (97.7% of DIN). The composition was more complex in the discharge area, where DIN was largely present as NH_4^+ in D1 (92% of DIN), and as NO_3^- in D2 (76.8% of DIN), indicating the impact of nitrification.

Table 6 Characteristics of nitrogen and phosphorus pollution in groundwater at Zhuhai Campus, Sun Yat-sen University

Area	Well	Depth (m)	NO_3^- (mg/L)			NH_4^+ (mg/L)			NO_2^- (mg/L)			PO_4^{3-} (mg/L)		
			Max	Ave	Min	Max	Ave	Min	Max	Ave	Min	Max	Ave	Min
Recharge	R1	4.2	4.11	2.98	2.08	0.70	0.19	–	0.018	0.009	–	0.027	0.017	–
	R2	10	0.13	0.09	–	0.23	0.17	0.11	0.004	0.003	–	0.019	0.018	–
	R3	13.2	0.30	0.11	–	0.32	0.20	0.11	0.007	0.005	–	0.021	0.016	–
Middle	M1	5	0.13	0.09	–	4.75	2.98	1.69	0.009	0.005	–	0.024	0.018	–
	M2	8.5	0.94	0.19	–	1.54	1.16	0.85	0.006	0.004	–	0.019	0.016	–
	M3	10	0.65	0.15	–	1.66	1.44	1.09	0.090	0.013	–	0.021	0.017	–
	M4	11.3	0.68	0.18	–	2.67	1.57	0.92	0.007	0.004	–	0.024	0.017	–
	M5	17.8	0.83	0.31	–	0.54	0.16	0.03	0.005	0.003	–	0.027	0.018	–
Discharge	D1	5	1.35	0.25	–	3.01	2.27	1.28	0.005	0.004	–	0.021	0.014	–
	D2	8	3.64	2.11	0.08	0.90	0.37	0.05	0.155	0.040	–	0.024	0.017	–
	D3	10	0.55	0.20	–	0.48	0.10	–	0.027	0.012	–	0.019	0.017	–
	D4	15	1.17	0.31	–	0.18	0.09	–	0.013	0.008	–	0.021	0.015	–
	D5	20	1.50	0.28	–	0.48	0.22	0.03	0.005	0.004	–	0.024	0.021	–
	D6	35	1.05	0.48	–	0.43	0.17	0.06	0.004	0.004	–	0.019	0.016	–

Note "–" indicates the value is below the thresholds of detection, which are: NO_3^-: <0.05 mg/L; NH_4^+: <0.02 mg/L, NO_2^-: <0.003 mg/L, PO_4^{3-}: <0.01 mg/L

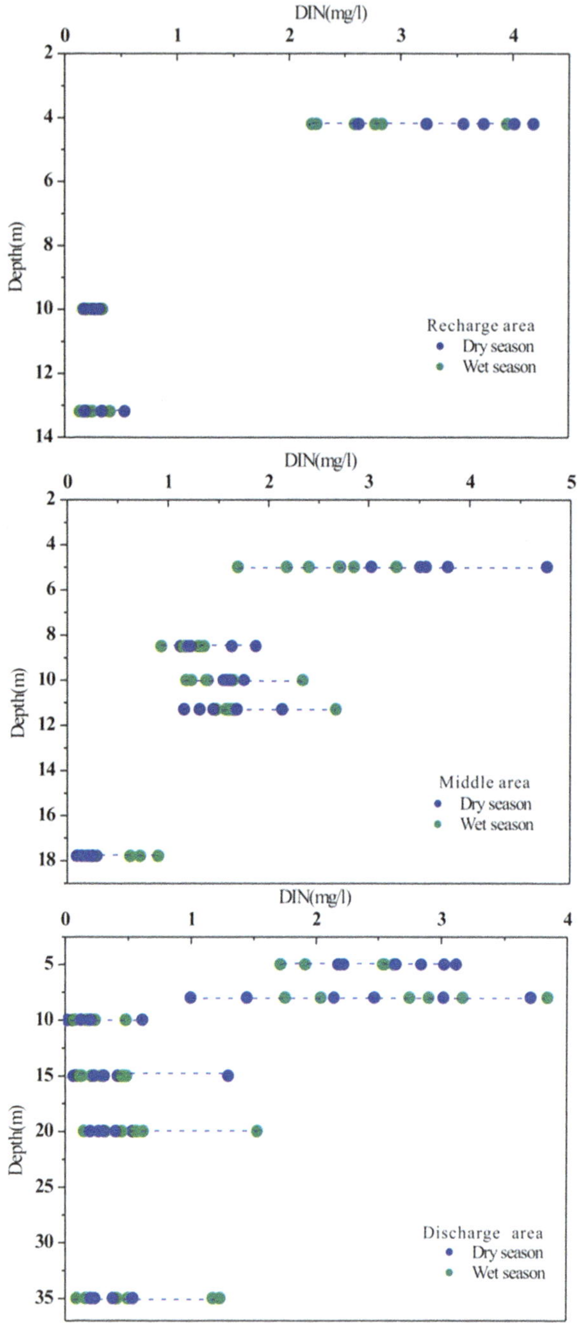

Fig. 11 Dissolved inorganic N concentrations in the groundwater sites of Zhuhai Campus

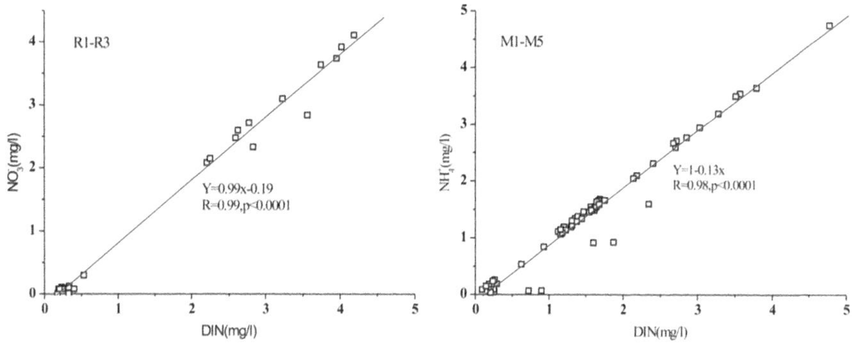

Fig. 12 Correlation between N forms and dissolved inorganic N in recharge and middle areas

3.3 Characteristics of Nitrogen and Phosphorus Pollution in Groundwater Under Conditions of Extensive Human Activity

Based on water samples collected from July 2009 to June 2010 (Table 7), the Tangjiawan residential area had significantly higher N and P contents in its groundwater than the ZCSYSU, with more N in particular. The NO_3^- concentration was between 9.51 and 151.3 mg/L, with an average as high as 87.6 mg/L. The NH_4^+ and NO_2^- contents were lower, and were not detected in some samples. The highest values of NH_4^+ and NO_2^- were 7.03 and 2.219 mg/L, respectively, which were both observed in well T04. The P content of the residential area was overall higher than that of the campus despite the similar geological settings, chiefly in wells T04 and T05, which had annual averages of 1.31 and 2.31 mg/L, respectively.

Due to the processes through which N migrates and transforms in saturated soils, N pollution of groundwater occurs largely in the form of NO_3^- (Chen et al. 2006a, b). At the Tangjiawan residential area, NO_3^- accounted for 83.0–99.0% of DIN. According to the World Health Organization, the maximum NO_3^- concentration allowed in domestic water is 10 mg N/L, and the value at all wells except T01 exceeded this threshold, indicating the status of severe N pollution. T01 is an old well built in 1599, located at the upper footslope within the local groundwater recharge area, which contributed to the lower degree of N pollution at this site. The NO_3^- content was much high downstream from T01. In wells T02, T04, and T05, the groundwater had NO_3^- concentrations of 33.5–88.2 mg/L. The concentrations in wells T03, T06, and T07 ranged from 103 to 151.3 mg/L. Large fluctuation in the rainy season was found compared to that in the dry season (Fig. 13). The Tangjiawan residential area is a historic area with a high building density, and a primitive and aging water drainage pipework that disposes human excreta mainly into cesspools. According to isotopic signals by Zhao et al. (2008) on groundwater NO_3^-, the main sources of N pollution in Tangjiawan were domestic water and seepage from cesspools. Complex variations in

Table 7 Groundwater nitrogen and phosphorus pollution in the Tangjiawan residential area

Well	NO_3^- (mg/L)			NH_4^+ (mg/L)			NO_2^- (mg/L)			PO_4^{3-} (mg/L)		
	Max	Ave	Min	Max	Ave	Min	Max	Ave	Min	Max	Ave	Min
T01	15.90	12.12	9.51	0.12	0.07	–	0.007	0.005	–	0.021	0.016	–
T02	85.50	74.61	59.80	0.35	0.15	–	0.017	0.011	–	0.030	0.022	–
T03	147.70	128.61	115.14	0.20	0.07	–	0.030	0.015	–	0.023	0.019	–
T04	83.00	59.12	33.50	7.03	2.32	0.09	2.219	0.263	0.007	2.308	1.310	0.292
T05	88.20	84.81	81.90	0.25	0.10	–	0.122	0.031	0.004	2.835	2.307	1.575
T06	151.30	135.38	121.40	0.37	0.11	0.02	0.202	0.050	0.004	0.196	0.098	0.047
T07	128.90	118.45	103.00	1.65	0.87	0.04	1.599	0.284	–	0.092	0.032	–

Note "–" indicates the value is below the thresholds of detection, which are: NO_3^-: <0.05 mg/L; NH_4^+: <0.02 mg/L, NO_2^-: <0.003 mg/L, PO_4^{3-}: <0.01 mg/L

Fig. 13 Seasonal variation of NO^{3-} in the groundwater of the Tangjiawan residential area

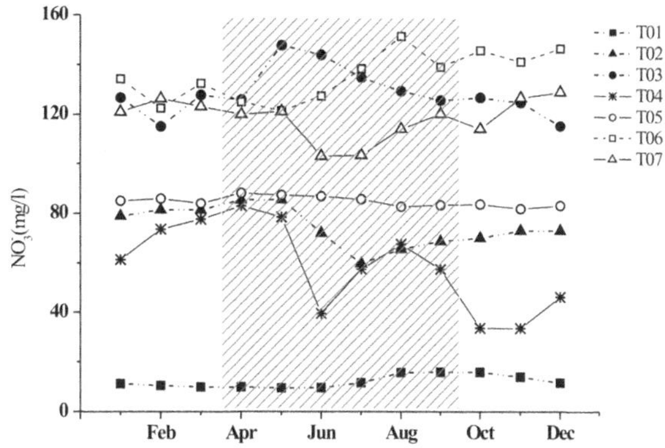

NO_3^- concentration could be explained by a combination of increased seepage from rainwater, denitrification, and dilution (Zhao et al. 2008).

In comparison, the P content was relatively low. Figure 14a shows the seasonal P content in the wells, which, together with Table 7, indicates that the PO_4^{3-} concentrations were highly variable among the wells. Upstream wells T01, T02, T07, T06, and T03 had less PO_4^{3-}, and none of them had a concentration over 0.1 mg/L, except T06 (0.047–0.196 mg/L). Downstream wells T04 and T05 were more severely polluted with P, with maximum values of 2.31 mg/L and 2.84 mg/L, respectively, which are significantly higher than those in the campus (maximum: 0.027 mg/L). Figure 14b shows the seasonal variation of PO_4^{3-}. PO_4^{3-} was not detected in wells T01, T02, T03, or T07 in May, possibly due to the dilution of groundwater by rainfalls. Being situated in a more polluted area, T04 showed great PO_4^{3-} fluctuations, with the lowest values in April and August. The PO_4^{3-} content was more stable in T05. Domestic wastewater and animal excreta contain high NH_4^+ and inorganic P concentrations (Xing et al. 2001). Since well T04 was located only 1–2 m from a cesspool, its high NH_4^+ and PO_4^{3-} concentrations could be attributed to seepage of domestic wastewater and excreta. The highest PO_4^{3-} content was found in well T05, which was located at the lowest altitude and situated 1–2 m from a cesspool in an area with frequent human activities (e.g., clothes washing).

As shown by the presence of numerous wells, groundwater used to be an important source of drinking water for Tangjiawan residents, who still collect water from some wells for washing food ingredients and clothes. The Campus catchment can be considered as a reasonable background site for Tangjiawan, representing the conditions of the local groundwater in earlier days. Therefore, it is likely that the local groundwater in Tangjiawan became heavily polluted with N and P due to human activities over hundreds of years.

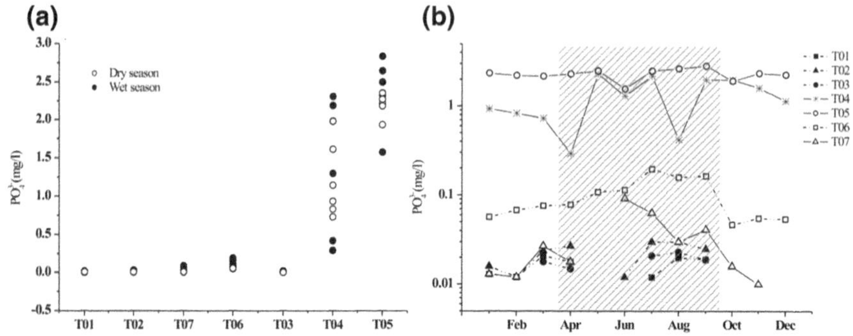

Fig. 14 Spatial distribution (**a**) and seasonal variation (**b**) of groundwater PO_4^{3-} in the Tangjiawan residential area

4 Case Study: Trace Metal Pollution in the Shima River

4.1 Study Area

Dongguan is located at the central Guangdong Province (113°31′–114°15′E, 22°39′–23°09′N) (Fig. 15), and has a subtropical monsoon climate with an annual average rainfall of 1,954 mm and an average temperature of 22.9 °C. Typhoons occur frequently from June to September, and rainstorms occur regularly. Dongguan has an area of 2,465 km^2, and higher and lower terrain is distributed in the southeast and northwest areas of the city, respectively. Dongguan is covered with unconsolidated quaternary clay soil that is thick from 0 to 18 m (Gao et al. 2015a). Hills and alluvial plains are the primary terrain, accounting for 88% of the total area of Dongguan (Wu et al. 2015). Dongguan underwent a transition from agricultural to an industrial and technological economy after Xiaoping Deng's Southern Tour in 1992. Since then, the residential population and the GDP have increased more than 3.7 and 57 times from 1992 to 2015 (Dongguan Bureau of Statistics 1978–2015), respectively.

The Shima River is the largest tributary of the Dong River in Dongguan (Fig. 15). Originating from the Baoan District in Shenzhen, the Shima River is 88 km in length with a watershed area of 1,249 km^2. Its original flow direction was from south to north, draining through an outlet located at Qiaotou town into the Dong River that is a crucial water source for the PRD region. Because of the demand of potable water from Hong Kong, eight pumping stations were installed along the Shima River as the major component of the diversion project in 1964 to supply freshwater from the Dong River to Hong Kong, resulting in the reversal of the river flow direction (Gao et al. 2017). The Shima River provided approximately 80% of Hong Kong's potable water use after the completion of the expansion project in 1987 (Xu 2006). Rapid industrial development and the population increase in Dongguan since 1990s have resulted in huge water demand. In 2003, the Quondam Diversion Project was substituted by a closed pipeline continually transporting water to Hong Kong, thus recovering its

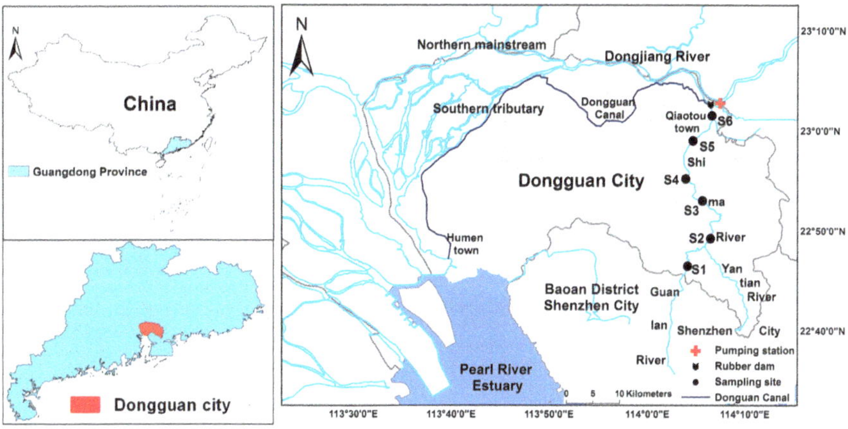

Fig. 15 Study area and sampling sites

original south-to-north flow direction. River water containing high concentrations of pollutants (e.g., trace metals) began to drain into the Dong River. Moreover, a rubber dam with a length of 92 m and height of 2.35 m was built at the outlet at Qiaotouto minimize contaminant discharge. In the dry season, the river water flows into the Dongguan Canal through an underground pipe, eventually draining through Humen into the estuary. However, the rubber dam may collapse during heavy rainfalls.

River water (S1–S6) was collected using the grab-sampling method in February, June, and November 2012. Water samples collected were stored in a refrigerator (4 °C) in the dark after being transported to the laboratory, and then were filtered through a polypropylene membrane with a diameter of 0.45 μm. Levels of metals or metalloid [zinc (Zn), cadmium (Cd), copper (Cu), chromium (Cr), nickel (Ni), lead (Pb), manganese (Mn), iron (Fe), and arsenic (As)] were determined using inductively coupled plasma–atomic emission spectrometry (ICP-AES; IRIS, Pleasant Prairie, WI, USA). The detection limits were 10 μg/L for Cu, Cr, Ni, and As, 5 μg/L for Zn, Mn, Pb, and Fe, and 1 μg/L for Cd.

Surface sediments were collected and stored at −40 °C in the dark after being transferred to the laboratory. Sediment sample was ground using an agate mortar and homogenized until all sediment particles passed through a 100 mesh nylon sieve after freeze-drying. Approximately 50 mg of sediment sample was totally digested in a HF-HNO_3 mixture, under high temperature and pressure conditions, in Teflon tubes for the determination of trace metal (Cd, Zn, Cu, Ni, Cr, Pb, As, scandium (Sc), and vanadium (V)) concentrations with inductively coupled plasma-mass spectrometry (ICP-MS, X-2; Thermo, Waltham, MA, USA). Fe, aluminum (Al), and Mn concentrations in the digestion mixture were measured using ICP-AES.

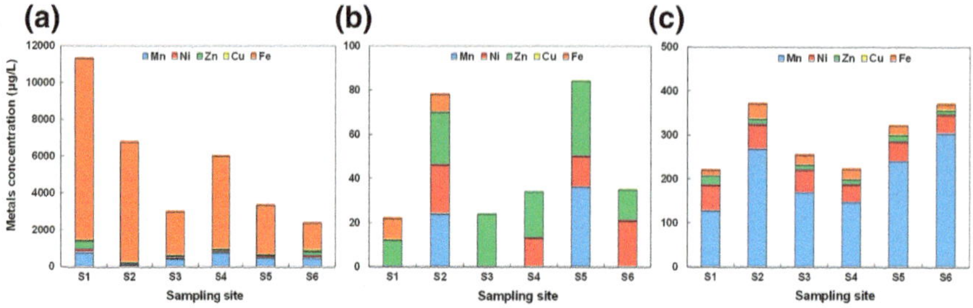

Fig. 16 Concentration of dissolved trace metals in river water in February (**a**), June (**b**), and November (**c**), 2012

4.2 Trace Metal Pollution in River Water

Rapid urbanization and industrialization processes have resulted in the release of large quantities of pollutants (e.g., trace metals, nutrients, etc.) into the environment, leading to severe pollution in surface water bodies in Dongguan (Xiong et al. 2010; Liu et al. 2011a, b). Moreover, the Shima River has suffered from intense human activities, resulting in a number of environmental issues (Gao et al. 2014, 2015b; Jiang et al. 2016). Total trace metal concentrations showed a clear spatiotemporal difference, decreasing from site S1 (11,302 μg/L) to S6 (2,386 μg/L) in February (Fig. 16a), which likely resulted from deposition following the adsorption of trace metals onto suspended particles (Varol and Şen 2012). Meanwhile, a slight increasing trend in metal concentrations was observed in June and November (Fig. 16b, c), which was likely associated with the confluence of river branches, such as the Yantian River, before site S2 and wastewater discharge along the river channel. Total trace metal concentrations in river water generally followed the descending order of February, November, and June, as more monthly precipitation was observed in June (256 mm) and November (168 mm) than February (51.2 mm). A dilution effect via precipitation was the main cause of the relatively lower concentration of trace metals in the rainy season in the Shima River.

In terms of individual metals, As, Cr, Cd and Pb, with detection limits of 10, 10, 1, and 5 μg/L, respectively, were not observed in the river water during the three investigated periods (Table 8). The highest observed concentrations were 748 μg/L for Mn, 151 μg/L for Ni, 494 μg/L for Zn, and 9,860 μg/L for Fe at site S1 in February, which substantially exceeded the national surface water quality standard (MEP 2002). This suggested that a large number of trace metals originated from upstream, in particular from the industrial wastewater discharge in Baoan District, Shenzhen city. In November, the trace metal concentrations, with the exception of Mn, of all water samples met the drinking water quality standards (MEP 2002; WHO 2011). Mn pollution likely promoted the occurrence of brown/black precipitate (Koukal et al. 2004), causing unsightly stains in the river water in February and November. Overall, river water pollution caused by trace metals might present a serious risk to urban

Table 8 Trace metal concentrations of the river water samples (μg/L)

Site	As	Cd	Cr	Pb	Mn	Ni	Zn	Cu	Fe
February									
S1	n.d.	n.d.	n.d.	n.d.	748	151	494	49.0	9860
S2	n.d.	n.d.	n.d.	n.d.	84.0	30.0	85.0	18.0	6550
S3	n.d.	n.d.	n.d.	n.d.	420	52.0	116	25.0	2390
S4	n.d.	n.d.	n.d.	n.d.	749	61.0	119	28.0	5050
S5	n.d.	n.d.	n.d.	n.d.	443	80.0	104	24.0	2700
S6	n.d.	n.d.	n.d.	n.d.	436	130	247	93.0	1480
June									
S1	n.d.	n.d.	n.d.	n.d.	n.d.	n.d.	12.0	n.d.	10.0
S2	n.d.	n.d.	n.d.	n.d.	24.0	22.0	24.0	n.d.	8.00
S3	n.d.	n.d.	n.d.	n.d.	n.d.	n.d.	24.0	n.d.	n.d.
S4	n.d.	n.d.	n.d.	n.d.	n.d.	13.0	21.0	n.d.	n.d.
S5	n.d.	n.d.	n.d.	n.d.	36.0	14.0	34.0	n.d.	n.d.
S6	n.d.	n.d.	n.d.	n.d.	n.d.	21.0	14.0	n.d.	n.d.
November									
S1	n.d.	n.d.	n.d.	n.d.	127	58.0	21.0	n.d.	14.0
S2	n.d.	n.d.	n.d.	n.d.	268	55.0	13.0	n.d.	35.0
S3	n.d.	n.d.	n.d.	n.d.	168	51.0	12.0	n.d.	24.0
S4	n.d.	n.d.	n.d.	n.d.	146	40.0	12.0	n.d.	24.0
S5	n.d.	n.d.	n.d.	n.d.	239	45.0	16.0	n.d.	21.0
S6	n.d.	n.d.	n.d.	n.d.	303	42.0	11.0	n.d.	14.0
MEPC[a]	50	1	10	10	100	n.i.	50	10	300
MEPC[b]	100	10	100	100	n.i.	n.i.	2000	1000	n.i.
WHO[c]									

n.d. not detected
[a,b]Environmental quality standard (Class I and V) for surface water of China (MEP 2002)
[c]Guidelines for drinking-water quality-Forth edition (WHO 2011)

residents by polluting the local water supply as a result of the short distance between the water source area and the drainage outlet of the Shima River (Fig. 15) (Table 9).

4.3 *Trace Metal Pollution in Riverine Sediment*

Riverine sediment, a primary component of river ecosystems, not only is a sink for suspended solid and soluble chemicals originated from both natural sources and anthropogenic inputs into watersheds, but also is an important source due to the resuspension of deposited substances into the overlaying water (Hou et al. 2013).

Table 9 Trace metal concentrations in sediment cores from the Shima River (mg/kg)

Site	V	Cr	Ni	Cu	Zn	As	Cd	Pb	Mn	Fe	Al	Sc
S1	41.4	101	100	127	972	6.62	1.35	65.3	626	40327	87631	9.502
S2	59.8	279	216	271	1310	11.1	1.88	71.9	893	47458	96945	10.8
S3	54.3	335	181	401	1370	12.8	1.64	131	849	40562	102183	10.7
S4	59.6	469	253	630	1704	14.3	1.91	109	803	41395	99618	10.2
S6	73.8	50.2	24.7	26.3	92.6	13.3	0.504	47.3	283	38389	91945	13.0
BGV	65.3	50.5	14.4	17	47.3	8.9	0.056	36	279	24200	72100	12.2

BGV environmental background value for soils of Guangdong Province (China National Environmental Monitoring Centre 1990)

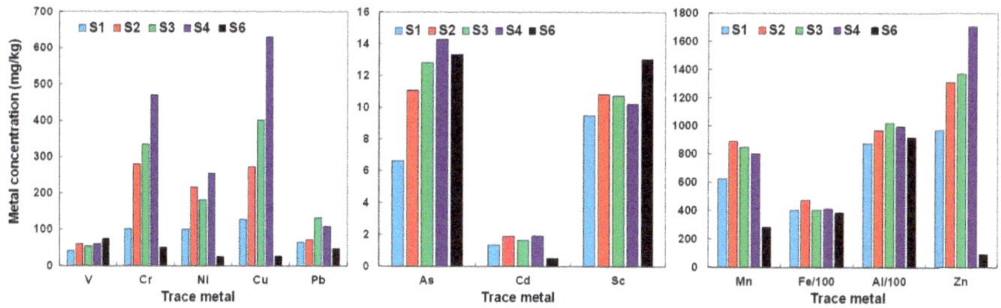

Fig. 17 Distribution of trace metal concentrations in riverine sediment of the Shima River

Generally, trace metals input into river systems tends to deposit on riverbeds after adsorption onto suspended particles and accumulation in sediments. In the dry season (February), the trace metal concentrations in the river water decreased from the upper reach site (S1) to the lower reach site (S6), whereas their concentrations in the surface sediment increased from site S1 to S4 along the river channel. This was in line with general sedimentation processes of trace metals in river systems. The metal concentrations were in the range of 41.4–73.8 mg/kg for V, 6.62–14.3 mg/kg for As, 9.50–13.0 mg/kg for Sc, 38.4–47.5 g/kg for Fe, and 87.6–102 g/kg for Al (Fig. 17), which were similar to the soil background values in Guangdong Province (China National Environmental Monitoring Centre 1990). Meanwhile, Cr, Ni, Cu, Zn, Cd, Pb, and Mn concentrations had mean values of 247, 155, 291, 1,089, 1.46, 85.0, and 690 mg/kg, respectively, which were substantially higher than local soil background values. Notably, Cr, Ni, Cu, Pb, Cd, Mn, and Zn concentrations decreased steeply at site S6, which was likely attributable to periodic dredging of sediment at the rubber dam.

To further assess the pollution level of each metal in the surface sediment, the geoaccumulation index (I_{geo}) was applied. The mean I_{geo} values were −0.786 for V, −0.179 for Al, −0.761 for Sc, and −0.250 for As, indicative of an uncontaminated status (Fig. 18). However, the mean I_{geo} values ranged from 0.194 to 0.559 for Pb, Mn, and Fe, indicating that they were uncontaminated-to-moderately contaminated elements. Cr had a mean I_{geo} value of 1.30, indicative of moderate contamination, and the mean I_{geo} values of Ni, Cu, Zn, and Cd fell in the range of 2–4, suggesting moderate-to-heavy contamination. The highest I_{geo} value was obtained for Cd (3.97), due mainly to a low background value and higher concentration in the sediments. Overall, the pollution status of individual metals based on the average I_{geo} values followed a descending order of Cd, Zn, Cu, Ni, Cr, Mn, Pb, Fe, Al, As, Sc, and V.

Hierarchical cluster analysis was conducted for the I_{geo} values of determined trace metals to preliminarily differentiate the sources of the sediment metals according to the pollution level and path. Cluster A included uncontaminated or slightly contaminated metals, including V, Sc, Al, As, and Fe. These metals in sediment were likely derived from parent materials due to the similarity between their concentrations and local soil background levels (Fig. 19). Trace metals clustered in groups B and C were the main pollution elements, and were identified as anthropogenic components (Bai

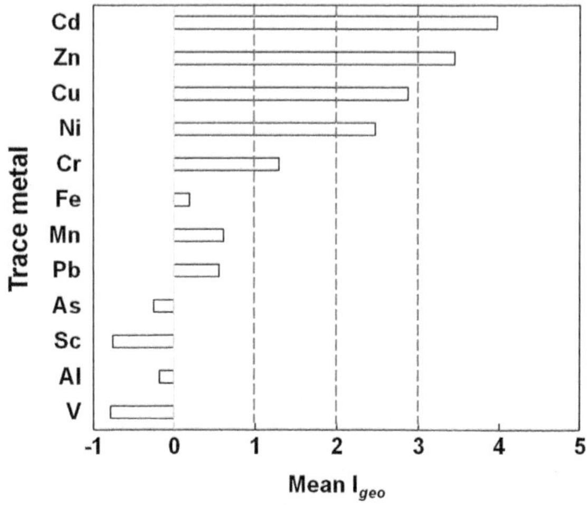

Fig. 18 Mean geoaccumulation index (Igeo) values of the trace metals in sediment

et al. 2011; Fu et al. 2014; Ma et al. 2016). Overall, coal and leaded-gasoline were found to be the major sources for sediment Pb in the highly urbanized catchment in China (Li et al. 2012), however, since the phase-out of leaded gasoline in 1999, coal was mainly responsible for Pb pollution in the aquatic environment (Zhang et al. 2016). Furthermore, Cr can be found in stainless and alloy steels manufacturing industries (Yeung et al. 2003). Whereas Mn might have originated from metal mining and smelting (Fu et al. 2014). In cluster C, Ni, Cu, Zn, and Cd showed moderate-to-heavy contamination in sediment. Effluents from machinery, electronics, and electroplating industries reportedly contain large quantities of Zn, Ni, and Cu in highly urbanized catchment (Li et al. 2013; Iqbal et al. 2015). This was in good agreement with the conditions in the Shima River catchment, where Cu, Ni, and Zn concentrations in river water decreased from sites S1 to S6 in February, suggesting that upstream industrial areas contributed to both river water and sediment pollution. However, Cd was likely associated with non-point pollution sources, because its concentrations were significantly lower in sediments than those in riparian soils ranging from 2.00 to 11.2 mg/kg along the Shima River (Gao et al. 2014), which was mainly indicative of agricultural runoff. Furthermore, soil Cd pollution was found to be associated mainly with the application of fertilizer (Cai et al. 2008). Consequently, the trace metals included in cluster B and C were mainly associated with industrial and agricultural activities.

5 Conclusions

The PRD region had experienced a significant increase in population and economic growth over the past several decades. To fully apprehend the effects of rapid development on the ecological environment in this region, three case studies regarding

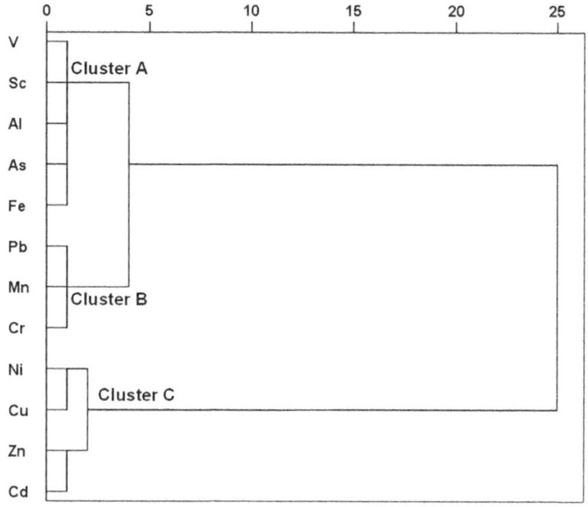

Fig. 19 Dendrogram of trace metals based on the geoaccumulation index (Igeo) in sediment samples

nutrients and trace metals pollution in different environmental settings have been conducted.

The total N input and output of the PRD showed little variations during the investigated periods, whereas the N surplus increased significantly. The rapid urbanization and intensive human activities exerted significant impacts on the evolution of nitrogen budgets during 2000–2010, indicating a severer N pollution in Guangzhou, Zhuhai, Shenzhen, Dongguan, and Foshan. On the contrary, the phosphorus input and output varied distinctly during 2000–2010. The major source of P input was found to be human and livestock excreta, which was followed by the application of fertilizer, and the crop harvests were mainly responsible for the P output. A high phosphorus surplus per unit area in Shenzhen, Dongguan, Foshan, Guangzhou, and Zhongshan city reflected a great risk of P-related pollution. In terms of nutrients pollution in the groundwater, DIN dominated by NO_3^- and/or NH_4^+ was highly concentrated in the shallow unconfined aquifer from the Zhuhai Campus of Sun Yat-sen University. It was notable that N and P concentrations in the groundwater were found to be higher in Tangjiawan residential area than in Zhuhai Campus, directly revealing the significant impacts of domestic wastewater on the groundwater systems.

The severe trace metals pollution had been observed in the highly urbanized Shima River catchment area. The river water pollution with high concentrations of trace metals in the dry season likely imposed obvious threats on the drinking water resource area. According to cluster analysis of metals, sediment V, Sc, Al, As, and Fe were derived from the parent materials. And industrial activities likely contributed significantly to sediment Pb, Cr, Ni, Cu, Zn, and Mn. However, Cd was supposed to be related to non-point pollution source. Riverine sediment polluted by Cr, Ni, Cu, Zn, and Cd likely resulted in the secondary pollution of river water due to its re-suspension during the flood processes.

It was evident that the rapid urbanization, industrialization and intensive human activities significantly affected surface water and groundwater, resulting in some hazards to the safety of urban water consumption in the PRD region. Thus, the improvement of water quality necessitates the implementation of pollution prevention measures, the control of agrochemical application, and the reduction of industrial wastewater discharge. Furthermore, we urge the local governments to carry out the dredging project in the polluted river channel near the water source areas.

References

Bai J, Xiao R, Cui B, Zhang K, Wang Q, Liu X, Gao H, Huang L (2011) Assessment of heavy metal pollution in wetland soils from the young and old reclaimed regions in the Pearl River Estuary, South China. Environ Pollut 159(3):817–824. https://doi.org/10.1016/j.envpol.2010.11.004

Cai LM, Jin M, Zhou YZ, Huang LC, Dou L, Zhang CB, Fu SM (2008) Multivariate geostatistics and GIS-based approach to study the spatial distribution and sources of heavy metals in agricultural soil in the Pearl River delta, China. Environ Sci 12:3496–3502

Cao N, Zhang YB, Chen XP (2009) Spatial-temporal change of phosphorus balance and the driving factors for agroecosystems in China. Chin Agric Sci Bull 25(13):220–225

Chen B, Duan JC, Mai BX, Luo XJ, Yang QS, Sheng GY, Fu JM (2006a) Distribution of alkylphenols in the Pearl River Delta and adjacent northern South China Sea, China. Chemosphere 63(4):652–661. https://doi.org/10.1016/j.chemosphere.2005.08.004

Chen JY, Wang Y, Zhang HB, Zhao XF (2006b) Overview on the studies of nitrate pollution in groundwater. Prog Geogr 25(1):34–44

Chen FJ, Jia GD (2009) Nitrogen budgets of the Beijiang river basin. Trop Geogr 29(1):11–15. https://doi.org/10.13284/j.cnki.rddl.001208

China National Environmental Monitoring Centre (1990) Background value of soil elements in China. China Environmental Science Press

Deng MH, Xie YX, Xiong ZQ, Xing GX, Yan XY (2007) Nitrogen budgets of the Yangtze delta region and their effect on the environment. Acta Sci Circum 27(10):1709–1716. https://doi.org/10.13671/j.hjkxxb.2007.10.015

Dongguan Bureau of Statistics (1978–2015) Statistical yearbook of Dongguan City

Fan ML, Wang XM, Wang Q, Lin WS, Jin H (2010) Atmospheric deposition of nitrogen and phosphorus into the Hengmen of Pearl River Estuary. J Trop Oceanogr 29(1):51–56

Fu J, Zhao C, Luo Y, Liu C, Kyzas GZ, Luo Y, Zhao D, An S, Zhu H (2014) Heavy metals in surface sediments of the Jialu River, China: their relations to environmental factors. J Hazard Mater 270:102–109. https://doi.org/10.1016/j.jhazmat.2014.01.044

Galloway JN, Howarth RW, Michaels AF, Nixon SW, Prospero JM, Dentener FJ (1996) Nitrogen and phosphorus budgets of the North Atlantic Ocean and its watershed. Biogeochemistry 35:3–25

Galloway JN (2005) The global nitrogen cycle: past, present and future. Sci China Ser C Life Sci 48:669–677. https://doi.org/10.1360/062005-261

Gao L, Chen JY, Wang J, Ke ZT, Yang XY, Shimizu Y, Zhu AP (2014) Heavy metal pollution of soils alongside Shima River, Dongguan and its ecotoxicity. Acta Pedol Sin 51(3):538–546

Gao L, Chen J, Tang C, Ke Z, Wang J, Shimizu Y, Zhu A (2015a) Distribution, migration and potential risk of heavy metals in the Shima River catchment area, South China. Environ Sci Process Impacts 17(10):1769–1782. https://doi.org/10.1039/c5em00156k

Gao L, Chen JY, Wang J, Ke ZT, Zhu AP, Xu K (2015b) Temporal-spatial variation and source identification of hydro-chemical characteristics in Shima catchment, Dongguan city. Environ Sci 36(5):61–69

Gao L, Wang Z, Shan J, Chen J, Tang C, Yi M (2017) Aquatic environmental changes and anthropogenic activities reflected by the sedimentary records of the Shima River, Southern China. Environ Pollut 224:70–81. https://doi.org/10.1016/j.envpol.2016.12.056

Hou D, He J, Lu C, Ren L, Fan Q, Wang J, Xie Z (2013) Distribution characteristics and potential ecological risk assessment of heavy metals (Cu, Pb, Zn, Cd) in water and sediments from Lake Dalinouer, China. Ecotoxicol Environ Saf 93:135–144. https://doi.org/10.1016/j.ecoenv.2013.03.012

Howarth RW, Billen G, Swaney D, Townsend A, Jaworski N, Lajtha K, Downing JA, Elmgren R, Caraco N, Jordan T, Berendse F, Freney J, Kudeyarov V, Murdoch P, Liang ZZ (1996) Regional nitrogen budgets and riverine N & P fluxes for the drainages to the North Atlantic Ocean: natural and human influences. Biogeochemistry 35:75–139. https://doi.org/10.1007/BF02179825

Huang D, Xu J, Zhang S (2012) Valuing the health risks of particulate air pollution in the Pearl River Delta, China. Environ Sci Policy 15(1):38–47. https://doi.org/10.1016/j.envsci.2011.09.007

Iqbal J, Shah MH, Shaheen N (2015) Distribution, source identification and risk assessment of selected metals in sediments from freshwater lake. Int J Sedim Res 30(3):241–249. https://doi.org/10.1016/j.ijsrc.2012.12.001

Jiang T, Zhong M, Zou LJ, LI K, Lin B, Zhu AP, Liu ZF (2016) Simulation and analysis of water quality in Dongjiang River based on the different scenarios of flood discharge from Shima River and the regulation of Dongjiang Hydro-Project. Acta Sci Nat Univ SunYatSen 55(2):117–123

Jin ZF, Wang FE, Chen YX, Qgura NR (2004) Nitrate pollution of groundwater in urban area. Acta Pedol Sin 41(2):252–258

Jonge VN, Elliott M, Orive E (2002) Causes, historical development, effects and future challenges of a common environmental problem: eutrophication. Hydrobiologia 475(476):1–19. https://doi.org/10.1023/A:1020366418295

Koukal B, Dominik J, Vignati D, Arpagaus P, Santiago S, Ouddane B, Benaabidate L (2004) Assessment of water quality and toxicity of polluted Rivers Fez and Sebou in the region of Fez (Morocco). Environ Pollut 131(1):163–172. https://doi.org/10.1016/j.envpol.2004.01.014

Li HB, Yu S, Li GL, Deng H (2012) Lead contamination and source in Shanghai in the past century using dated sediment cores from urban park lakes. Chemosphere 88:1161–1169. https://doi.org/10.1016/j.chemosphere.2012.03.061

Li F, Huang J, Zeng G, Yuan X, Li X, Liang J, Wang X, Tang X, Bai B (2013) Spatial risk assessment and sources identification of heavy metals in surface sediments from the Dongting Lake, Middle China. J Geochem Explor 132:75–83. https://doi.org/10.1016/j.gexplo.2013.05.007

Liu J, Guo JS, Fang F, Li Z, Chen YP (2011a) Phosphorus input-output budgets for a small watershed in purple soil sloped farmland in the Three Gorges area and its pollution potential. Res Environ Sci 24(12):1385–1392. https://doi.org/10.13198/j.res.2011.12.57.liuj.010

Liu YD, Wu DL, Li MM, Liang M (2011b) Pollution situation of Dongguan reaches of Dongjiang River and its relevant countermeasures. J Dongguan Univ Technol 18(5):6–9

Liu JF, Song ZG, Xu T (2006) Study on ionic composition of rainwater at Guangzhou and the primary factors of rainwater acidity. Acta Sci Circum 27(10):1998–2002. https://doi.org/10.13227/j.hjkx.2006.10.013

Ma X, Zuo H, Tian M, Zhang L, Meng J, Zhou X, Min N, Chang X, Liu Y (2016) Assessment of heavy metals contamination in sediments from three adjacent regions of the Yellow River using metal chemical fractions and multivariate analysis techniques. Chemosphere 144:264–272. https://doi.org/10.1016/j.chemosphere.2015.08.026

MEP MoEPotPsRoC (2002) Environmental quality standards for surface water

Nancy BD, Carly JS (2005) Nitrogen deposition and reduction of terrestrial biodiversity: Evidence from temperate grasslands. Sci China Ser C Life Sci 35:720–728. https://doi.org/10.1360/062005-265

Russell MJ, Weller DE, Jordan TE, Sigwart KJ, Sullivan KJ (2008) Net anthropogenic phosphorus inputs: spatial and temporal variability in the Chesapeake Bay region. Biogeochemistry 88:285–304. https://doi.org/10.1007/s10533-008-9212-9

Valiela I, Boynton W, Hollibaugh JT, Jay D, Kemp WM, Kremer J et al (1992) Understanding changes in coastal environments: the LMER Program. EOS 73:481–485. https://doi.org/10.1029/91EO00354

Varol M, Şen B (2012) Assessment of nutrient and heavy metal contamination in surface water and sediments of the upper Tigris River, Turkey. Catena 92:1–10. https://doi.org/10.1016/j.catena.2011.11.011

Wang XY, Yan ES, Qu Y (2009) Material flow analysis of the phosphorus cycle in the upper watershed of the Miyun Reservoir in Beijing. Acta Sci Circum 29(7):1549–1560. https://doi.org/10.13671/j.hjkxxb.2009.07.003

WHO (2011) Guidelines for drinking-water quality, 4th edn.

Wu S, Peng S, Zhang X, Wu D, Luo W, Zhang T, Zhou S, Yang G, Wan H, Wu L (2015) Levels and health risk assessments of heavy metals in urban soils in Dongguan, China. J Geochem Explor 148:71–78. https://doi.org/10.1016/j.gexplo.2014.08.009

Xing GX, Cao YC, Shi SL, Sun GQ, Du LJ, Zhu JG (2001) N pollution sources and denitrification in waterbodies in Taihu lake region. Sci China Ser B Chem 31(2):130–137. https://doi.org/10.3321/j.issn:1006-9240.2001.02.006

Xing GX, Yan XY (1999) Direct nitrous oxide emissions from agricultural field in China estimated by the revised 1996 IPPC guidelines for national greenhouse gases. Environ Sci Policy 2:355–361. https://doi.org/10.1016/S1462-9011(99)00025-8

Xing GX, Zhu ZL (2000) Analysis and estimation of nitrogen sources and outputs in China Watershed. Acta Pedol Sin 37(supplement):72–82

Xing GX, Zhu ZL (2002) Regional nitrogen budgets for China and its major watersheds. Biogeochemistry 57(58):405–427. https://doi.org/10.1023/A:1016508323200

Xiong X, Jiang Y, Ren FP, Dong MY, Tian YH, Lei YM (2010) Land-use change and its ecological effect to the water in downstream river network area of Dongjiang basin. J Nat Resour 25(8):1321–1331

Xu Q (2006) Water conservancy encyclopedia China. China Water Power Press, pp 256–257

Yang ST, Cheng HG, Bu QS, Zhang JY, Shi XX (2006) Estimation of soil erosion and its application in assessment of the absorbed nitrogen and phosphorus load in China. Acta Sci Circum 26(3):366–374. https://doi.org/10.13671/j.hjkxxb.2006.03.003

Yeung ZLL, Kwok RCW, Yu KN (2003) Determination of multi-element profiles of street dust using energy dispersive X-ray fluorescence (EDXRF). Appl Radiat Isot 58(3):339–346. https://doi.org/10.1016/s0969-8043(02)00351-2

Zhang R, Guan ML, Shu YJ, Shen LY, Chen XX, Zhang F, Li TG (2016) Historical record of lead accumulation and source in the tidal flat of Haizhou Bay, Yellow Sea: insights from lead isotopes. Mar Pollut Bull 106:383–387. https://doi.org/10.1016/j.marpolbul.2016.02.046

Zhao XF, Chen FJ, Chen JY, Tang CY, Luo YL, Jia GD (2008) Using nitrogen isotope to identify the sources of nitrate contamination in urban groundwater-a case study in Zhuhai city. Hydrogeol Eng Geol 3:87–92. https://doi.org/10.16030/j.cnki.issn.1000-3665.2008.03.011

Zhu ZL (1997) Nitrogen balance and cycling in agroecosystems of China. Kluwer Academic Publishers, Dordrecht, pp 323–330

Dr. Lichun Xie is currently a Lecturer of Physical Geography at the Guangdong University of Finance and Economics. Her current research interests include fluxes and transport mechanisms of nutrients in groundwater and surface water under the influence of urbanization in coastal areas, and changes of hydrologic processes in response to urbanization. Xie has authored or co-authored over 20 articles in these areas.

Dr. Lei Gao is currently an Associate Research Fellow at the School of Geography and Planning, Sun Yat-Sen University, China. His current research interests include migration, transformation regularity, and effect of trace metals in watershed, historical reconstruction of aquatic environment, and geochemical cycling of phosphorus at the overlaying water-sediment interface. Dr. Gao has authored or co-authored 20 articles in the field of environmental and hydrological science.

Dr. Jianyao Chen is a Full Professor of Hydrology at the School of Geography and Planning in the Sun Yat-sen University, China. His research interests include hydrological processes in coastal watershed, nutrients in groundwater, and paleohydrology. Chen has authored or co-authored one book and over 100 articles in these areas.

Flooding Hazards and Risk Analysis in the Pearl River Delta, China

Xianwei Wang

Abstract The Pearl River Delta (PRD) of southern China is known for its complex river networks and low lying terrain. The rapid population growth and economic development make this area vulnerable to flood events. This chapter briefly describes the flood hazards, assess their flood risk, and illustrate the local defense strategies and key infrastructures. The first section is a general introduction to the geographic feature and climatic setting. The second section illustrates the heavy precipitation, the river fluvial floods and the flood defense infrastructures in the Bei Jiang catchment. The third section describes the urban pluvial floods in Guangzhou, including the main urban flood types, storm features, drainage modeling, and flood hazards. The fourth section discusses storm surges and coastal floods, including flood types, storm surge flood hazards, a typical coastal flood defense in Zhongshan, and four extreme typhoon and storm surge events occurred in the 21st century. The last section is a summary. Overall, most cities in the PRD region are facing severe flood hazards under the protection of levees. More thorough and quantitative analyses are required to better understand the local specific flood hazards and risk.

Keywords Heavy storms · Storm surges · Flood · Risk analysis
Pearl River Delta (PRD)

X. Wang (✉)
School of Geography and Planning, Sun Yat-sen University, Guangzhou 510275, Guangdong, China
e-mail: Wangxw8@mail.sysu.edu.cn

X. Wang
Guangdong Key Laboratory for Urbanization and Geo-simulation, 135 West Xingang Road, Guangzhou 510275, Guangdong, China

X. Yang and S. Jiang (eds.), *Challenges Towards Ecological Sustainability in China*, https://doi.org/10.1007/978-3-030-03484-9_9

1 Introduction

The Pearl River ranks the third in length and the second by volume in China. The Pearl River Delta (PRD) is located in the Pearl River estuary (Fig. 1) and consists of two alluvial deltas (East and West) that are formed by three major tributaries of the Pearl River, i.e., Xi Jiang (West), Bei Jiang (North) and Dong Jiang (East). Xi Jiang originates from Yunan-Guizhou Plateau in the northwest of Guangdong province, and is the dominant tributary in length (2400 km), drainage area and streamflow of the Pearl River. Bei Jiang collects runoffs from the Lingnan Mountains in the North of PRD and converges into Xi Jiang at Sanshui in Foshan, where the west PRD starts. The minor eastern PRD is formed by the Dong Jiang (below Shilong in Dongguan) and separated from the western PRD by the Pearl River's core branch. The low lying deltaic plains are crisscrossed by main watercourses and numerous local channels, which pour into the South China Sea through eight outlets from south to north, i.e., Yamen, Hutiaomen, Jitimen, Modaomen, Hengmen, Hongqimen, Jiaomen, and Humen (Fig. 1), and are protected by thousands of closed river levee and coastal seawall systems.

The PRD region has a sub-tropic climate controlled by the South China Sea Monsoon. Typically, its summers are warm and wet and its winters are dry, with a mean annual air temperature of 22 °C and an annual mean precipitation of 1700 mm,

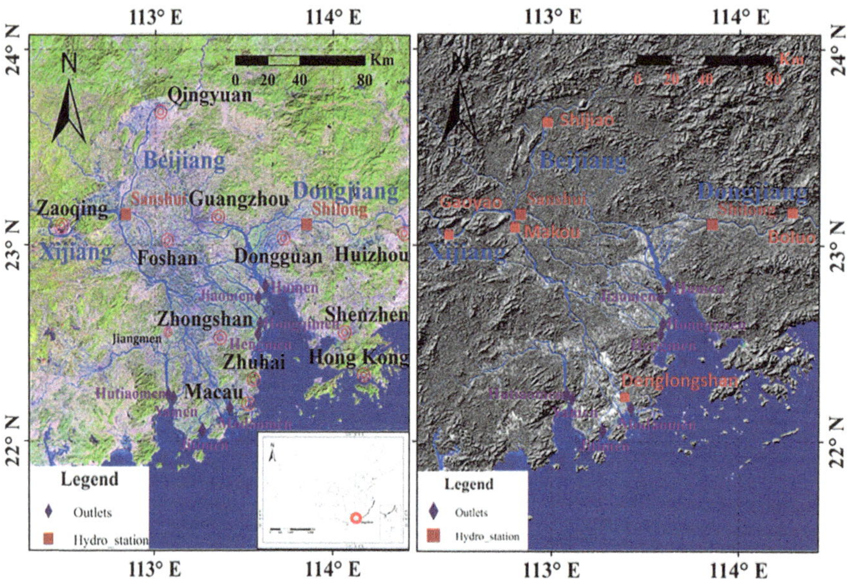

Fig. 1 The geographic features of the Pearl River Delta and its surrounding area. The background (left) is Landsat images in 2015, and the right is a shaded map derived from digital elevation model (DEM) data (1: 50000, 25 m grid) provided by the Department of Land and Resources of Guangdong Province

ranging from 1200 to 2500 mm (GBWA 2014; Liu et al. 2014). Near 85% of the annual precipitation falls during the rainy season from April to September (GBWA 2014; Yang et al. 2015). During the flooding/rainy season, the front and convective storm events are dominated in the first rainy season from April to June, while the local convective thunderstorms and the typhoon-brought storms are dominant during the second rainy season from July to September (Pan et al. 2017). An average of 2.8 typhoons per year have struck this region during the recent years, contributing to 20–30% of the annual rainfall (Ren et al. 2006). The typhoon-induced storm surges severely impact the low lying delta plains (Zhang et al. 2009b).

The PRD is an emerging metropolitan area with ten large cities including Guangzhou, Dongguan, Shenzhen, Zhuhai, Jiangmen, Zhongshan, Foshan, Zhaoqing, Qingyuan and Huizhou, plus Hong Kong and Macau (Fig. 1). They form the Guangdong-Hong Kong-Macau Greater Bay Area or the Yuegang'ao Greater Bay Area, which is likely to become the world's largest urban area in area, economy and population according to the World Bank Group (WBG 2015). The three cities of Zhaoqing, Qingyuan and Huizhou are adjacent to the PRD region geographically, but are counted into the Greater Bay Area economically. The seven cities (excluding Zhaoqing, Qingyuan and Huizhou) in PRD had a population of 52.46 million and GDP by 4224.8 billion Yuan at the end of 2015, comprising 48.4 and 64.1% of the provincial total count (Xing et al. 2016). The rapid population growth and economic development make this area more vulnerable to flooding hazards, such as the river fluvial flooding, urban storm pluvial flooding, and coastal storm surge flooding.

2 River Fluvial Flooding in the Bei Jiang Catchment

2.1 Precipitation

The rising terrains in the north of PRD uplift the warm and humid air blown from the South China Sea and cause heavy rainfall within the Bei Jiang catchment (Fig. 1). The regional mean annual rainfall in the southern China (Fig. 2a) is 1673 ± 339 mm during 2008–2015, 46% of which fell in the first rainy season from April to June (Fig. 2b), and 33% in the second rainy season from July to September (Fig. 2c). Besides the coastal areas, the Bei Jiang catchment and the PRD areas are often the storm centers, especially in the first rainy season when front storms and localized thunderstorms are dominated. The rainfall has large temporal variations, and the maximum rainfall is almost doubled compared to the mean rainfall (Fig. 2d–f), leading to severe flash-type flooding hazards along the Bei Jiang and the PRD.

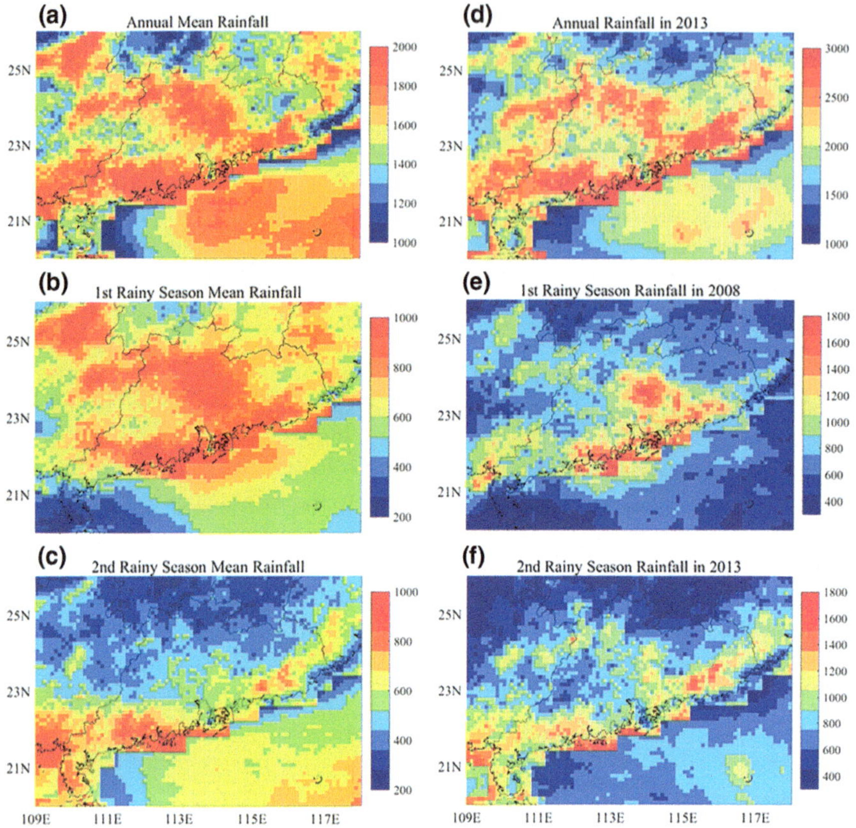

Fig. 2 Spatial distribution of mean rainfall (mm) for **a** annual, **b** first rainy season, **c** second rainy season during 2008–2015, and the maximum rainfall for **a** annual in 2013, **b** first rainy season in 2008, **c** second rainy season in 2013 from CMPA precipitation product

2.2 Catastrophic Flooding Events

There were catastrophic flooding events in the downstream of the Pearl River during the last century (Table 1). The most devastating one occurred in 1915, when both the Bei Jiang and Xi Jiang rivers recorded the largest flood peak which encountered at Sanshui at the same time (Fig. 1). Most of river embankments and levees in the Bei Jiang and in PRD rivers failed. Two thirds of Guangzhou was inundated for a whole week, and many villages in PRD were in deep water over a month. Near four million people were seriously affected, and more than ten thousands of fatalities were reported.

Another catastrophic event happened in June and July of 1994, when again both the Bei Jiang and Xi Jiang rivers recorded the largest flood peak which encountered at Sanshui at the same time (Table 1). Over 12 million people were seriously impacted,

Table 1 Major recorded flooding events in the mid and downstream of Xi Jiang and Bei Jiang during 1915–2015. The flooding information was published on the website of Guangdong Provincial Bureau of Beijiang River Basin Administration (GPB-BRBA), http://www.gdbjdd.com.cn/

Year month	Basin	Impact areas (10^4 hm^2)	Impact pop. (10^6)	Fatality	Economy loss 10^8 Yuan	Damage description	*Q1 (m^3/s)	*Q2 (m^3/s)
1915 Jun–Jul	Xi, Bei Jiang	93.33	3.78	>10 k	n.a.	Two thirds of Guangzhou were inundated for a week	54500	21000
1949 July	Xi Jiang	39.33	3.70	n.a.	n.a.	Guilin, Liuzhou and Nanning were inundated over two weeks, depth 5–6 m, 90% of houses were in water	48900	n.a.
1968 June	Xi, Bei Jiang	12.73	n.a.	n.a.	n.a.	Water levels at Sanshui were above warning level for 37 days. Bei Jiang levees were in dangerous conditions	42600	n.a.
1982 May	Bei Jiang	13.2	2.29	493	4.4	1:20 year flooding, damaged 60000 houses, 201 levees and 758 small dams, 71 hydropower plants, the main Beijing-Guangzhou railway were suspended for 13 days.	n.a.	15200
1994 June	Xi, Bei Jiang	125	13.19	371	n.a.	1:50 year storm, flooding peak from Xi and Bei Jiang encountered at Sansui, where water level were beyond warning level for 10 days; 1502 km levees, 2060 sluice gates and 332 small dams failed	52600	16700
1994 July	Xi Jiang	n.a.	12.83	n.a.	**632	1:100 year storm, most dike and levees in PRD were overtopped and failed, Zhuhai, Foshan were inundated for a few meters deep, and Zhongshan was in water over one month.	55000	19700
2005 June	Xi, Bei Jiang	n.a.	3.71	54	28.2	Guangzhou, Foshan, Dongguan, Zhongshan, Zhuhai were in water.	n.a.	13500
2006 July	Bei Jiang	25.1	0.34	111	30.7	100a storm hit the Beijiang and PRD, 248 km road were destroyed, Beijing-Guangzhou Railway were suspended, electricity, telephone, television, water pipes were stopped	n.a.	17500

Note *Q1 were reported at hydrologic stations of Wuzhou or Gaoyao in Xi Jiang, while Q2 were reported at Hengshi or Shijiao in Bei Jiang. **632: total direct economy loss for both June and July flooding events in 1994 while near four hundred fatalities were reported. The economy loss was the largest one on record, 63.2 billion Yuan (RMB). Fighting against flooding is always the first strategy for the local government.

2.3 Reservoirs

The major fluvial flooding hazard for PRD now comes from the Bei Jiang river although the streamflow from the Xi Jiang river is over three times of Bei Jiang. The main river channel along the Bei Jiang is 468 km long, and its drainage area above Sanshui (Sixianjiao hydrologic station) is 46,710 km^2, with the mean slope of 0.26‰. The annual mean discharge is 428.8×10^8 m^3, ranging from 163.7 to 722.2×10^8 m^3. Because of the storms centers in the mid and downstream of Bei Jiang river (Fig. 2), the flooding peak routines and retreats very fast like flash flooding, leading to severe challenges for flooding forecast and management.

A series of flooding defense infrastructures had been constructed to reduce the impact of flooding on the lower part of Bei Jiang catchment and on the PRD. Large reservoirs had been built in the upper streams, cascaded dams were constructed across the main channels, and large dike systems were piled up along the downstream main channels below the Feilai Gorge Reservoir in Qingyuan and along watercourses in the PRD (Table 2). The Nanshui Reservoir was the largest one (12.84×10^8 m^3) constructed in 1960s, when most reservoirs were constructed in the catchment head of tributaries. The second wave of reservoir construction was in 1990s, and all cascaded the main channel of Bei Jiang. The Feilai Gorge Reservoir in the downstream of Bei Jiang in Qingyuan is the largest one (19.04×10^8 m^3), while the Lechang Gorge Reservoir is the only large one (3.44×10^8 m^3) completed in the 21st century (2013) in the upper stream (Wujiang, main tributary) of the Bei Jiang. Geographically, the Liuxihe reservoir lies in the Liuxihe catchment, adjacent to the east of Bei Jiang catchment and drains to the Pearl River that flows through the downtown Guangzhou. The total volume of the ten large reservoirs is 49.38×10^8 m^3, together having a storage index (reservoir volume/annual discharge) of 0.11 and maximum flooding control volume by 33×10^8 m^3. They dramatically reduce the flooding risk in the entire catchments, especially in the downstream and metropolitan areas of PRD.

2.4 Levees

Reservoirs in the upstream and across the main river channels are the first buffers to reduce the flash flooding, while the levees are the last and most crucial barriers to defend the low lying floodplain. The Bei Jiang Levee is one of the seven state key levee systems in China and the most crucial one in Guangdong Province. Bei Jiang Levees were initially built in the Song Dynasty (960–1279) and were strengthened in the Ming Dynasty (1368–1644), the Qing Dynasty (1644–1911), and the Republic of

Table 2 The ten large reservoirs (>1.0×10^8 m^3) in the Bei Jiang catchment

No.	Reservoir name	River/location	Start year	Finish year	Total volume (10^8 m^3)	Flooding volume (10^8 m^3)	Drainage areas (km^2)
1	Tanling	Lianjiang (head)	1962	1967	1.80		142
2	Xiaokeng	Zhenjiang (head)	1964	1969	1.13		139
3	Liuxihe	Lixihe (head)	1956	1958	3.76	2.39	539
4	Nanshui	Nanshui (Head)	1958	1969	12.84	9.37	608
5	Jinjiang	Jinjiang (Upper)	1990	1993	1.89		1500
6	Changhu	Wengjiang (Tributary)	1969	1974	1.55	0.55	4800
7	Lechang Gorge	Wujiang (Upper)	2007	2013	3.44	2.11	4988
8	Mengbazhou	Up Bei Jiang	1992	1998	2.04		14720
9	Baishijiao	Mid Bei Jiang	1992	1999	1.89	0.85	17740
10	Feilai Gorge	Down Bei Jiang	1992	1999	19.04	13.16	34100
Total					49.38	33	

China (1912–1949). The present Bei Jiang Levees were completed during 2003–2007 after several times of constructions during 1955–1957, 1970–1972, and 1983–1987, and were designed for a 1:100 year storm flooding event, plus 1.5 m height of uncertainty (GPB-BRBA 2007).

The Bei Jiang Levees begin at Shijiao in Qingyuan along the left side of Bei Jiang although there are other levees above Shijiao, and end at Shishan in Nanhai, Foshan, with a total length of 64 km (Fig. 3). They protect Qingyuan, Foshan, Guangzhou and the low lying PRD, with total population over 20 million, GDP of 2000 billion Yuan (RMB) and many key civil infrastructures, such as the new Baiyuan International Airport, Beijing-Guangzhou railways/high-speed railways, express ways, and so on.

The present Bei Jiang Levees consist of three sessions (Fig. 3). The first main session (46 km) distributes from Shijiao to Sanshui in the downstream of Bei Jiang, and the second session (18 km) is from Sanshui to Shishan of Foshan in PRD, where Bei Jiang meets Xi Jiang at Sanshui, Foshan. The third session is a 10-km long dual levee in the mid of the first session at Datang, where the levee foundation is soft. Another two sessions with soft foundation are in Shijiao and Lubao. During the 1:100 year flooding event in June 1994 in Bei Jiang, a severe piping event occurred under the levee foundation at Shijiao. On 17 June 1994, the flooding water level within Bei Jiang at Shijiao rose to 14.5, 3.85 m beyond the warning level. At 3:00 pm on 19, a 3-m (diameter) pipe formed at Xialingzhou (Shijiao) levee foundation, and water pumped out 1.5 m high above ground. The levee and the behind mega cities were in dangerous situation. An emergency remediation was carried out immediately, and near 600 rescuers including experts, officers, soldiers, military students and local residents, had fought continuously over 17 h to stopped the piping accident (Fig. 4).

Fig. 3 A Bei Jiang levee distribution map manually marked based on GoogleEarth images

After this piping event, numerous investigations had been done to analyze the geologic features on the levee foundations up to 30 m below at several weak and soft foundation sessions. After a 10-year investigation and a series of discussions, a full strengthening and repairing project with a budget of 2.5 billion yuan was proposed and had been completed during 2005–2007. Together with the ten large reservoirs, they are expected to control a 1:300 year storm flooding event in the Bei Jiang catchment. Those flooding defense infrastructures in the Bei Jiang Catchment greatly reduce the flooding hazards to the mega cities of the Guangzhou-Foshan and PRD.

Fig. 4 Four Bei Jiang levee photos at Shijiao, Qiangyuan: **a** Upstream view of waterside, **b** Downstream view, **c** Text describing the 94.6 levee piping emergency remediation, and **d** Shijiao Monument erected in 2014 to honor those in the 94.6 flooding rescue. Photos were taken by Xianwei Wang on 16 October 2016

2.5 Flooding Stage Variations in PRD

Gaoyao, Shijiao and Boluo are three key hydrologic stations used to monitor the water discharge from Xi (west) Jiang, Bei (north) Jiang and Dong (east) Jiang catchments, respectively (Fig. 1b). Above each of them, their drainage areas are 351500 (84.7%), 38400 (9.2%) and 25300 (6.1%) km^2, collecting annual mean runoff by 219.9 (77.1%), 41.7 (14.6%) and 23.0 (6.1%) km^3 during 1954–2004 (Zhang et al. 2008). The annual water discharge at Gaoyao in Xi Jiang and at Sijiao in Bei Jiang were mainly controlled by the catchment precipitation variations and did not show a significant change during 1954–2004 because of their relatively small reservoir storage index (~0.1), a ratio of total reservoir storage capacity to the annual mean streamflow, while it showed a significant decrease trend at Boluo in Do Jiang since it has a high storage index by 0.78 due to numerous large reservoirs constructed within the catchment (Zhang et al. 2008; Chen et al. 2010). Such a large reservoir storage capacity in Dong Jiang, such as Xinfengjiang and Fengshuba Reservoirs, the two largest reservoirs in Guangdong Province, are beneficial for flood mitigation and greatly reduce the fluvial flooding hazards along the Dong Jiang catchment and in the east PRD in the flooding season, and also provide crucial fresh water resources in the dry season for over thirty million people in PRD like Shenzhen, Hong Kong, Dongguan, Huizhou and so on.

The major fresh water in PRD is poured down from Xi Jiang via Gaoyao. Although the annual mean water discharge did not show a significant change during 1954–2004 (Zhang et al. 2008, 2009a), the annual maximum streamflow at Gaoyao and Shijiao had an upward trend (Fig. 5). However, their annual maximum water levels instead showed a decline trend. Accordingly, the annual maximum water level at stations of Makou and Sanshui in the upper PRD also showed a decrease trend (Zhang et al. 2009b). For instance, the flood stage at Sanshui during the catastrophic flood event in June 1994 (simplified as "94.6") was 1.28 m lower compared with previous similar flood discharges (Luo et al. 2007). Meanwhile, in the middle and lower portion of the delta, the flood stages were much elevated mainly due to the backwater effect of the spring high flooding tide and storm surges (Zhang et al. 2008).

The lower water levels at the upper PRD for a similar water discharge are attributed to the downcutting of river beds, which is contributed by less sediment filling from upstreams, inchannel sediment mining, navigation dredging and river bed erosion due to faster flow velocity (Luo et al. 2007; Zhang et al. 2009b, 2010). For instance, the annual sediment load from the Pearl River to PRD had declined by one third from 80.4×10^6 t during 1957–1995 to 54.0×10^6 t during1996–2004 (Zhang et al. 2010). Field surveys found that over 8.7×10^8 m^3 of sands were excavated from 1986 to 2003, causing an average downcutting of 0.59–1.73 m, 0.34–4.43 m, and 1.77–6.48 m in the main channels of the West River Delta, North River Delta, and East River Delta, respectively (Luo et al. 2007). The deeper channels along the Bei Jiang waterways in PRD altered the streamflow allocation of Xi Jiang (Makou) and Bei Jiang (Sanshui) when they meet at Sanshui. The flow ratio at station Sanshui was ~18% of the total flow (Sanshui and Makou) prior to 1991, and increased to 26.2% in 1997. In other words, their annual mean streamflow had doubled, increasing from 1240 m^3/s during 1960s–1980s to 2587 m^3/s during 1996–1998 (Luo et al. 2007).

The deeper river bed and the resulted lower water levels in the upper PRD indicate that the river freshwater flooding hazards from the Xi Jiang are greatly reduced. The deeper river bed and the rising sea level also suggest larger tidal prisms, and the spring high tides could propagate further. The tidal regime changes significantly throughout the PRD region. The tidal limit in Dong Jiang increased from just above Shilong in 1980s to near Boluo in 2000s, from above Sanshui in 1980s along Bei Jiang to near Shijiao in 2000s, and from above Gaoyao in 1980s along Xi Jiang to the middle between Gaoyao and Wuzhou in 2000s (Fig. 1). The tidal range and durations in the upper PRD like Makou and Sanshui had greatly increased from 1950s to 2005 (Zhang et al. 2010). With the defense of reservoirs and levee systems, the river fluvial flooding is not a major hazard to PRD, and the coastal flooding from spring high tide and storm surges are most dangerous sources for the low lying delta plain. More details will be discussed in the session of Coastal Flooding.

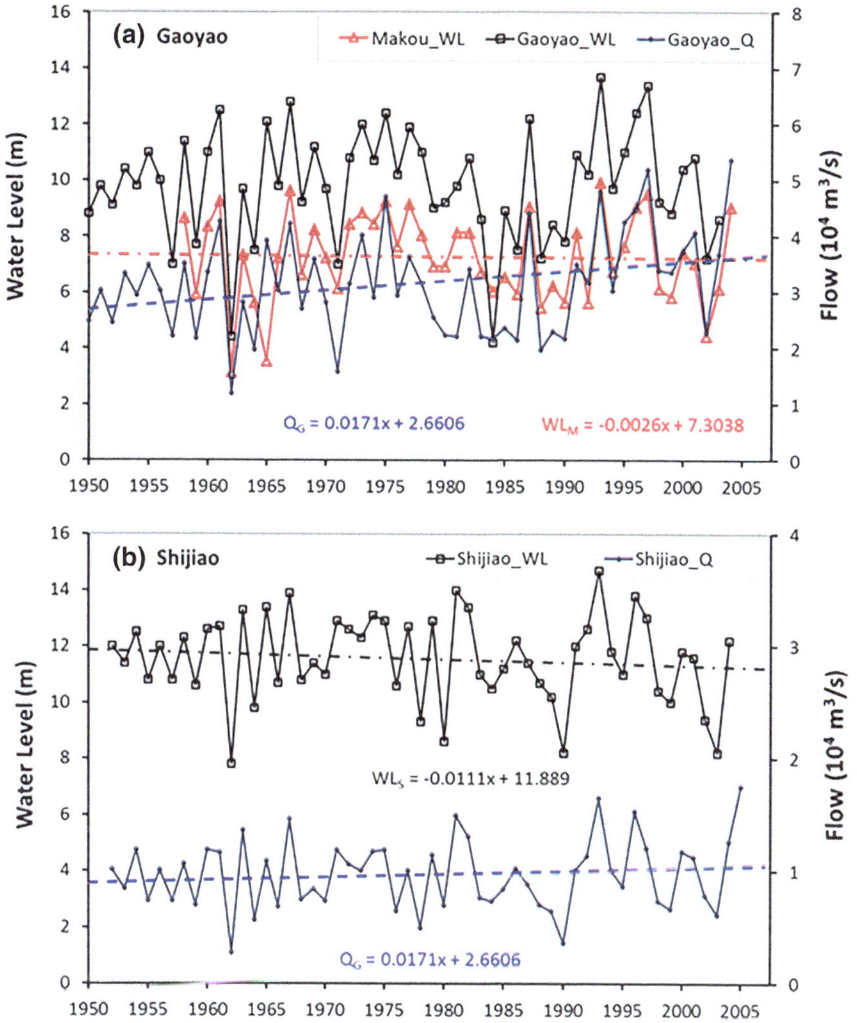

Fig. 5 The annual maximum flow at Gaoyao of Xi Jiang (**a**) and Shijiao of Bei Jiang (**b**), and their water level including at Makou, up which Xi Jiang and Bei Jiang converge and divert. Replotted from Zhang et al. (2009b)

3 Urban Pluvial Flooding in Guangzhou

3.1 Main Urban Flooding Types

Urbanization alters natural scape and disturbs the local energy balance and the atmosphere-ecology-hydrology system. Urbanization affects the rainfall diurnal distribution and increases the afternoon rainfall occurrence and amounts (Burian and

Shepherd 2005). The natural hydrological cycle is modified by removing vegetation and soil layers and building impervious layers and drainage infrastructure (Collier 2016). Both changes in rainfall pattern and hydrological cycle have important implications for urban water management, flood defend engineering and emergency responders (Burian and Shepherd 2005).

The PRD area has experienced rapid urbanization during the past 30 years. For example, in the Tianhe and Panyu Districts of Guangzhou, the remote sensing-derived impervious surface ratio increased from 16% to 71% and from 2% to 40% from 1990 to 2013, respectively (Pan et al. 2017). However, most (about 83%) of the drainage pipes were designed according to the standards for storms with a one-year return period, and less than 10% of the pipes with a two-year return period (Wu et al. 2016). Clearly, the return period is very low because a 25-year return period is practiced in many countries. Consequently, the core city area of Guangzhou has been frequently flooded due to the low drainage capacity of pipes and channels. This is a very common type of urban pluvial flooding in the highly urbanized area in PRD, such as Guangzhou, Shenzhen and Dongguan.

Another type of urban pluvial flooding is common in the low-lying delta plains, such as Zhongshan (like Zhongshan, Xiaolan, Guzhen), and Foshan (like Shunde, Nanhai). These cities are located in the middle PRD and surrounded by the Pearl River waterways. They were protected by closed high river levee systems. Some of these areas are lower than the spring high tide and river flooding peaks. Thus, the local storm runoffs could not drain out by gravity during the high tide or river flooding. Pumping is required to keep these areas dry when there are high water levels in the surrounding waterways. This type of urban flooding will be analyzed in the coastal flooding session. This session takes Guangzhou as an example to demonstrate the first type of urban pluvial flooding in PRD.

3.2 Urban Storm Features

Heavy storms are the direct cause of urban pluvial flooding. Most storms occur in the first (April, May and June) and second (July, August and September) rainy seasons, when it has 72% of annual rainfall during 2008–2012 in Guangzhou (Fig. 6a). It had a rainfall peak from 12:00 to 7:00 pm in the afternoon, and about 43% of rainfall occurred in this period (Fig. 6b). About 75% of those heavy storm (>15 mm/h) durations were shorter than 3 h at both sites in Guangzhou (Table 3). The downtown Site 1 at Tianhe District had 46% more storms than the suburban Site 2 at Panyu District (Table 3) during 2008–2012. The peak rainfall time of 54% storm events at Site 2 concentrated in the second quartile, while 44% of them were in the first quartile at Site 1 (Table 4). Similar patterns were also observed in the urban area of Beijing (Yang et al. 2014).

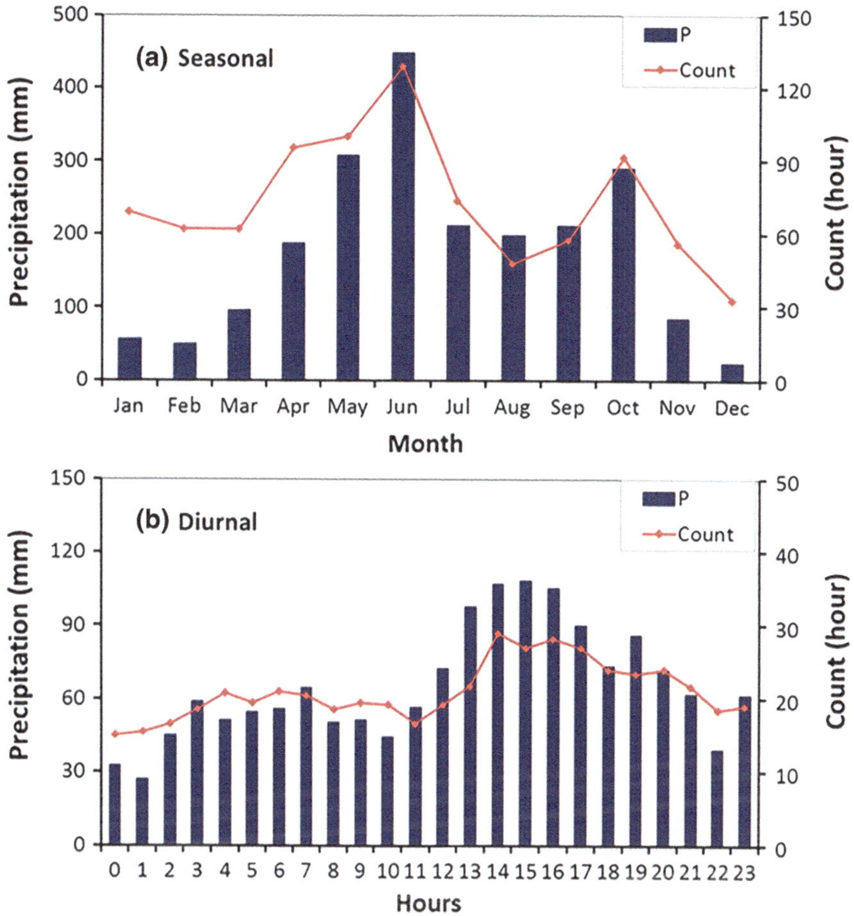

Fig. 6 Monthly (**a**) and hourly (**b**) mean rainfall and rain hours during 2008–2012 based on records of the standard meteorological station (Site ID: 59287) in Guangzhou

However, both sites exhibit a similar distribution pattern in rainfall depth, time of peak intensity, and mean intensity in the rising and falling limbs after normalizing the rainfall duration and depth (Pan et al. 2017). The normalized time of peak rainfall is $33 \pm 5\%$ of the mean storm duration. The rising limb falls $43 \pm 5\%$ of the total rainfall. This indicates that the storm hyetograph is similar at both sites (25 km apart), and thus the storm curve developed at one site could characterize the generic rainfall temporal distribution in the study area.

Table 3 The number of storms with different durations at two sites (Tianhe and Panyu district of Guangzhou) for the period 2008–2012 (*Source* Pan et al. 2017)

Rain Gauges	Duration (h)							
	<1	1–2	2–3	3–4	4–5	5–6	>6	Total
Site 1 (Suburban)	12	27	15	3	6	5	3	71
Site 2 (Urban)	25	37	16	11	6	5	4	104
Total	37(21%)	64(36%)	31(18%)	14(8%)	12(7%)	10(6%)	7(4%)	175

Table 4 The number of storms in each quartile defined according to the occurrence of peak rainfall considering a normalized time at two sites (Tianhe and Panyu district of Guangzhou) for the period 2008–2012 (*Source* Pan et al. 2017)

Quartiles	1st	2nd	3rd	4th	Total
Site 1 (Suburban)	31(44%)	27(38%)	9(13%)	4(6%)	71
Site 2 (Urban)	33(32%)	56(54%)	8(8%)	7(7%)	104
Total	64(37%)	83(47%)	17(10%)	11(6%)	175

3.3 Urban Drainage Modeling

It is quite challenging to accurately predict the urban pluvial flooding hazards because of the complicated artificial drainage systems, such as the typical dual or triple drainage system (Collier 2016). It was exacerbated by the unknown information for the underground drainage infrastructures for many old cities like in the old downtown of Guangzhou.

Two dimensional (2D) hydrodynamic models are often used to simulate the urban drainage at regular grids, such as the Mike 11, FloodMap, LISFlood-FP etc. (Bates and De Roo 2000; Liu et al. 2015; Collier 2016). These 2D models require prohibitive computing resources when they are applied to large areas or with fine resolutions. Computational stability and efficiency are important factors to be considered in urban flooding practice. Thus, cellular automaton (CA)-based 2D models are developed to simulate the urban water dynamics at higher computing efficiency. Its surface water flow is governed by a set of gravitational diverging rules in a rectangular template of three cells by three cells of a raster layer (Parsons and Fonstad 2007). Such a CA model had also been developed and tested in the downtown Shiqiao, Panyu District of Guangzhou (Liu et al. 2015). The CA model could simulate the dynamic flooding extent and depth with an accuracy of 90% or 4 cm in water depth, which is comparable to a physically based 2D model (e.g., FloodMap). Additionally, the CA model has much lower cost for model setting up and higher computing efficiency that can meet the needs of city emergency management (Liu et al. 2015). However, the underground pipe drainages are often simplified or even neglected in most 2D hydrodynamic models, including the CA model.

In contrast, the Storm Water Management Model (SWMM) developed by the American Environmental Protection Agency is good at modeling underground pipe drainages, while the surface runoff is simulated with lump models in different sub-catchments (USEPA 2009). SWMM has been widely used for urban drainage management, flood-control facility design, and water quality modeling (Croci et al. 2014; Li et al. 2015). SWMM was also established in the Shiqiao Street, downtown of Panyu District, Guangzhou (Pan et al. 2017). The total SWMM modeling area is 15.53 km^2. Each sub-catchment was delineated from the main pipe network and high-resolution airborne LiDAR DEM data, and was augmented by repeated field visits and validation. There are 194 main pipes, 230 notes and 258 sub-catchments in the SWMM modeling (Fig. 7). The dynamics of water flow (depth and velocity) within each pipe can be explicitly simulated, and the flooding sites are also predicted. The inundation depth and areas are further computed based on the SWMM predicted inundation volume and the fine DEM data.

SWMM can simulate the flow dynamics quite well once all required data are collected, such as the underground pipe information (radius, height) and the fine topography used to divide the sub-catchment, which is crucial to correctly model pipe flow dynamics and flooding sites. It is time-consuming to establish a sound SWMM modeling frame and thus is difficult to be applied for real-time flooding management in megacities like Guangzhou since many underground pipes were not correctly recorded and the pipe design adopted a low standards of 1 to 5-year return periods. SWMM can still predict the overland flow caused by intense rainfall, while 2D hydrodynamic models may work better by assuming a maximum piping drainage. Meanwhile, SWMM is quite useful for sub-catchment drainage planning, drainage infrastructure designs, and pinpointing drainage working status.

3.4 Urban Flooding Hazards

The short-term and intense storms are the direct driving force of urban pluvial flooding, while the local micro topography and pipe/pump capacity determine the flooding hazards and frequency. It's difficult to map the urban flooding hazards in a mega city such as Guangzhou considering its complicated artificial hydrologic system, which is frequently changing with rapid urbanization and timely drainage updating. For example, there were severe flooding sites (water clogging) each year before 2011 in the old downtown area of Yuexiu District of Guangzhou. After a series of drainage-improving projects for several years, its flooding stage had been greatly improved, while new flooding hotspots appeared in the lately emerging business center of Tianhe District, where the drainage improvement in the upstream even resulted in more flooding sites in the downstream areas, especially when the water level of Pearl River is in high stage.

The short-term and intense thunder storms cause sever hazards to the metropolitan areas of Guangzhou and PRD. For instance, in the early morning on 7 May 2017, a localized thunderstorm initially occurred at Huadu and Bei Yun District, the northern

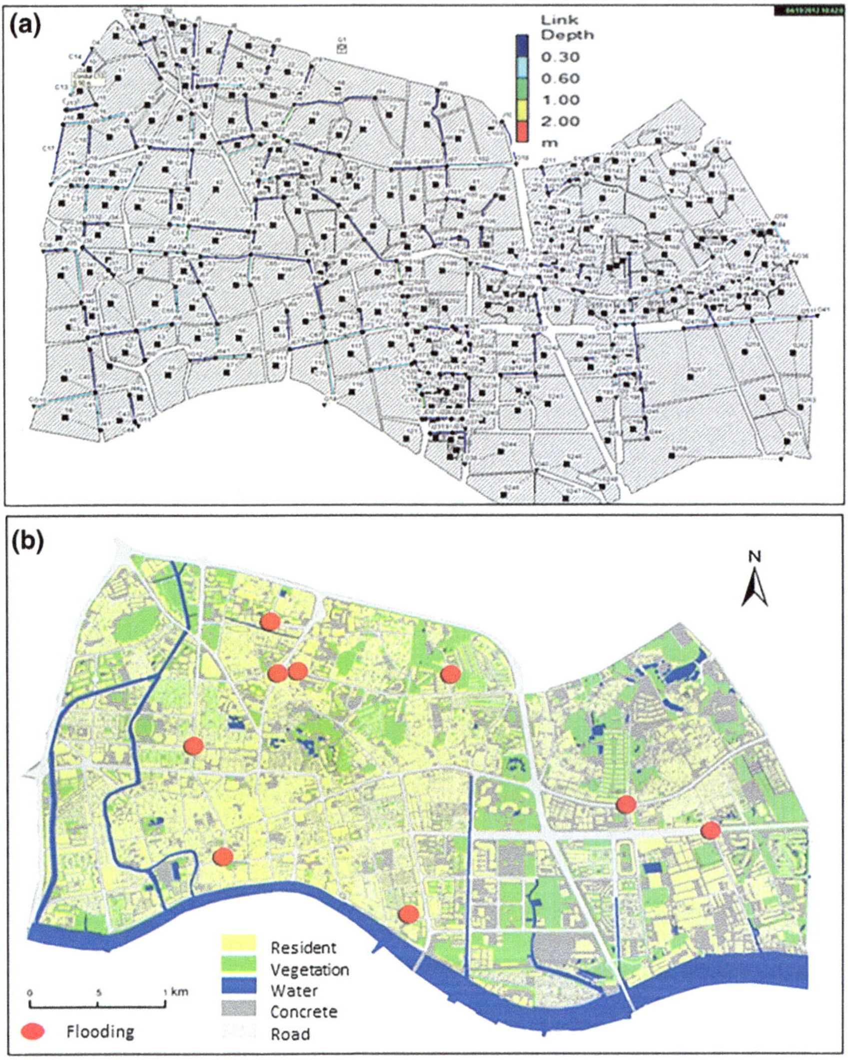

Fig. 7 SWMM modeling of pipe water depth (**a**) and surface flooding sites (**b**) for different storm events at the downtown Shiqiao, Panyu District, southern Guangzhou, China

Guangzhou, moved eastward to Zengcheng and Huangpu, then turned to west at Tianhe and south at Haizhu and Panyu district (Fig. 8). The storm centers were located in the suburban areas with rising topography, such as Huadong, Jiulong, and Xintang. The maximum event-total precipitation was recorded at Jiulong (545 mm). The maximum 3-h (382 mm) and 24-h (524 mm) accumulation at Xintang broke the records of Guangzhou, while its hourly accumulation (184 mm) was just a few mm less than the maximum record (188 mm). Figure 9 illustrates the hourly precipitation

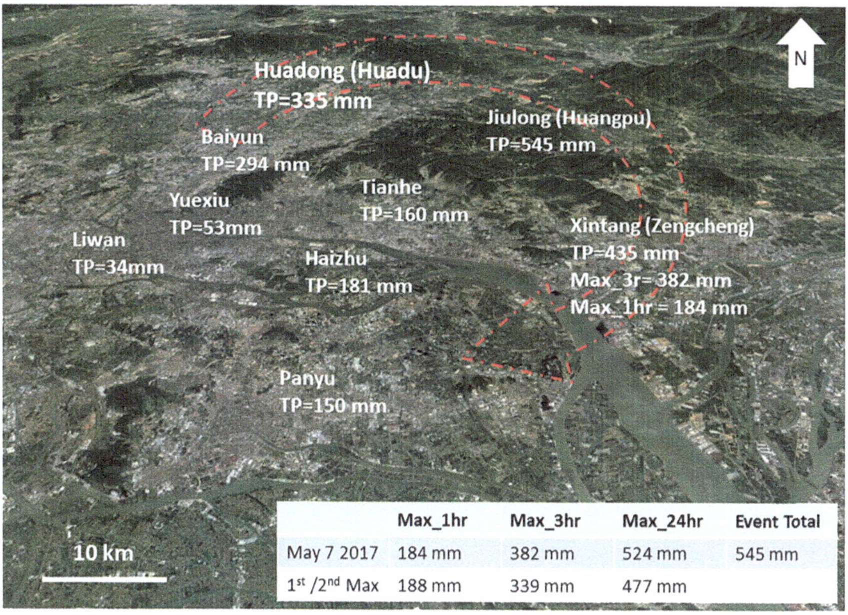

Fig. 8 The maximum event total precipitation at each district of Guangzhou on 7 May 2017. The red dashed arrow shows the storm movement track. The inset table lists the maximum 1, 3, 24-h and event total precipitation for this event and for historic records in Guangzhou

distribution at four sites. Most of the storm rainfall occurred within a few hours and had a sharp peak intensity, which brought severe hazards for storm water draining. Almost every street under those storm centers was inundated for a few hours (Fig. 10). Many main roads had been inundated over 50 cm depth, acting as main drainage channels and boat-evacuating routes.

Such a localized thunderstorm had large spatial heterogeneity, which makes it a severe challenge to weather forecast and urban hydrologic modeling. The huge amount of rainfall concentrated in a small area is also a big problem for urban drainage design. The current design standards in Guangzhou and other cities in PRD are far below the demand to instantly drain out such amount of heavy rainfall. There is still a long way to adapt to the heavy summer rainfall for urban drainage planning and management. In order to achieve a better and safer living environment, flooding knowledge and education are significant for all classes of the society in PRD, such as urban planners, government officers, drainage management staff and the local community. Good flooding risk perception and full preparedness can greatly mitigate the impact and damages of flooding events.

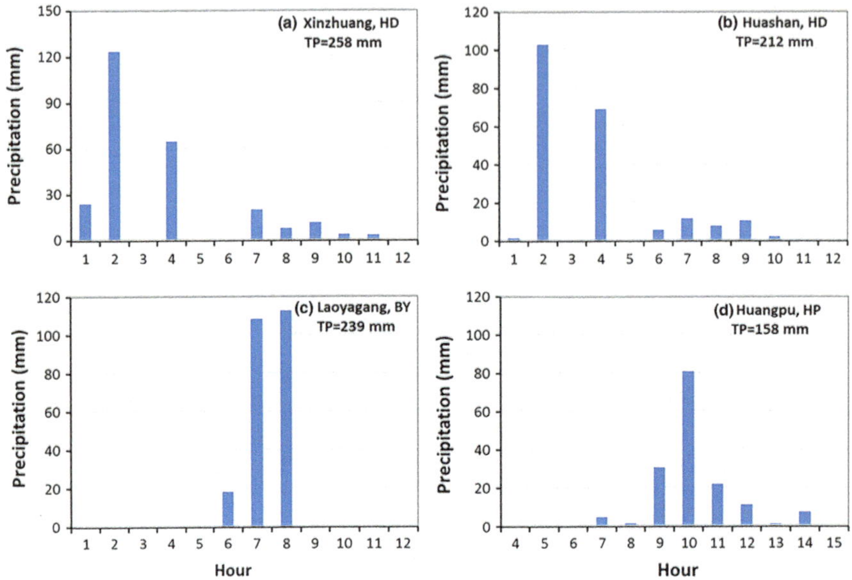

Fig. 9 The intense thunderstorm rainfall recorded at meteorological sites of **a** Xinzhuang, **b** Huashan of Huadu (HD) district, **c** Laoyagang of Baiyun (BY) district and **d** Huangpu (HP) district of Guangzhou on 7 May 2017

4 Storm Surges and Coastal Flooding

4.1 Coast Flooding Types

Typhoon and storm surges can abruptly and severely affect coastal areas, causing sea water intrusion, coastal erosion, civil infrastructure damage and even large death tolls (Menendez and Woodworth 2010). PRD is in sub-tropic and tropic climate settings and had 2–3 landfall typhoons on average per year (Ren et al. 2006). PRD bears a large proportion of population and economy of the Guangdong Province and is exposed to frequent typhoon and storm surges. For instance, the super typhoon Rammasun swept Guangdong from July 17 to 18, 2014. It reported a death toll of 18 in China, caused direct economic losses of 13.12 billion Yuan (RMB) and affected 2.56 million people in the Guangdong province (Deng et al. 2014). The storm surge is mainly responsible for the total economy loss and death toll.

Storm surges (SS) are the unusual rise of sea surface related with low pressure weather systems, such as tropical cyclones and strong extratropical cyclones (Peek and Young 2013). Their impacts can be exacerbated by riverine flood, waves and spring high Astronomical Tides (AT). AT is the rise and fall of sea surface compelled by the joint gravitational force exercised by the Moon, the Sun and the Earth and the Earth's rotation. The combined effects of storm surges, waves and AT are also

Fig. 10 Three flooding pictures at **a** the town hall of Zhongxin, Zengcheng, **b** Guangzhou Business College campus, Tianhe, **c** Guangyuan Road, Huangpu district of Guangzhou on 7 May 2017

called as storm tide, which formulates the Total Water Level (TWL) height. When large storm surges encounter with a high AT, they may generate an extremely high water level, resulting in disastrous impacts on the low lying coastal areas (Wang et al. 2017).

Coastal flooding is usually a natural process and could not be prevented completely. Several common pathways of coastal flooding are possible in PRD: direct inundation, where water level exceeds the elevation of natural land without defense infrastructures (Collier 2016); overtopping the height of seawall and river embankments; failure of flooding defenses, such as barrier, levees, seawalls and embankments etc.; backwater effects, which reduce the river discharge rate and result in abnormal water level (fresh water piles up in river channels), or prevent the storm water from draining out under gravity from the low lying terrains protected by blocked levees.

The backwater effects frequently occur and show different severity at different section of PRD each year. For instance, the downtown of Zhongshan was inundated for almost a month during the June-July flooding season of 1994; the river-front streets/roads in Haizhu, Huangpu, Tianhe and Yuexiu districts of Guangzhou along the Pearl River were inundated for a few hours during the spring high tide and

Hato (1713) typhoon landfall on 23 August 2017. Accompanying typhoon landfall is heavy rainfall (Fig. 2c, f), which causes severe challenges to urban storm water management and urban life in sea front cities like Shenzhen, Zhuhai and Zhongshan in PRD.

4.2 Storm Surge Flooding Hazards

The low lying terrains of PRD are susceptible to typhoon-induced floods (Zhang et al. 2013). Hallgatte et al. (2013) even ranked Guangzhou as the highest risk city among the 136 largest coastal cities in the world because of the potential coastal flooding losses, although their analysis neglected the mitigation effect of the huge flooding defense infrastructures and greatly overestimate the actual flooding risk in Guangzhou (Feng et al. 2013). In the Great Bay area of PRD, there is an urgent need to conduct comprehensive flood risk assessment that take into account catastrophic impacts of extreme riverine flood, super typhoons and storm surges, and to offer flood risk maps for the local governments and public for better flooding risk management and development planning (Wang et al. 2017).

Figure 11a shows that elevations in most PRD area are less than 5 m, while the mean sea level is around 80 cm referring to the 1985 Yellow Sea Geodatum of China, and the water levels of spring high tide ranges from 1.5 m to 3 m in PRD. In other words, a large part of areas in PRD are lower than the spring high tide, such as in the downtown of Zhongshan and Shunde, where storm water could not drain out under gravity and artificial pumping is in great demand. These low lying areas are protected by complicated river embankment systems (Fig. 11b). For example, there are 1167 km of levees within the administrative areas of Guangzhou, among which 303 km has protect criteria of 1:200 year, 173 km for 1:100 year criteria, 574 km for 1:50 year, and 109 km for 1:20 year (GBWA 2014).

The levees play a crucial role in defending the coast flooding and protect the low lying delta plains. The entire PRD is almost fully inundated by a 1:100 year storm

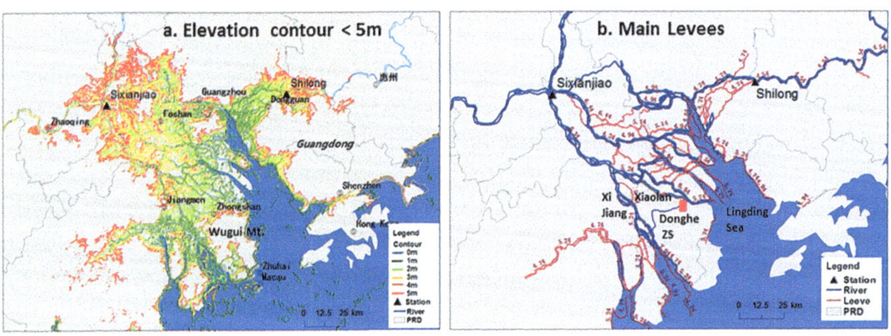

Fig. 11 The elevation contour (**a**) and main levees (**b**) in PRD

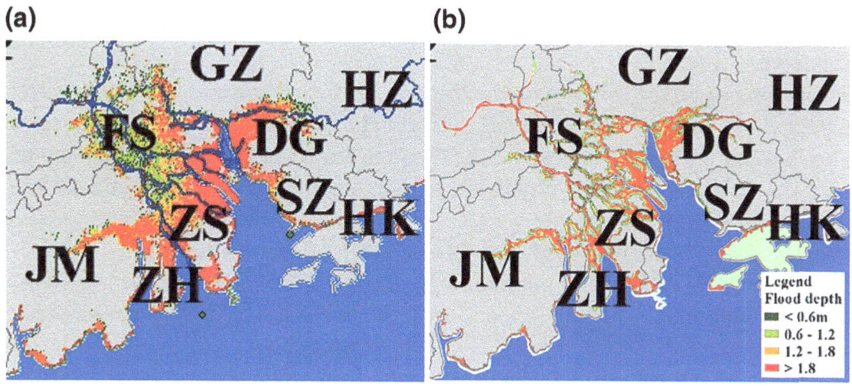

Fig. 12 The storm surge inundation depth by **a** a static model without considering levees and **b** a 2D hydraulic model considering the levees' height in PRD for a 1:100 year storm surge event

surge event plus the spring high tide without the protection of levees (Fig. 12a), and its impact is dramatically reduced when levees protection is considered (Fig. 12b). Dongguan still faces severe storm surge hazards due to the low lying terrains, relatively higher storm surges and lower design standards of the levee system in the east PRD. Under the protection of large levee systems, the flooding impact is greatly reduced, resulting in fast development in the low lying delta for residents, industry and commercial malls. The concentration of population and economy makes these areas more vulnerable to flooding events, especially to super typhoon and severe storm surges. That's a well-known double-edged sword for flood protection provided by levees. Unlike a natural flood, levee failure flood is often rapid, forceful, extremely damaging, and occurs with little or no warning. Given enough time, levees either will be overtopped or even failed, resulting in severe impacts on population and properties at unprecedented scales (ASFPM 2007).

4.3 *Coastal Flooding Defense in Zhongshan*

The City of Zhongshan is a typical region that is severely impacted by both fresh flooding and storm surge flooding. It is located in the lower part of PRD (Fig. 11). The main area (simplified as Zhongshan) is a V-shaped island surrounded by Xi Jiang in west, Xiaolan watercourse in north, and Lingding Sea in east. The Wugui Mountain (Peak = 531 m a.s.l) stands in the center, and its surrounding areas are delta plains. The largest northwest plain is protected from west and north by the Zhong Shun Da Wei levees, which are expected to defend the 1:100 year storm surge events according to the latest repairing standards. The western terrain is generally higher than the east, and the north is higher than the south above Qijiang River. There are many river channels inside the main levees. Qijiang River is the largest one, which crosses the island from west to east and connects Xi Jiang and the Xiaolan-Hengmen watercourse (Fig. 11b).

The Qijiang River valley is the lowest part of Zhongshan and had been severely impacted by both fresh flooding and storm surge flooding. One of the most severe ones in recent 30 years occurred during the June-July flooding season of 1994 when this area was inundated for a month due to overtopping and backwater effects of the high water level in the outside watercourses. After the event, a series of flooding defense projects were proposed. The Donghe Water Control Project was and is still the largest integrated flooding defense engineering in Zhongshan (Fig. 13a). There is another gate in the west of Qijiang River, the Xihe (west) Tide Gate, and several other gates along the levees. Besides flooding defense, both gates work together like a giant shiplock for flushing the internal rivers, increasing navigation and irrigation by controlling the water levels inside and outside both gates during the diurnal tidal level variations (Fig. 13b). During a normal flooding tide, the Donghe (east) gate is closed, and the Xihe (west) gate is open to allow the fresh water in Xijiang flow into the Qijiang River; during an ebb tide, the Xihe gate is closed, and the Donghe gate is open, allowing the internal water in Qijiang River to discharge into the Hengmen watercourse. Together with levees, reservoirs, ponds and channel storage and other pumping stations, the flood defense in Zhongshan is expected to stand for a 1:100 year local heavy storm event.

A large part of terrains inside the Zhong Shun Da Wei levees is less than 2 m a.s.l., while the high tide is around 1.5 m in the Xiaolan watercourse. The water in the Qijiang River (and other internal rivers) cannot drain out to the watercourse under gravity during high tide, about half of a day (Fig. 13b). Once local heavy storm runoff exceeds the internal storage capacity, artificial pumping is demanded. During the 1713 Typhoon Hato landfall (on 23 August 2017), the water level outside Donghe gate (in Xiaolan-Hengmen Watercourse) exceeded the warning level over three hours, and the internal water in Qijiang River could not drain out under gravity over 20 h, when the internal water level continued increasing due to local rainfall. Pumping operation was started to pump the internal water out for a few hours in order to secure maximum internal channel storage capacity in the Qijiang River before the typhoon landfall. Lucky enough, there was no heavy rainfall accompanying this typhoon, and the pumping operation was shortly stopped. The Donghe and Xihe tide gates and the levees successfully defended the storm surge, while the outside water level (2.88 m) almost reached the constraining level (2.97 m) of the Donghe gate. The maximum water head residual ($H_{out} - H_{in}$) for the Donghe gate is 2.67 m, indicating the internal water level has to increase with the outside water level once it reaches the maximum residual head, resulting in less internal channel storage capacity. Accurate weather and flow forecasting and leading pumping operation would greatly improve the performance of flooding defense infrastructures and the public safety inside the levees.

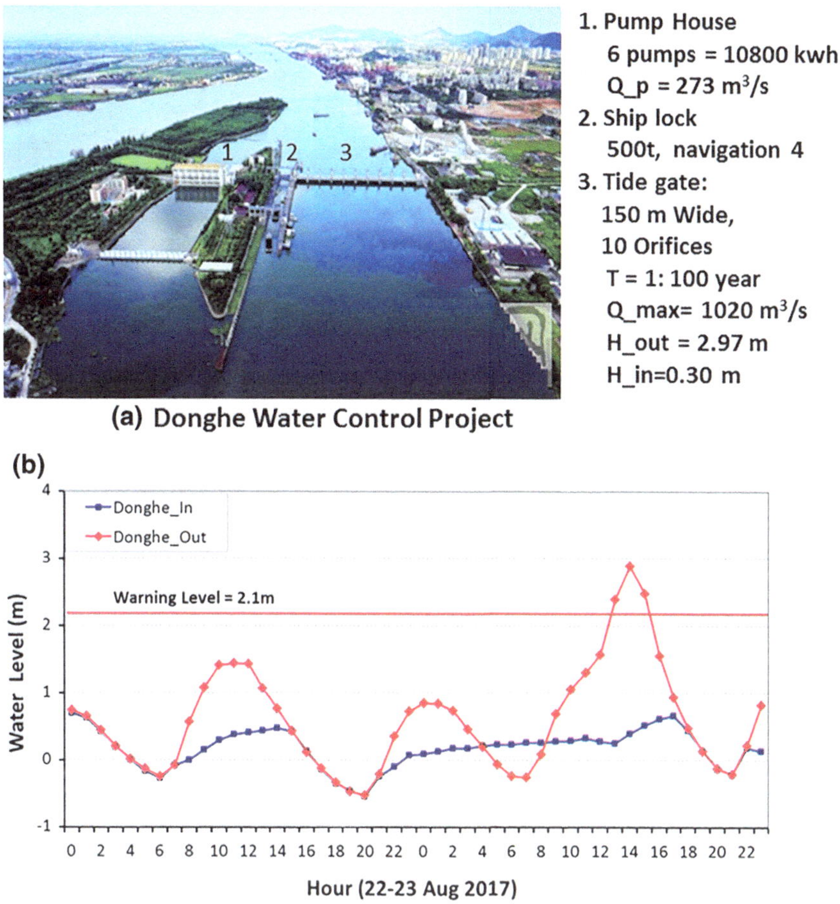

Fig. 13 **a** The Donghe Water Control Project in Zhongshan and **b** the water level variations Inside and Outside the tide gate during the 1713 Typhoon Hato on 22–23 August 2017. Water levels are above the mean sea level referring to the Pearl River Geodatum, which is 78 cm lower than the 1985 Yellow Sea Geodatum of China

4.4 Extreme Storm Surge Events

The low-lying PRD areas are prone to the impacts of typhoons and storm surges. Figure 14 illustrates the paths and landfall information of four typhoons that severely affected PRD in the 21 century, i.e., 0814 Hagupit, 1319 Usagi, 1604 Nida and 1713 Hato. All were formed in the Philippine Sea, and made landfall in the Guangdong province, China. They moved from southeast to northwest and produced the maximum storm surges in the Pearl River Estuary because of their strong anti-clock wind and coincident flooding tide.

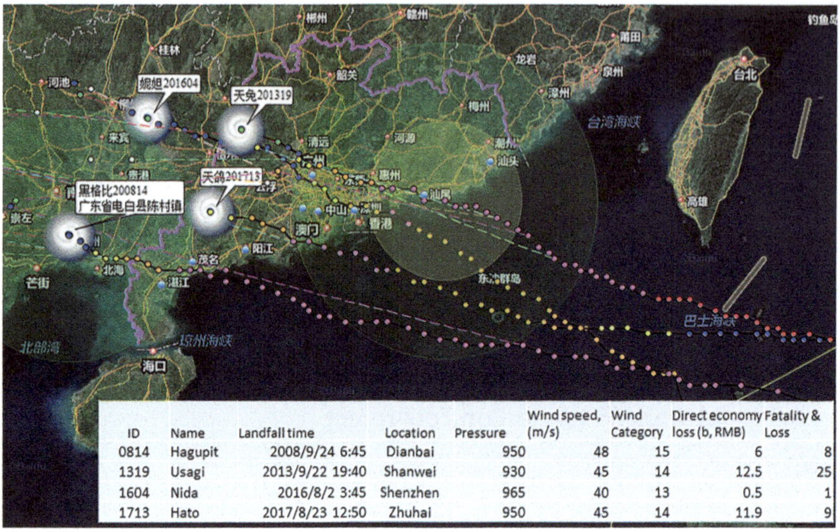

ID	Name	Landfall time	Location	Pressure	Wind speed, (m/s)	Wind Category	Direct economy loss (b, RMB)	Fatality & Loss
0814	Hagupit	2008/9/24 6:45	Dianbai	950	48	15	6	8
1319	Usagi	2013/9/22 19:40	Shanwei	930	45	14	12.5	25
1604	Nida	2016/8/2 3:45	Shenzhen	965	40	13	0.5	1
1713	Hato	2017/8/23 12:50	Zhuhai	950	45	14	11.9	9

Fig. 14 Four typhoons that severely affected PRD in the 21 century. The direct economy loss (Billion Yuan) and fatality numbers are reported in the Guangdong Province, China

4.4.1 0814 Hagupit

Hagupit made landfall at 6:45 on 24 September (lunar: 25 August, neap tide) 2008 in Dianbai County, Maoming (Fig. 14). Although its landfall location was about 200 km south of PRD, the shortest distance from typhoon centers to PRD before landfall was only 100 km and during flooding tide, thus generating large storm surges in the PRD. The total economy loss was 6 billion yuan and the death toll was 8 in Guangdong Province. Zhuhai were severely damaged, including levee failure, overtopping, road collapsed, crop and four deaths/lost.

4.4.2 1319 Usagi

Usagi was the strongest typhoon formed in 2013 on Earth, being called Monster Super Typhoon. It made landfall at 19:40 on 22 September 2013 (lunar: 18 August, spring tide) in Shanwei, about 180 km north of the Pearl River Estuary (Fig. 14). The record-broke low pressures and strong wind caused 25 deaths and 12.5 billion yuan direct economy loss in Guangdong province. The cities of Shanwei were severely damaged, reporting 16 deaths and 11.8 billion yuan direct economy loss. The sea-front roads and streets in Zhuhai were inundated by the storm surge and wave setup. Many water level stations in PRD exceeded their warning levels.

(a) The Fishery Lady in Zhuhai on 2 Aug 2016 **(b)** The Fishery Lady before Nida

(c) The White Swan Hotel in Guangzhou on 23 Aug 2017 **(d)** The White Swan Hotel after Hato

Fig. 15 The storm surge inundation and dry land pictures of the Fishery Lady (**a**, **b**) in Zhuhai during typhoon Nida landfall on 2 August 2016 (lunar: June 30, near spring high tide), and the front parking lot of the White Swan Hotel (**c**, **d**) in Guangzhou during typhoon Hato landfall on 23 August 2017 (lunar: July 1, near spring high tide)

4.4.3 1604 Nidan

Nidan was a normal scale typhoon, but it was only one that made landfall near the Daya Bay nuclear power plant in Shenzhen at 3:40 am on 2 August 2016 (lunar: 30 June, near spring tide), and then passed through the center of PRD (Fig. 14). Red warning was issued for the entire PRD cities, and schools and business were closed. Many water level stations exceeded their warning levels, especially in the Lingding Sea and the upstream of Humen outlet. Storm surges also inundated the sea-front streets in Zhuhai, e.g., about one meter in the park of Fishery Lady (Fig. 15a, b).

4.4.4 1713 Hato

The severe typhoon Hato made landfall in Zhuhai (near Macao) at 12:50 on 23 August 2017 (lunar: 1 July, near spring tide) (Fig. 14). Hato swept PRD and caused direct economy loss of 11.9 billion yuan and 12 deaths, especially in Zhuhai, Zhongshan, Jiangmen, Dongguan, Foshan and Guangzhou. Macao reported 8 deaths and 153 injuries, and the chief officer of the Meteorological Administration was resigned, partially due to the tardy operation of issuing warning information. The sea-front streets in Macao and Zhuhai were severely inundated. The spring high tide and storm surges caused most of water level stations in PRD exceed the warning levels for a few hours (Fig. 13b). The water-front streets in Guangzhou (the upper PRD) were even inundated by over 30 cm for a few hours, such as the parking lot of the most prestigious White Swan Hotel, in Shamian, Yuexiu District (Fig. 15c, d).

5 Summary

The Yuegang'ao Greater Bay Area has been developed from rural towns into a cluster of megacities during the past 40 years, and is very likely to become one of the largest urban areas of the world in the near future. Many cities in PRD lie in low lying terrains or even are reclaimed from the sea. They are under protection by complicated levee systems and artificial pumping stations. Levee protection is a double-edged sword. Given enough time or an extreme event, levees either will be overtopped or even failed, resulting in catastrophic impacts on population and damage to properties behind levees at unprecedented scales. It is quite controversial to assess the flooding hazards and potential risk in the low lying terrains behind those large levee systems.

As far as the flooding hazards are concerned in PRD, several pathways are possible, i.e., the river fluvial flooding, local urban pluvial flooding, and the coastal storm surge flooding. The reservoirs in the Bei Jiang catchment cut off the peak runoff at the first stage. Then the huge levee systems play a crucial role in protecting these cities in PRD. The Bei Jiang Levees constrain the roaring flooding flow generated in the steep terrains of the Bei Jiang catchment, and escort it into the Lingding Sea via watercourses blocked by high embankments in the PRD, thus protecting a huge number of population, property and key civil infrastructures behind the embankments.

Within the megacities, such as Guangzhou, local storm water drainage is a severe problem. On one hand, many natural open channels are narrowed down and covered by concrete due to urbanization, while the drainage pipes are constructed at low designed criteria, most in 1:1 year and a few in 1:5 years. On the other hand, the pluvial flooding is exacerbated by the backwater effects due to the high water level in the internal river channels, main outer watercourses and the sea.

In the sea-front cities in the lower part of PRD, such as Zhuhai, Shenzhen and Zhongshan, coastal storm surge flooding is most dangerous. The upstream flooding discharge and the local heavy rainfall aggravated the flooding hazards. Huge artificial pumping capacity is in great demand in order to drain out the internal runoff within the blocked levees during high tide, especially in cases of outer watercourse fresh flooding discharge, spring high tide and large storm surges.

This is just a generic analysis on the flooding hazards facing by the cities in PRD. More thorough and quantitative analysis is required to better understand the local specific flooding hazards and risk. Meanwhile, those factors must be addressed in a synoptic way when local detailed flooding risk assessment and management is carried out.

Acknowledgements The research reported in this chapter was partially funded by the Water Resource Science and Technology Innovation Program of Guangdong Province, China (#2016-19). The edition from Dr. Xiaojun Yang greatly improves this chapter and is highly appreciated.

References

ASFPM (2007) National flood policy challenges levees: the double-edged sword. ASFPM White Paper, 1–10. Online article. http://www.floods.org/. Accessed 30 Oct 2017

Bates PD, De Roo AP (2000) A simple raster-based model for flood inundation simulation. J Hydrol 236:54–77

Burian SJ, Shepherd JM (2005) Effects of urbanization on the diurnal rainfall pattern in Houston. Hydrol Process 19:1089–1103

Chen Y, Zhang Q, Xu CY, Lu X, Zhang S (2010) Multiscale streamflow variations of the Pearl River basin and possible implications for the water resource management within the Pearl River Delta, China. Quat Int 226:44–53

Collier CG (2016) Hydrometeorology. Wiley, Hoboken, NJ, USA

Croci S, Paoletti A, Tabellini P (2014) URBFEP model for basin scale simulation of urban floods constrained by sewerage's size limitations. Procedia Eng 70:389–398

Deng SY, Li Q, Yue MX (2014). Rammasun Guangdong losses of more than 13.1 billion. Nan Fang Daily, 2014-07-20001 (In Chinese)

Feng W, Zhuang J, You D, Xu W (2013) A discussion on reasons for impact of over-estimated sea level rise on storm surge disaster forecasting of the Pearl River Estuary. Trop Geogr 33(5):640–645

Guangzhou Bureau of Water Authority (GBWA) (2014) 2013 Guangzhou Water White Book. Online article. http://www.gzwater.gov.cn. Accessed 6 Oct 2017

Guangdong Provincial Bureau of Beijiang River Basin Administration (GPB-BRBA) (2007) Online materials. http://www.gdbjdd.com.cn/. Accessed 12 Oct 2017

Hallegatte S, Green C, Nicholls RJ, Corfee-Morlot J (2013) Future flood losses in major coastal cities. Nat Clim Change 3(9):802–806

Liu T, Zhang YH, Xu YJ, Lin HL, Xu XJ, Luo Y, Xiao J, Zeng WL, Zhang WF, Chu C (2014) The effects of dust–haze on mortality are modified by seasons and individual characteristics in Guangzhou, China. Environ Pollut 187:116–123

Liu L, Liu Y, Wang X, Yu D, Liu K, Huang H, Hu G (2015) Developing an effective 2-D urban flood inundation model for city emergency management based on cellular automata. Nat Hazards Earth Syst Sci 15:381–391. https://doi.org/10.5194/nhessd-2-6173-2014

Li F, Duan H, Yan H, Tao T (2015) Multi-objective optimal design of detention tanks in the urban stormwater drainage system: framework development and case study. Water Resour Manag 29:2125–2137

Luo X, Zeng E, Ji RY, Wang CP (2007) Effects of in-channel sand excavation on the hydrology of the Pearl River Delta, China. J Hydrol 343:230–239

Menendez M, Woodworth PL (2010) Changes in extreme high water levels based on a quasi-global tide-gauge data set. J Geophys Res 115:C10011. https://doi.org/10.1029/2009JC005997

Pan C, Wang X, Liu L, Huang H, Wang DS (2017) Improvement to the Huff curve for design storms and urban flooding simulations in Guangzhou, China. Water 9(6):411. https://doi.org/10.3390/w9060411

Parsons JA, Fonstad MA (2007) Cellular automata model of surface water flow. Hydrol Process 21:2189–2195

Peek KM, Young RS (2013) Understanding the controls on storm surge through the building of a national storm surge database. J Coast Res 29(6A):17–24

Ren F, Wu G, Dong W, Wang X, Wang Y, Ai W, Li W (2006) Changes in tropical cyclone precipitation over China. Geophys Res Lett 33:L20702

United States Environmental Protection Agency (USEPA) (2009) Storm water management model applications manual, United States Environmental Protection Agency, Washington, DC, USA
Wang XN, Wang X, Zhai J, Li X, Huang H, Li C, Zheng J, Sun H (2017) Improvement to flooding risk assessment of storm surges by residual interpolation in the coastal areas of Guangdong Province, China. Quat Int. https://doi.org/10.1016/j.quaint.2016.12.025
World Bank Group (WBG) (2015) World Bank report provides new data to help ensure urban growth benefits the poor, 26 Jan 2015
Wu H, Huang G, Meng Q, Zhang M, Li L (2016) Deep Tunnel for regulating combined sewer overflow pollution and flood disaster: a case study in Guangzhou City, China. Water 8:329–336
Xing X, Ye J et al (eds) (2016) Guangdong Statistic Yearbook 2016. Online data downloaded on 10 Oct 2017 from http://www.gdstats.gov.cn/
Yang L, Scheffran J, Qin H, You Q (2015) Climate-related flood risks and urban responses in the Pearl River Delta, China. Reg Environ Change 15:379–391
Yang L, Tian F, Smith JA, Hu H (2014) Urban signatures in the spatial clustering of summer heavy rainfall events over the Beijing metropolitan region. J Geophys Res 119:1203–1217
Zhang SR, Lu XX, Higgitt DL, Chen CT, Han J, Sun H (2008) Recent changes of water discharge and sediment load in the Zhujiang (Pearl River) Basin, China. Global Planet Change 60:365–380
Zhang Q, Xu C, Chen Y, Jiang J (2009a) Abrupt behaviors of the streamflow of the Pearl River basin and implications for hydrological alterations across the Pearl River Delta, China. J Hydrol 377:274–283
Zhang W, Yan YX, Zheng JH, Li L, Dong X, Cai HJ (2009b) Temporal and spatial variability of annual extreme water level in the Pearl River Delta, China. Global Planet Change 69:35–47
Zhang W, Ruan X, Zheng J, Zhu Y, Wu H (2010) Long-term change in tidal dynamics and its cause in the Pearl River Delta, China. Geomorphology 120:209–223
Zhang K, Li Y, Liu H, Xu H, Shen J (2013) Comparison of three methods for estimating the sea level rise effect on storm surge flooding. Clim Change 118(2):487–500

Dr. Xianwei Wang is a Full Professor in the School of Geography and Planning at the Sun Yat-sen University, China. His research interests include remote sensing in hydrology and engineering, hydrologic modeling and flooding risk management. Wang has authored or co-authored three book chapters and over 40 peer-reviewed journal articles in these areas.

Part III
Challenges Towards Ecological Sustainability in Other Areas of China

The last section includes five chapters discussing various issues challenging the transition towards ecological sustainability in other areas of China. Chapter "Potential Impacts of Urban Sprawl on the Thermal Environment in the Nanjing Metropolitan Area Based on the SLEUTH and WRF Models" discusses the potential impacts of urban growth on the thermal environment in China's Nanjing metropolitan area, home of more than 8 million people by using a dynamic spatial model and the Weather Research and Forecasting (WRF) model. Chapter "Linking Land Use with Water Pollution in Coastal Watersheds of China" examines the land use/cover-water pollution relationship with two coastal watersheds in China's Fujian Province as the study site. Chapter "Water and Ecological Security at the Heart of China's Silk Road Economic Belt" discusses recent research progress on water security and ecosystem stability in the central Asia, the core region of the Silk Road economic belt. Chapter "The Spatial Network of Beibu Gulf Urban Agglomeration and Its Challenges for Coordinated Integral Development" discusses the urban networks and their inter-city interactions of the Beibu Gulf Urban Agglomeration (BGUA), a newly planned urban agglomeration in Southern China that is expected to become a strategic focus to support future regional economic development. Chapter "Strategies to Promote Green Development in Coal Resource-Based Regions with China's Shanxi Province as an Example" included in the book discusses some strategies to promote the green development in coal resource-based regions with China's Shanxi Province as an example.

Potential Impacts of Urban Sprawl on the Thermal Environment in the Nanjing Metropolitan Area Based on the SLEUTH and WRF Models

Fanhua Kong, Haiwei Yin, Fei Jiang and Jiayu Chen

Abstract The urban heat island (UHI) effect threatens the livability and sustainability in many cities. In this study, taking the Nanjing metropolitan area as the study area, we first simulated and predicted urban growth by 2040 under three ecological security scenarios (i.e., low, moderate, and high) by using the Slope, Land use, Exclusion, Urban extent, Transportation, and Hillshade (SLEUTH) model. Then, we investigated the impacts of future urbanization on the urban thermal environment by using the Weather Research and Forecasting (WRF) model under the future urban growth scenarios. Lastly, we analyzed the thermal environment characteristics corresponding to the changes in land use and urban sprawl. The results indicate that in 2040, the urban built-up land area would increase while the ecological protection grade would decrease. This means that urban sprawl would obviously affect the thermal environment and the urban heat island would intensify. The results also show that from the present to the three future scenarios (S1, S2, and S3), the areal extent of Zone 5 (pixels with air temperature >34.8 °C) would increase from 697.22 to 856.19 km^2, 1236.08 km^2, and 1434.78 km^2, respectively, which would lead to prominent diurnal variation in the UHI intensity with higher values during nighttime. These results should be helpful for urban planners and managers to better understand urban growth impacts upon the thermal environment in the future and the corresponding measures need to be taken to ensure a livable and sustainable city.

F. Kong (✉) · F. Jiang · J. Chen
International Institute for Earth System Science (ESSI), Nanjing University, Xianlin Ave.163, Nanjiang 210023, Jiangsu, China
e-mail: fanhuakong@163.com

F. Jiang
e-mail: jiangf@nju.edu.cn

J. Chen
e-mail: imbamiss@163.com

H. Yin
Department of Urban Planning and Design, Nanjing University, No. 22, Hankou Road, Nanjing 210093, Jiangsu, China
e-mail: qzyinhaiwei@163.com

X. Yang and S. Jiang (eds.), *Challenges Towards Ecological Sustainability in China*, https://doi.org/10.1007/978-3-030-03484-9_10

Keywords Urban sprawl · Thermal environment · SLEUTH model · WRF model Nanjing metropolitan area

1 Introduction

As urbanization progresses, vegetation and natural surfaces are gradually replaced by impervious surfaces, leading to alterations in the energy balance of an area, such as decreases in latent heat and increases in the sensible heat flux (Weng and Larson 2005; Morris et al. 2016a, b; Wang et al. 2016). At the same time, tall buildings and narrow streets in cities can also trap heated air between them. Ultimately, the urbanization conversion process creates an urban canopy layer (UCL) of buildings, which will alter radiation and thermal properties within a given urban space (Morris et al. 2017). A study conducted by Chen and Frauenfeld (2016) projected that rapid urbanization will lead to a temperature increase up to1.9 °C in various regions of China. Many cities are now suffering from intensifying urban heat island (UHI) effects, which are usually accompanied by frequent heat waves and outdoor thermal discomfort (Poumadere et al. 2005; Liu and Zhang 2011; Morris 2016; Morris et al. 2016a). These UHIs have severely affected the human living environment and increased energy consumption; consequently, UHIs threaten the livability and sustainability of cities (Weng and Larson 2005).

Surface UHI characteristics can be captured easily through the retrieval of land surface temperatures (LSTs) with remote sensor data; however, it is much more difficult to obtain atmospheric UHI characteristics at the city or larger scale, as field observations are typically necessary. The Weather Research and Forecasting (WRF) model, which is a mesoscale numerical weather prediction system (Skamarock et al. 2008), has been widely employed to study the urban climate in recent years and represents an alternative approach to studying atmospheric UHIs (Chen et al. 2011; Morris et al. 2016b; Jandaghian et al. 2018). The model has been also applied to investigate the impacts of urbanization and variations in UHIs in different cities and metropolitan areas (Kusaka et al. 2012; Conry et al. 2015; Morris et al. 2016a; Papangelis et al. 2012; Wang et al. 2016). However, most of the previous studies were centered on the impacts of present urbanization on UHIs, and few on possible impacts of future urbanization scenarios (i.e., future urban sprawl scenarios) on urban thermal environments or climate change.

Urban expansion is driven by multiple biophysical and socio-economic factors (Irwin et al. 2009; Akın et al. 2014; Yin et al. 2018). Dynamic spatial models can be used to examine land use changes and urban expansion, which can help urban resource managers and planners and designers to better understand urban sprawl and their associated consequences under different development scenarios (Barredo et al. 2003; Yang and Lo 2003; Santé et al. 2010; Vermeiren et al. 2012). Because of its flexibility, openness, non-linearity, and ability to self-adapt, the SLEUTH (Slope, Land use, Exclusion, Urban extent, Transportation, and Hillshade) model, developed by Clarke et al. (1997), has been widely used in land use research (Silva and Clarke

2002; Berling-Wolff and Wu 2004; Yin et al. 2018). Specifically, the SLEUTH model has become increasingly popular in urban growth simulations and predictions (Herold et al. 2003; Jantz et al. 2010; Mahiny and Clarke 2012; Norman et al. 2012; Vermeiren et al. 2012; Onsted and Chowdhury 2014). However, few studies have been conducted to find out how the simulated urban growth under different scenarios will impact on the urban thermal environment in the future.

The study aimed to examine the potential impacts of future urban growth scenarios on the thermal environment by using the urban growth model SLEUTH coupled with the WRF model. The Nanjing metropolitan area, Jiangsu Province, China, was selected as the study site largely because this region has experienced rapid growth during the past several decades. Three different urban development scenarios were investigated based on the grade of ecological security patterns (i.e., low, moderate, and high). The SLEUTH model was used to simulate the urban growth in 2040 under these three scenarios. At the same time, we used the WRF model to simulate the thermal environment under the different urban development scenarios, and then, we analyzed the potential impacts on the urban thermal environment under these different development scenarios. More specifically, the following two questions were addressed: (1) how would the urban extent of the study area grow under different development scenarios and (2) how much would the urban thermal environment be affected under these three development scenarios? These results should be helpful for urban planners to better understand the possible impacts of urban growth on the thermal environment, and accordingly, mitigation measures can be developed to ensure a livable and sustainable city.

2 Study Area

Nanjing, Jiangsu Province's capital city, is located in the eastern coastal region of China (Fig. 1). It is characterized by a subtropical monsoon climate with four distinct seasons including a hot, wet summer. According to Nanjing meteorological data for 1951–2010, the summer (daily air temperature higher than 22 °C over five consecutive days) lasts for 119 days on average. The average annual air temperature is 15.4 °C, the average daily maximum air temperature is 31.9 °C, and in July, the daily peak is 39.7 °C (Yin et al. 2017). The mean annual precipitation is ca. 1100 mm, with approximately 80% falling in the wet season (April to September).

Nanjing has experienced rapid economic growth in the last few decades with the Gross Domestic Product (GDP) increasing from 3.5 to 1045 billion RMB during 1978–2016 (Nanjing Statistics Bureau 2016; Kong et al. 2012). Along with this dramatic economic growth, Nanjing has also experienced intense rural migration to the city. From 1952 to 2016, the population has increased from 1.1 to 8.3 million (Nanjing Statistics Bureau 2016). The built-up land has also grown from 116 km^2 in 1978 to 734 km^2 in 2016 (Statistical Yearbook of Nanjing 2016). The urban growth pattern is impacted by physical geography environment. Eastward growth is constrained by a steep topography, and westward and northward growth is restricted

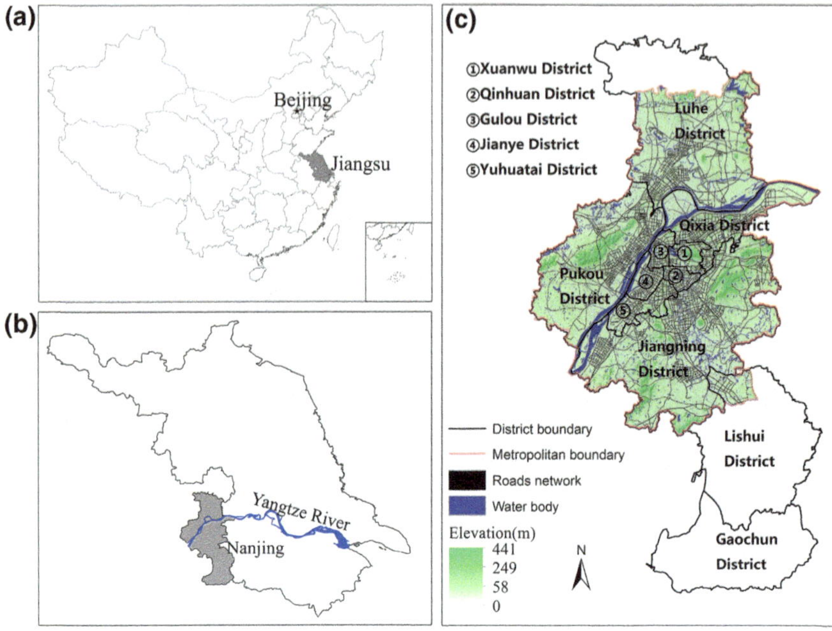

Fig. 1 Location of the study area, road network and topography

by the Yangtze River (Fig. 1). In the Nanjing 2006–2020 Master Plan, the urban development is expected to expand further to the east and south, and there is a proposal to build a new city in the area north of the Yangtze River (Nanjing Planning Bureau 2006). In this paper, we take the Nanjing metropolitan area as the study area, which consists of eight districts (shown in Fig. 1) covering an area of about 4388 km^2 (Fig. 1c).

3 Research Methods

3.1 Data

The primary data used in this study include images acquired by three sensors from Landsat satellites, i.e., TM (Thematic Mapper), ETM+(Enhanced Thematic Mapper Plus), and OLI (Operational Land Imager), in 1988, 2009, 2000, and 2017, respectively. Other data included topographic maps (1:50,000) and various urban planning documents (e.g., Nanjing master planning reports of 2011–2020 and Ecological Red Line planning of 2016–2030). Geometric correction was done for the remotely sensed data with the root mean square errors (RMSEs) of less than 1 pixel (30 m). The land use maps of 1988, 2000, 2009, and 2017 were generated using a

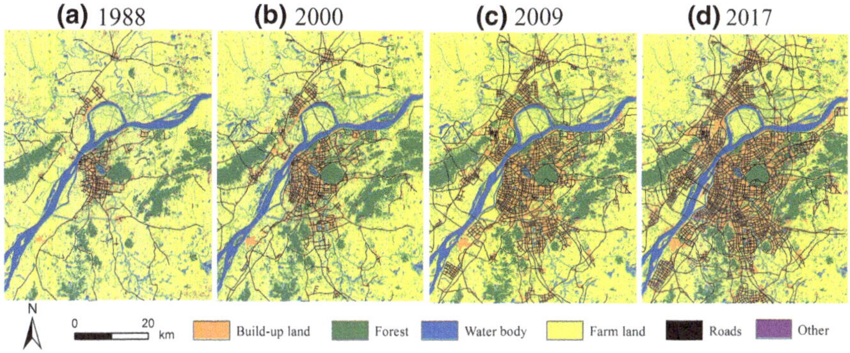

Fig. 2 Land use maps of the study area for the years **a** 1988, **b** 2000, **c** 2009, and **d** 2017

supervised classification method, and urban built-up land was mapped (Fig. 2). The overall classification accuracy was 94.64% according to the error matrixes method, and the Kappa index was 0.93. In addition, a road network dataset was developed through manually digitization (Figs. 1 and 2).

3.2 *Methods*

3.2.1 Simulation of Urban Sprawl Based on the SLEUTH Model

(1) Data input preparation. Each cell state in the SLEUTH model can be set as urban or non-urban land. The state of each cell is determined by the interaction among the five coefficients, namely, dispersion, breed, spread, slope, and road gravity. The SLEUTH model considers four types of urban growth, i.e., spontaneous growth, road-influenced growth, edge growth, and new center spreading (see Clarke and Gaydos 1998).

Five input data layers are needed in order to run the SLEUTH model: slope, exclusion layer, urban extent, transportation, and hillshade. In this study, the urban extent layer was derived from the remote sensor data, and it is a binary raster with urban and non-urban only (Fig. 3a). The roads (Fig. 3b) were not weighted because it has been reported that there was no significant difference noted as a result of weighting roads (Chaudhuri and Clarke 2013). The digital elevation model (DEM) data was used to generate both the hillshade and slope layers (Fig. 3c).

We defined the slope here as a percentage of the vertical to horizontal, and if a cell with a slope greater than 100%, it was assigned a slope value of 100 (Yin et al. 2018). The exclusion layer to be used for model calibration only included urban green space and large water bodies, while the exclusion layer used in the model simulation stage was defined according to the different ecological security scenarios, which will be explained in more detail later during each scenario design described in the sections

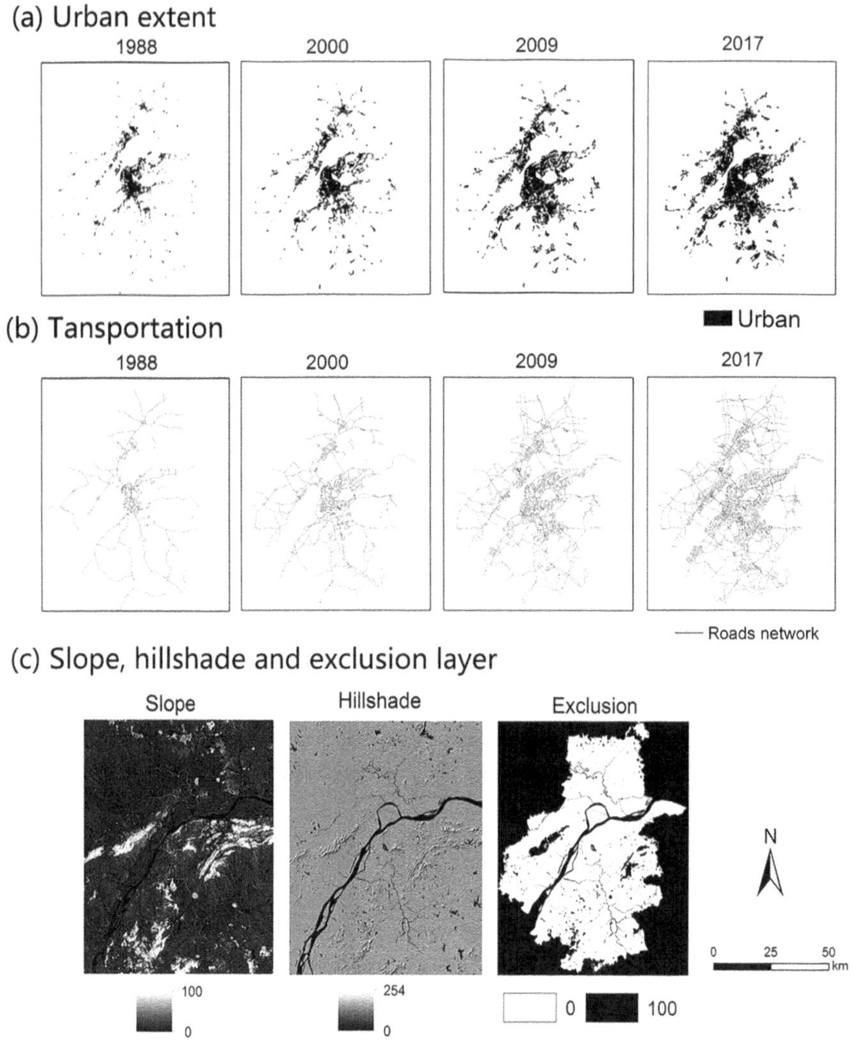

Fig. 3 The input layers for the SLEUTH model (**a** urban extent; **b** transportation; **c** slope, hillshade and exclusion layer)

below. Last, all data used in the model were converted to the Graphics Interchange Format (GIF), with a resolution of 60 m.

(2) SLEUTH model calibration. The SLEUTH model calibration involved finding the optimal combination of the five coefficients that can be used to simulate future urban sprawl. The model calibration is achieved by using the brute force Monte Carlo method. This procedure involved the following three stages: coarse, fine, and final calibrations. In each stage, the model was calibrated against historical data. Several

Table 1 Summary of the coefficients for each of the SLEUTH model's calibration stages

Growth coefficients	Coarse calibration		Fine calibration		Final calibration		Final coefficients
	Range	Step size	Range	Step size	Range	Step size	
Diffusion	1–100	25	50–100	10	60–80	4	79
Breed	1–100	25	75–100	5	90–100	2	100
Spread	1–100	25	50–100	10	70–80	2	100
Slope	1–100	25	25–75	10	35–60	5	1
Road gravity	1–100	25	25–100	15	70–90	5	97

metrics were used to quantify the calibration performance and further narrow down the coefficient value ranges (see Clarke and Gaydos 1998).

In this study, the 1988 data layer was used as the initial layer, and the 2000, 2009, and 2017 data layers were used for model calibration. In the coarse and fine calibration stages, we used five and seven Monte Carlo iterations, respectively. The product of seven metrics including Compare, Pop, Edges, Cluster, Slope, Xmean, and Ymean (Yin et al. 2018) is considered to be the Optimal SLEUTH Metric (OSM), which can be used to evaluate the goodness of fit and then select the optimal parameter combination (Onsted and Clarke 2012). Last, nine iterations were used to determine the final optimal coefficient combinations (Tables 1 and 2).

(3) Determination of the built-up land suitability pattern under different ecological security scenarios. Through adjusting the coefficient values and designating the exclusion layers, the SLEUTH model can accommodate different scenarios for future urban sprawl (Clarke et al. 1997; Clarke and Gaydos 1998; Yin et al. 2018). Based on the current urban land use pattern and the urban development program, the following three ecological security scenarios were accordingly designed: high ecological security development (HESD, S1), moderate ecological security development (MESD, S2), and low ecological security development (LESD, S3) (Yin et al. 2018). In this study, we used the potential-constraint discriminant matrix method (Yin et al. 2013) to design the development scenarios.

First, we calculated the land development potential by pixels according to Tobler's first law of geography (Eq. 1) (Tobler 1970). The PLAND index of the urban built-up land in 2017 was calculated by using the moving window method with a radius of 300 m in Fragstats 4.2, and then, the regions with ≥70% PLAND values and ≥3 km^2 were extracted as the urban growth poles with assigned value of 100. Accessibility can easily be calculated by using the Cost-distance method in ArcGIS (see Kong et al. 2012). The travel speed was set as 30 km/h, and the cost-distance was defined as 20 min/10 km; specially, in places with no roads the cost values were assigned according to following three categories: 1000 for rivers and water, 500 for mountains, and 120 for others. A GIS (geographic information system)-based maximum value overlay analysis was performed for summing up the land development potential at

Table 2 Summary of the metrics for each of the SLEUTH model's calibration stages

Metrics	Definition of the metrics	Coarse calibration	Fine calibration	Final calibration
Compare	Comparison of modeled final urban extent to real final urban extent	0.3667	0.3565	0.3622
Pop	r^2 Population: Least square regression score of modeled urbanization compared with actual urbanization for control years	0.9960	0.9946	0.9931
Edges	Edge r^2: Least square regression score for modeled urban edge count compared with actual urban edge count for control years	0.9824	0.9819	0.9861
Cluster	R^2 cluster: Least square regression score for modeled urban clustering compared with known urban clustering for control years	0.9386	0.9998	0.9983
Slope	Average slope r^2: Least square regression of average slope for modeled urbanized cells compared with average slope of known urban cells for control years	0.9883	0.9894	0.9802
Xmean	$X - r^2$; center of gravity [X]: Least square regression of average X values for modeled urbanized cells compared with average X values of known urban cells for control years	0.9939	0.9776	0.9877
Ymean	$Y - r^2$; center of gravity [Y]: Least square regression of average Y values for modeled urbanized cells compared with average Y values of known urban cells for control years	0.9929	0.9975	0.9933
OSM	The optimal SLEUTH Metric, above seven indices multiplied	0.3284	0.3358	0.3405

the pixel scale for all selected growth poles, and then, the results were divided into five grades by using the natural breakpoint method (Fig. 4a). The equation used was as follows:

$$P = S \times \frac{1}{1 + 0.01 * \exp^{(0.15*T)}} \quad (1)$$

where P is the land development potential at the pixel scale (0–100), S is the comprehensive strength of the urban growth pole (0–100), and T is the accessibility to the urban growth poles (min).

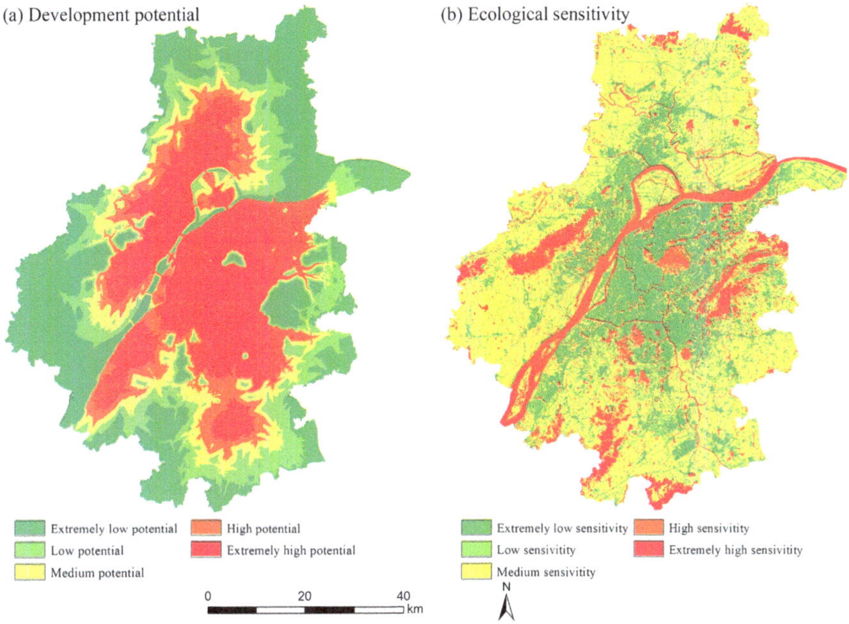

Fig. 4 Grades of the **a** development potential and **b** ecological sensitivity at the pixel level

Second, we calculated the ecological constraint by pixels based on the ecological sensitivity analysis (Yin et al. 2013). Ecological sensitivity refers to the sensitivity of the ecological system to external disturbance, and it reflects the occurrence possibility of potential ecological problems (Li et al. 2013). Ecological sensitivity analysis can be used to evaluate the ecological risk and ecological resource security (Yin et al. 2015). In this paper, we selected four sensitivity factors (vegetation, slope, water, and agricultural land) to build a hierarchy evaluation system to conduct the ecological sensitivity analysis (Table 3). A GIS-based maximum value overlay analysis was used to make the multifactor comprehensive evaluation, and then, the ecological sensitivity zonation was accordingly obtained (Yin et al. 2015) (Fig. 4b).

Third, we analyzed the built-up land suitability based on the potential-constraint discriminant matrix method (Yin et al. 2013). According to the results for the development potential and ecological constraints, a potential-constraint discriminant matrix was developed for each of the three (high, moderate, and low) ecological security scenarios (Tables 4, 5 and 6). Under the high ecological security development scenario (S1), ecological security was the priority in order to protect the ecological environment. The ecological security and development potential were both equally important under the moderate ecological security development scenario (S2), while under the low ecological security development scenario (S3), development potential was the priority and ecological security was just regarded as the bottom line (i.e., the minimum amount necessary). A GIS-based recoding analysis was applied, and

Table 3 Ecological sensitivity factors and their grades and values

Sensitivity factor	Classification (buffer)	Value	Ecological sensitivity grade
Vegetation	Park	9	Extremely high sensitivity
	100 m buffer	7	High sensitivity
	Forest (NDVI ≥ 0.6)	7	High sensitivity
	Forest (NDVI < 0.6)	5	Moderate sensitivity
	Lawn	3	Low sensitivity
Slope	<5%	1	Non-sensitivity
	5–10%	3	Low sensitivity
	10–15%	5	Moderate sensitivity
	15–25%	7	High sensitivity
	>25%	9	Extremely high sensitivity
Water	Yangtze river	9	Extremely high sensitivity
	200 m buffer	7	High sensitivity
	Large reservoirs and river	7	High sensitivity
	100 m buffer	5	Moderate sensitivity
	Small reservoirs, river, pond	5	Moderate sensitivity
Agricultural land	–	3	Low sensitivity

Table 4 Discriminant matrix of built-up land suitability under the high ecological security scenario

	Non-sensitivity	Low sensitivity	Moderate sensitivity	High sensitivity	Extremely high sensitivity
Non-potential	3	1	1	1	1
Low potential	3	3	1	1	1
Moderate potential	5	5	3	3	1
High potential	7	7	3	3	1
Extremely high potential	9	9	5	3	1

the built-up land suitability zonation was obtained (Fig. 5). The values assigned to different land suitability types were as follows: extremely high suitability (9), high suitability (7), moderate suitability (5), low suitability (3), not suitable (1).

Finally, the weighted values of the built-up land suitability layer were created through a user-defined option. Therefore, the highest suitability was assigned with the lowest value in the exclusion layer, and the probability of a cell not being urbanized was scored as 90, 75, 50, 25, and 0% according to the increasing degree of built-up land suitability. Then, the exclusion layers for S1, S2, and S3 were generated (Fig. 6).

(4) Urban growth simulation. We use the slope, urban extent, transportation, hill-shade, and exclusion layer for each of various scenarios as the initial input data of

Table 5 Discriminant matrix of built-up land suitability under the moderate ecological security scenario

	Non-sensitivity	Low sensitivity	Moderate sensitivity	High sensitivity	Extremely high sensitivity
Non-potential	3	3	1	1	1
Low potential	3	3	3	1	1
Moderate potential	5	5	3	3	1
High potential	9	7	5	3	1
Extremely high potential	9	9	7	3	1

Table 6 Discriminant matrix of built-up land suitability under the low ecological security scenario

	Non-sensitivity	Low sensitivity	Moderate sensitivity	High sensitivity	Extremely high sensitivity
Non-potential	3	3	1	1	1
Low potential	5	5	3	1	1
Moderate potential	7	7	5	3	1
High potential	9	9	7	3	1
Extremely high potential	9	9	9	5	1

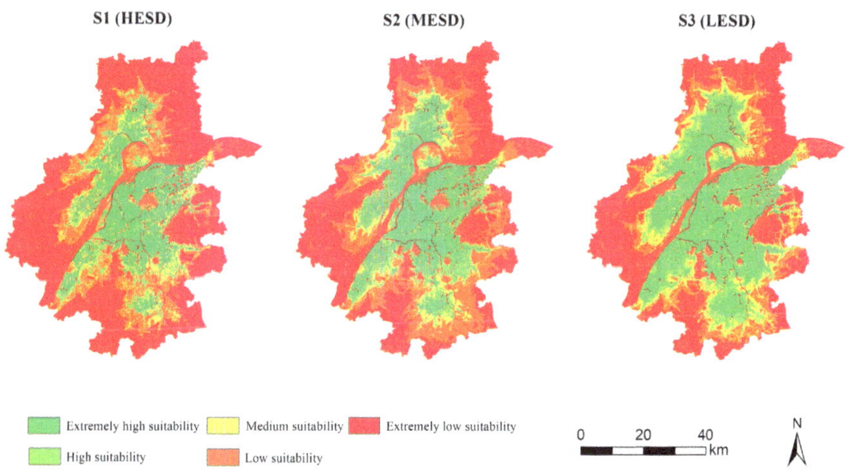

Fig. 5 Built-up land suitability pattern under the high (S1), moderate (S2), and low ecological security scenario (S3)

2017, and then 100 Monte Carlo iterations were performed in the model's prediction mode under the three ecological security scenarios (Yin et al. 2015, 2018). The

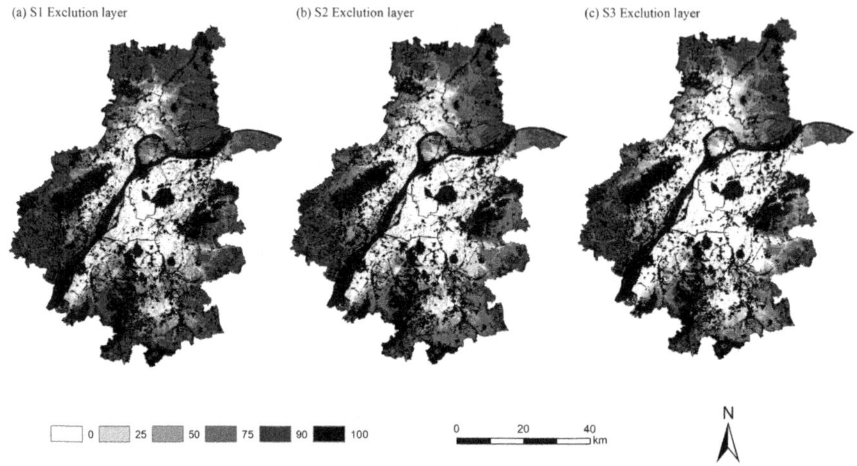

Fig. 6 The exclusion layers used for the simulations under scenarios 1–3. Note that attribute values of 100 indicate the completely excluded areas

Table 7 Statistics on land use changes from 2017 to 2040 under the three scenarios (S1, S2 & S3) considered

Land use type	S1-2040 in km^2	S2-2040 in km^2	S3-2040 in km^2
Urban built-up land	+201.6	+299.9	+391.9
Green space	−47.3	−66.5	−82.2
Agricultural land	−93.5	−153.5	−217.7
Water	−1.6	−3.2	−4.5
Other	−59.2	−76.7	−87.5

Note + means an increase in area and—means a decrease

thresholds used in the urbanization probability maps under the three scenarios were the same as in the model calibration stage. Based on these thresholds, the urban extent in 2040 was then reconstructed (Yin et al. 2018) (Table 7 and Fig. 7). The 2040 development probability was used as the threshold (50%), and if a cell with probability was larger than this threshold value, it was supposed to be the future urban area.

3.2.2 Simulation of the Urban Thermal Environment Based on the WRF Model

(1) WRF model setting and experimental scheme design. We used the WRF model (Version 3.5) to explore the impacts of three urban development scenarios on the thermal environment of the Nanjing metropolitan area. WRF is a mesoscale numerical weather prediction system designed to serve both operational forecasting and

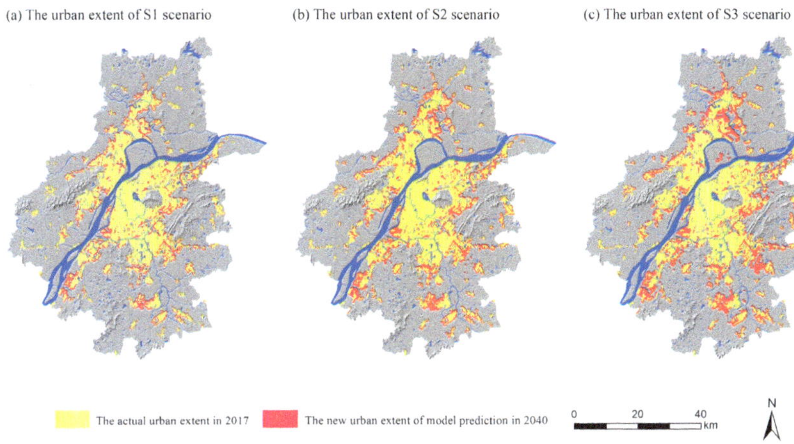

Fig. 7 The simulated urban extents in 2040 under the three scenarios (S1, S2, and S3)

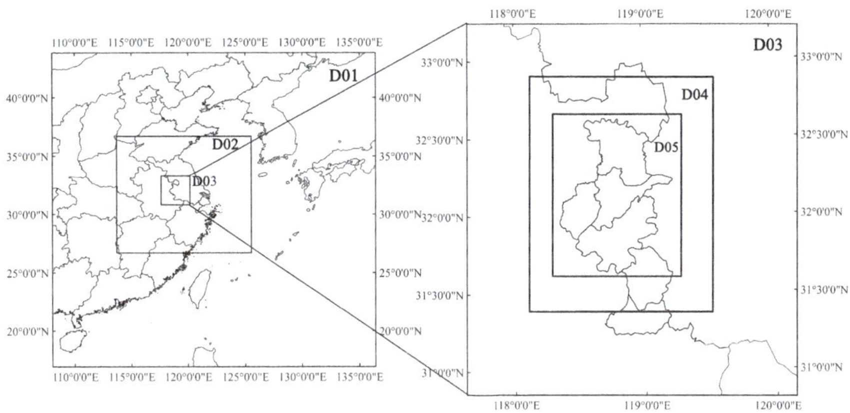

Fig. 8 The nested domains in the WRF model

atmospheric research needs, it can be used in a broad scales ranging from meters to thousands of kilometers (www.wrf-model.org). In this study, a five nested domain system (i.e., domain D01 to domain D05, with corresponding spatial resolutions of 27 km, 9 km, 3 km, 1 km, and 333.33 m, respectively, Fig. 8) was used to simulate the thermal environment of Nanjing at a resolution of 333.33 m. Both two-way and one-way nesting techniques were applied. Domains D01 to D03 were run using two-way method, and then the D04 and D05 domains were downscaled using one-way technique. The spatial resolutions of the static terrain data were 10 m, 5 m, 2 m, 30 s, and 30 s, respectively. There were 35 vertical layers with the model top fixed at 50 hPa (Feng et al. 2018). The settings of the simulation configurations are shown in Table 8.

Table 8 The WRF model configuration and physics schemes used in this study

Number of domains	5
Resolution	27 km (domain D01), 9 km (domain D02), 3 km (domain D03), 1 km (domain D04) and 333.33 m (domain D05)
Number of grid points	114 × 99 (D01), 123 × 126 (D02), 90 × 81 (D03), 144 × 120 (D04), 315 × 270 (D05)
Microphysics scheme	Single-Moment 5-class microphysical scheme
Land-surface layer scheme	Noah land surface model
Radiation scheme (long wave)	Rapid radiation transfer model (RRTM)
Radiation scheme (short wave)	Goddard shortwave
Boundary layer physics	Yon-sei University scheme
Cumulus convection scheme	Kain-Fritsch scheme

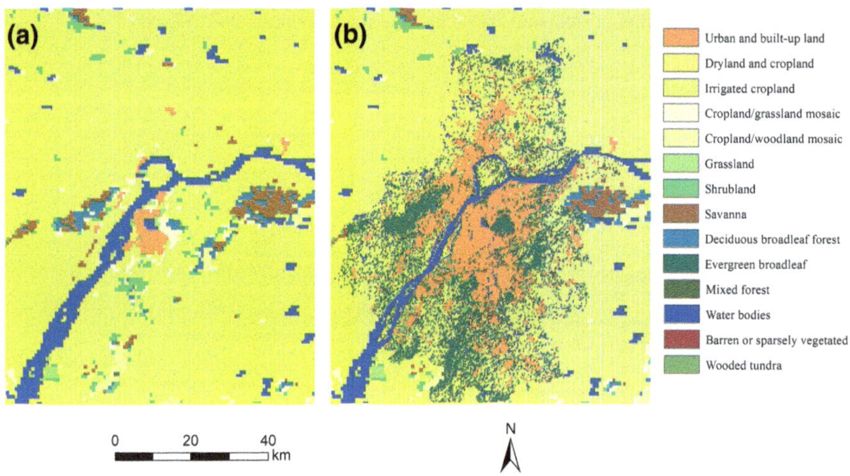

Fig. 9 The difference between **a** the land use data used in the WRF model (Noah land surface model) and **b** the actual situation within the study area in 2017 (**b**)

The WRF model was first applied to simulate the urban thermal environment of the Nanjing metropolitan area for 23–24 July 2017. The land use data inputs in the WRF model in domains D04 and D05 were replaced by the classified land use data from the remote sensing imagery of 2017. The results for the mean air temperature at a height of 2 m were then compared with the observation data (Figs. 9 and 10).

(2) WRF model evaluation. In this study, we used the near-surface 2 m height air temperature to study the thermal environment changes under the different urban sprawl scenarios; this height was chosen because ambient air temperatures at this height are much more directly related to public health than air temperatures at other heights (Basara et al. 2010; Hu et al. 2016; Oswald et al. 2012). First, we extracted the air temperatures at a 2 m height at the Xianlin campus of Nanjing University

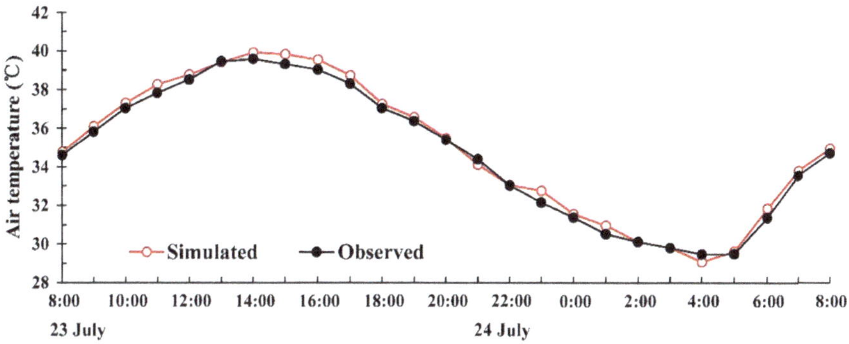

Fig. 10 Comparison of the simulated air temperature at 2 m height with the observed air temperature on the same day (23–24 July)

from the simulation results of the WRF model, and then, we compared the data with the observed air temperatures at the same location. We used the mean error (MB, Eq. 2), root mean square error (RMSE, Eq. 3), and correlation coefficient (r, Eq. 4) to evaluate the thermal environment simulation accuracy (Fig. 10):

$MB = \frac{1}{N}\sum_{1}^{N}(model - obs)$	(2)
$RMSE = \sqrt{\frac{1}{N}\sum_{1}^{N}(model - obs)^2}$	(3)
$r = \frac{\sum_{1}^{N}(model-\overline{model})(obs-\overline{obs})}{\sqrt{(\sum_{1}^{N}(model-\overline{model})^2)\sum_{1}^{N}(obs-\overline{obs})^2}}$	(4)

where *model* is the simulated air temperature, *obs* is the observed air temperature, $\overline{model}$ and $\overline{obs}$ are the mean of the simulated and observed air temperature, respectively, and N is the number of analyzed data points.

(3) WRF model simulation. By using the same parameter settings as the calibration process for the WRF model, the urban thermal environment in 2040 under the three development scenarios predicted by the SLEUTH model was simulated. Then, we extracted the simulated 2 m height air temperatures at four time points (0:00, 6:00, 12:00, 18:00) under the three urban development scenarios. Finally, the daily average temperature spatial patterns under the three scenarios were captured (Table 9 and Fig. 11). At the same time, the spatio-temporal variation of the urban heat island intensity (UHII), which was defined by the difference between the mean air temperature and the simulated air temperature in the farmland, was also analyzed (Table 10, Figs. 12 and 13).

Table 9 Composition of area and percentage of different thermal environment zone type (km^2, %)

Zone type	2017	Scenarios		
		S1	S2	S3
Zone 1 (<33 °C)	349.40 (7.96)	358.06 (8.15)	335.23 (7.63)	333.86 (7.60)
Zone 2 (33–34 °C)	701.26 (15.97)	689.55 (15.70)	95.42 (2.17)	88.73 (2.02)
Zone 3 (34–34.4 °C)	1587.64 (36.16)	1442.26 (32.85)	1249.46 (28.46)	1151.68 (26.23)
Zone 4 (34.4–34.8 °C)	1054.98 (24.03)	1044.75 (23.79)	1474.62 (33.58)	1381.76 (31.47)
Zone 5 (>34.8 °C)	697.22 (15.88)	856.19 (19.50)	1236.08 (28.15)	1434.78 (32.68)

Table 10 Compositions Areas of thermal environment landscapes (km^2, %)

Zone type	Present (2017)	Scenarios		
		S1	S2	S3
Zone 1 (<−0.5 °C)	398.98 (9.09)	379.79 (8.65)	347.83 (7.92)	345.07 (7.86)
Zone 2 (−0.5–0 °C)	993.11 (22.62)	840.15 (19.13)	646.07 (14.71)	595.71 (13.57)
Zone 3 (0–0.5 °C)	1401.04 (31.91)	1319.89 (30.06)	1322.84 (30.13)	1222.90 (27.85)
Zone 4 (0.5–1.0 °C)	717.59 (16.34)	770.41 (17.55)	907.43 (20.67)	895.53 (20.40)
Zone 5 (1.0–2.0 °C)	726.24 (16.54)	799.92 (18.22)	777.89 (17.72)	841.33 (19.16)
Zone 6 (>2.0 °C)	153.45 (3.50)	280.64 (6.39)	388.75 (8.85)	490.26 (11.17)

4 Results

4.1 SLEUTH and WRF Model Calibrations

The final values of the five coefficients were diffusion (79), breed (100), spread (100), slope (1), and road gravity (97) (Table 1). The diffusion coefficient was high (79), which indicated that there was evident spontaneous growth. At the same time, the spread coefficient was the largest (100), which indicated that urban development was dominated by urban edge sprawl. These results also suggest that edge sprawl was the overwhelming trend for urban sprawl, however, the spontaneous growth was a comparatively minor factor. The breed coefficient was the largest (100), which indicated that a high possibility of a new city center was developing in the study area. The road gravity coefficient was high (97) indicating that the road accessibility had a very important effect on urban sprawl. The slope coefficient was 1, which indicated

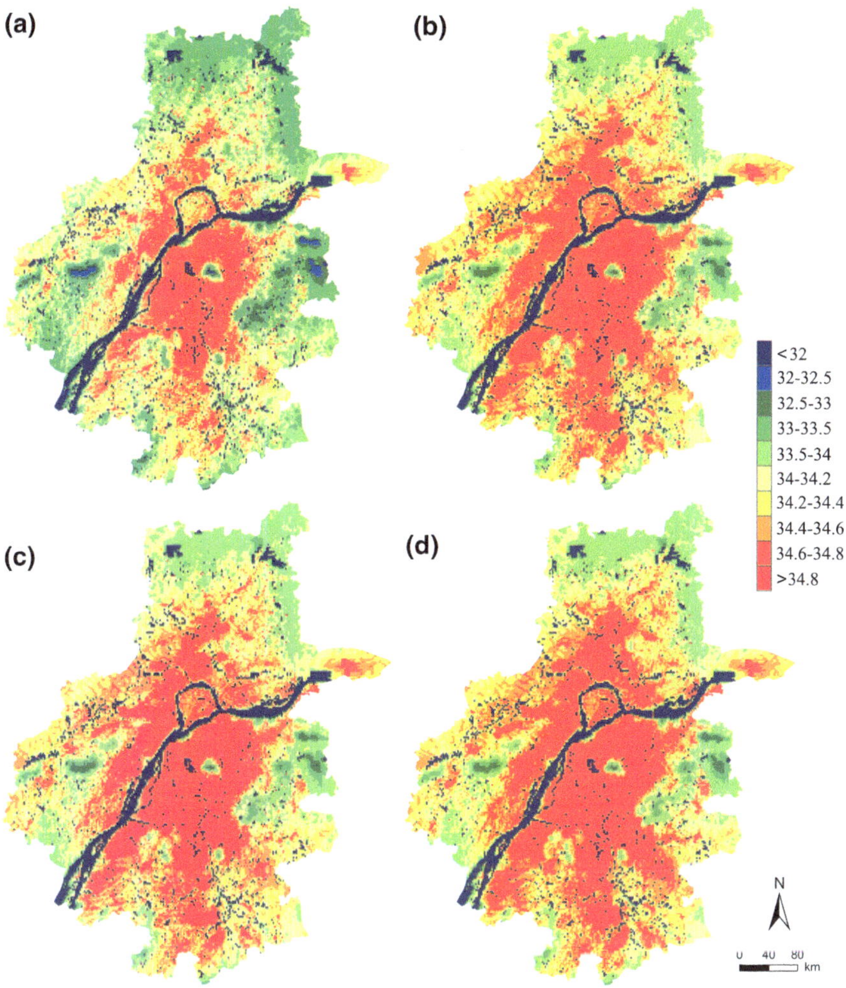

Fig. 11 The spatial patterns of daily average air temperature under different urban sprawl scenarios (**a** 2017, **b** S1, **c** S2, **d** S3)

that terrain conditions weakly restricted the urban growth. According to the results, the urban sprawl of Nanjing City was mainly affected by the spread, breed, and road gravity coefficients, and it happened primarily on the fringe or high road accessibility areas.

The products of the selected seven metrics indicated that the simulation accuracy of the SLEUTH model increased from 0.3284 during the coarse calibration stage to 0.3405 during the final calibration stage, and the metric values were all more than 0.93 except for the compare metric (Table 2). These results demonstrate that the

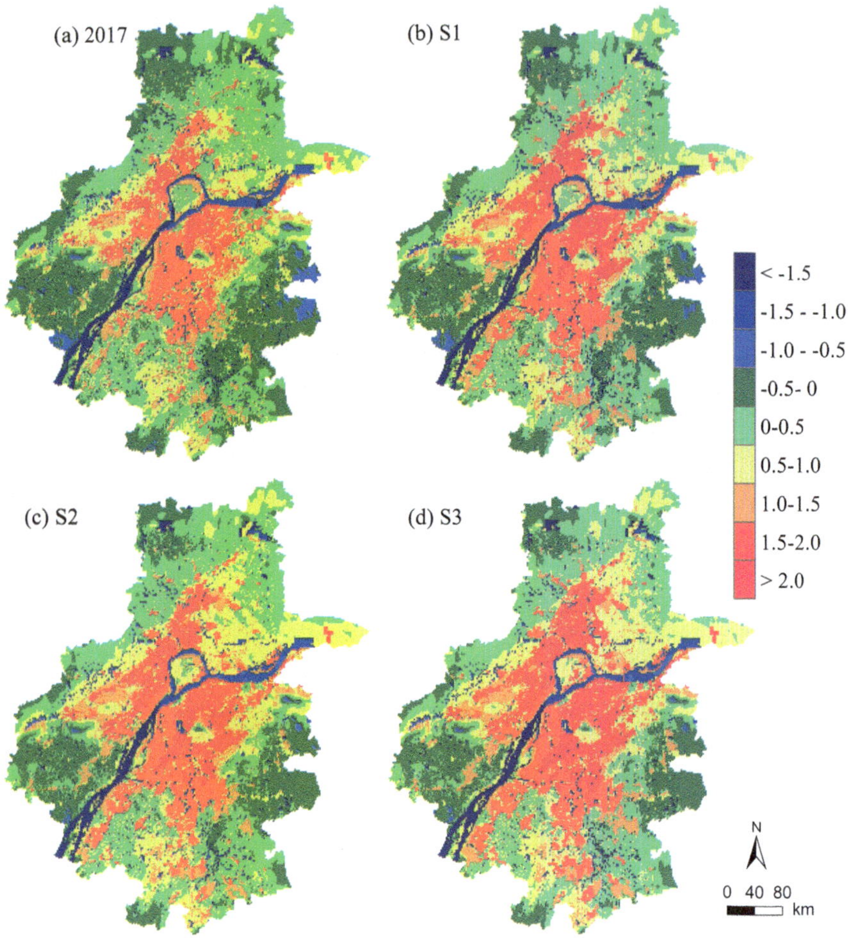

Fig. 12 The spatial variation of the heat island intensity in 2017 and under the three different scenarios

SLEUTH model was able to capture much of the urban growth, and it was also able to simulate the urban forms successfully.

The urban air temperatures at a height of 2 m on 23–24 July were compared against the observations collected at the Xianlin campus of Nanjing University, and these results revealed a satisfactory performance of the modeling system, even though around 15:00, which is the hottest time of a day, the simulated air temperature was slightly higher than the observed one (Fig. 10). The statistical analysis showed that the MB, RMSE, and r were 0.21 °C, 0.42 °C, and 0.94, respectively.

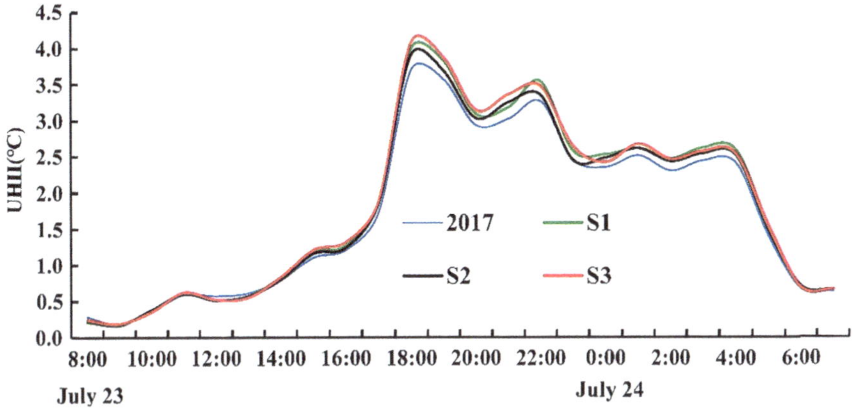

Fig. 13 Mean diurnal variation of UHII in 2017 and under the three different scenarios

4.2 Scenario Analysis of Urban Sprawl

According to the simulation results (Table 7 and Fig. 7), it can be found that the urban area is projected to grow fast under the three ecological security scenarios tested from 2018 to 2040; meanwhile, the simulated urban built-up land area in 2040 is projected to increase with decreases of the ecological protection grade. Because of the consideration given to the ecological environment, the urbanized area under the HESD scenario (S1) will only increase by 201.6 km^2, while ecological spaces such as woodlands and agricultural lands will decrease by only 47.3 km^2 and 93.5 km^2, respectively. In comparison, the urbanized area under the LESD scenario (S3) will increase by 391.9 km^2, while woodlands and agricultural lands will decrease by 82.2 km^2 and 217.7 km^2, respectively, which is related to the fact that in this scenario the urban built-up land development potential is given the priority and only minimum consideration is given to the protection of the ecological environment. The projected urban growth area in the moderate ecological security development (MESD) scenario (S2) is 299.9 km^2, which is a value that falls between those obtained for the two other scenarios. Generally, the development concepts and applied policies will have a significant impact on the growth of urban built-up land in the future.

4.3 Impacts of Urban Sprawl on the Thermal Environment Under Three Scenarios

4.3.1 Spatial Pattern of Daily Mean Air Temperature in Response to the Urban Sprawl Scenarios

The simulated urban thermal environment under the land use pattern of 2017 and those projected under the three urban growth scenarios were classified into five zone types, and the results are shown in Fig. 11 and Table 9.

The results showed that urban sprawl will have a very obvious impact on the thermal environment pattern. The daily mean air temperature at a height of 2 m over the urbanized area was higher than that over the other un-urbanized areas, and it was also higher downtown than in the suburbs, but it was relatively low in the mountains or areas of water (Figs. 2, 7, and 11). Compared with the thermal environment of 2017, the area covered by high air temperatures ($\geq$34.8 °C, Zone 5) under the three development scenarios of 2040 will grow with increase in the built-up land (Fig. 11 and Table 9). The results showed that the area of Zone 5 (pixels with air temperatures >34.8 °C) from the present (2017) to S1, S2, and S3 increased from 697.22 km^2 to 856.19 km^2, 1236.08 km^2, and 1434.78 km^2 respectively, and the corresponding percentages of the study area increased from 15.88% to 19.50%, 28.15%, and 32.68%, respectively. In S3, the area of Zone 5 reached twice that of 2017, and almost the whole downtown area experienced the urban heat island effect (Table 9 and Fig. 11). The area and percentage of Zone 1 (pixels with air temperatures $\leq$33 °C) changed very little (Table 9), and it was mainly distributed in the mountains and water areas (Fig. 11). In 2017, the area and percentage of Zone 3 (pixels with 34 °C < air temperatures $\leq$ 34.4 °C) was the biggest, but it decreased gradually from 1587.64 km^2 and 36.16% in 2017 to 1151.68 km^2 and 26.23%, respectively, in 2040. The reductions in the area of Zone 3 were transferred to Zones 4 and 5. All of these results indicate that the land use change, especially the change in built-up land, will have remarkable impacts on the urban thermal environment patterns. The different urban sprawl scenarios will produce different thermal environment patterns, but all of them show an increase in the area of high temperatures.

It was also found that in the LESD scenario (S3), the increase of the high air temperature areas in the northern and southern regions was in remarkable accordance with the increases of the urbanized areas (Figs. 7 and 11). Meanwhile, the high temperature areas continued expanding to the south and east, and the original isolated high temperature areas gradually integrated and merged into a larger high-temperature region (Fig. 11); this was indicative of a larger and much more intensified UHI.

4.3.2 Spatio-Temporal Variations of the Urban Heat Island Intensity (UHII)

The daily UHII under the land use pattern of 2017 and those projected under the three urban growth scenarios were classified into six zone types, and the results were shown in Fig. 12 and Table 10.

From the results, we can find that from the present in 2017 to S1, S2, and S3, the UHI will increase with the urban sprawl (Table 10 and Fig. 12). Zone 1 (cool island region), where the UHII is less than 0.5 °C (this zone is mainly made up of water bodies), will gradually decrease. Zones 2 and 3, where the UHII is between −0.5 and 0.5 °C (these zones are mainly located in farmland or at the edge of the urbanized area) will also gradually decrease because of the transformation from farmland to built-up land. Conversely, Zones 4, 5, and 6 (heat island region), where the UHII is more than 0.5 °C (these zones are mainly located downtown and in its surrounding areas) will gradually increase. Specially, Zone 6 (strong heat island region), where the UHII is more than 2 °C (this zone is located at the center of the built-up land area) displayed a clear spreading pattern. By comparing S1 with S3, we found that the urban sprawl led to the connection of formerly isolated built-up land areas, which resulted in the increase of Zone 6.

The mean diurnal variations of the UHII are shown in Fig. 13. As can be seen in this figure, the UHIIs at different times under the S1, S2, and S3 scenarios were obviously higher than those in 2017, and the UHII of S3 was higher than that of S1 and S2. These results indicate that urban sprawl will cause a change in the urban thermal environment and an increase in the UHII. The results also show that there will be obvious diurnal variation in the UHII with higher values during nighttime (Fig. 13), which is consistent with previous studies in other cities (Barlow et al. 2015; Yow and Carbone 2006).

5 Discussion and Conclusions

With the Nanjing metropolitan area as the study area, we have used the SLEUTH model to predict the urban built-up land growth in 2040 under three ecological security scenarios (i.e., low, moderate, and high). Furthermore, we have used the WRF model to simulate and analyze the impacts of urban sprawl on the urban thermal environment under these development scenarios. The main findings and recommendations are two-fold. First, through reconstructing the exclusion/attraction layers, the different development policies and strategies can be incorporated into the SLEUTH, which will be contributed to improving the model simulation accuracy model. Here, we found that different development concepts can significantly affect the future urban growth. Specifically, the area and rate of urban built-up land growth during 2018–2040 increased from the HESD (S1) to the LESD (S3) scenario, and this was mostly due to the prioritization of different development concepts (ecological protection priority (S1), balance between development and protection (S2), development priority (S3)) in the future. On the other hand, land use change is an important factor that can cause changes in the urban thermal environment. Urbanization led to a significant increase in the temperature, especially in newly developed areas, and the intensity and duration of the UHI are expected to increase in response to the expansion of built-up land under the three different urban development scenarios. These results will contributed to a better understanding by the urban planners and managers on the potential impacts of urban sprawl on the thermal environment, and accordingly, intelligent planning approaches can be devised to ensure a livable and sustainable city in the future.

Although we successfully coupled the urban growth model SLEUTH and the WRF model in order to simulate the impacts of urban sprawl under different development scenarios on the thermal environment, some aspects related to this study could be further improved in the future. First, the U.S. Geological Survey (USGS) Land Use/Land Cover data were applied as the default land cover inputs in the WRF model, but these data may not be good enough to accurately represent the status quo of land use in the study area. This could have led to uncertainties in the WRF simulations and inaccuracies in the estimates of urban sprawl impacts on the thermal environment. In this study, to increase the accuracy of the simulation, we used the land use data of 2017 to rebuild the USGS Land Use/Land Cover data in domains 4 and 5 (D04 and D05, Fig. 9). However, the different nested structures interacted with each other. Therefore, future studies will need to use improved land use data in all selected domains (D01–D05) to better quantify the influence of urban growth on the thermal environment. Second, the WRF model is very sensitive to indices such as the fractional vegetation coverage (FVC), leaf area index (LAI), and albedo; these indices, however, vary from place to place because of the different built-up land densities, average building heights, and green space ratios. While we used the recommended default parameters to simulate the urban thermal environment in this study, this may have had an influence on the simulation accuracy to a certain extent. Hence, in the future it will be very essential to improve the acquisition of high spatial resolution data in order to set the parameters (FVC, LAI, and albedo) in the WRF model and reduce the uncertainty of the simulation.

References

Akın A, Clarke KC, Berberoglu S (2014) The impact of historical exclusion on the calibration of the SLEUTH urban growth model. Int J Appl Earth Obs Geoinf 27:156–168

Barlow JF, Halios CH, Lane SE, Wood CR (2015) Observations of urban boundary layer structure during a strong urban heat island event. Environ Fluid Mech 15(2):373–398

Barredo JI, Kasanko M, McCormick N, Lavalle C (2003) Modelling dynamic spatial processes: simulation of urban future scenarios through cellular automata. Landsc Urban Plan 64(3):145–160

Basara JB, Basara HG, Illston BG, Crawford KC (2010) The impact of the urban heat island during an intense heat wave in Oklahoma City. Adv Meteorol 2010:230365. https://doi.org/10.1155/2010/230365

Berardi U (2016) The outdoor microclimate benefits and energy saving resulting from green roofs retrofits. Energy Build 121:217–229

Berardi U, GhaffarianHoseini A, GhaffarianHoseini A (2014) State-of-the-art analysis of the environmental benefits of green roofs. Appl Energy 115:411–428

Berling-Wolff S, Wu J (2004) Modeling urban landscape dynamics: a case study in Phoenix USA. Urban Ecosyst 7(3):215–240

Chaudhuri G, Clarke K (2013) The SLEUTH land use change model: a review. Environl Resour Res 1(1):88–105

Chen F, Kusaka H, Bornstein R, Ching J, Grimmond CSB, Grossman-Clarke S et al (2011) The integrated WRF/urban modelling system: development, evaluation, and applications to urban environmental problems. Int J Climatol 31(2):273–288

Chen L, Frauenfeld OW (2016) Impacts of urbanization on future climate in China. Clim Dyn 47:345–357

Clarke KC, Gaydos LJ (1998) Loose-coupling a cellular automaton model and GIS: long-term urban growth prediction for San Francisco and Washington/Baltimore. Int J Geogr Inf Sci 12(7):699–714

Clarke KC, Hoppen S, Gaydos L (1997) A self-modifying cellular automaton model of historical urbanization in the San Francisco Bay area. Environ Plan B: Plan Des 24(2):247–261

Conry P, Sharma A, Potosnak MJ, Leo LS, Bensman E, Hellmann JJ et al (2015) Chicago's heat island and climate change: bridging the scales via dynamical downscaling. J Appl Meteorol Climatol 54(7):1430–1448

Feng J, Wang J, Yan Z (2014) New advances in the research of climatic effect of urbanization. Adv Meteorol Sci Technol 4(5):21–29

Feng S, Jiang F, Jiang Z, Wang H, Cai Z, Zhang L (2018) Impact of 3DVAR assimilation of surface PM 2.5 observations on PM 2.5 forecasts over China during wintertime. Atmos Environ

Herold M, Goldstein NC, Clarke KC (2003) The spatiotemporal form of urban growth: measurement, analysis and modeling. Remote Sens Environ 86(3):286–302

Hu XM, Xue M, Klein PM, Illston BG, Chen S (2016) Analysis of urban effects in Oklahoma City using a dense surface observing network. J Appl Meteorol Climatol 55(3):723–741

Irwin EG, Bell KP, Bockstael NE, Newburn DA, Partridge MD, Wu J (2009) The economics of urban-rural space. Ann Rev Resour Econ 1(1):435–459

Jandaghian Z, Touchaei AG, Akbari H (2018) Sensitivity analysis of physical parameterizations in WRF for urban climate simulations and heat island mitigation in Montreal. Urban Clim 24:577–599

Jantz CA, Goetz SJ, Donato D, Claggett P (2010) Designing and implementing a regional urban modeling system using the SLEUTH cellular urban model. Comput Environ Urban Syst 34(1):1–16

Kong F, Yin H, Nakagoshi N, James P (2012) Simulating urban growth processes incorporating a potential model with spatial metrics. Ecol Ind 20:82–91

Kusaka H, Chen F, Tewari M, Dudhia J, Gill DO, Duda MG et al (2012) Numerical simulation of Urban Heat Island effect by the WRF Model with 4-km grid increment: an inter-comparison study between the urban canopy model and slab model. J Meteorol Soc Jpn Ser II 90:33–45

Li XX, Koh TY, Entekhabi D, Roth M, Panda J, Norford LK (2013) A multi-resolution ensemble study of a tropical urban environment and its interactions with the background regional atmosphere. J Geophys Res: Atmos 118(17):9804–9818

Liu L, Zhang Y (2011) Urban heat island analysis using the Landsat TM data and ASTER data: a case study in Hong Kong. Remote Sens 3(7):1535–1552

Mahiny AS, Clarke KC (2012) Guiding SLEUTH land-use/land-cover change modeling using multicriteria evaluation: towards dynamic sustainable land-use planning. Environ Plan B: Plan Des 39(5):925–944

Morris KI (2016) Computational study of Klang Valley's Urban Climatology, and Urbanisation of Putrajaya City, Malaysia

Morris KI, Chan A, Morris KJK, Ooi MC, Oozeer MY, Abakr YA et al (2017) Urbanisation and urban climate of a tropical conurbation, Klang Valley, Malaysia. Urban Climate 19:54–71

Morris KI, Chan A, Ooi MC, Oozeer MY, Abakr YA, Morris KJK (2016a) Effect of vegetation and waterbody on the garden city concept: an evaluation study using a newly developed city, Putrajaya, Malaysia. Comput Environ Urban Syst 58:39–51

Morris KI, Chan A, Salleh SA, Ooi MCG, Oozeer MY, Abakr YA (2016b) Numerical study on the urbanisation of Putrajaya and its interaction with the local climate, over a decade. Urban Climate 16:1–24

Nanjing Statistics Bureau (2016) Nanjing Statistical Yearbook. China Statistics Press, Beijing, (in Chinese)

Nanjing Planning Bureau 2006–2020 (2006). Nanjing City Master Planning. Nanjing Press, Jinan (in Chinese)

Norman LM, Feller M, Villarreal ML (2012) Developing spatially explicit footprints of plausible land-use scenarios in the Santa Cruz Watershed, Arizona and Sonora. Landsc Urban Plan 107(3):225–235

Onsted JA, Chowdhury RR (2014) Does zoning matter? A comparative analysis of landscape change in Redland, Florida using cellular automata. Landsc Urban Plan 121:1–18

Onsted J, Clarke KC (2012) The inclusion of differentially assessed lands in urban growth model calibration: a comparison of two approaches using SLEUTH. Int J Geogr Inf Sci 26(5):881–898

Oswald EM, Rood RB, Zhang K, Gronlund CJ, O'Neill MS, White-Newsome JL et al (2012) An investigation into the spatial variability of near-surface air temperatures in the Detroit, Michigan, metropolitan region. J Appl Meteorol Climatol 51:1290–1304

Papangelis G, Tombrou M, Dandou A, Kontos T (2012) An urban "green planning" approach utilizing the Weather Research and Forecasting (WRF) modeling system. A case study of Athens, Greece. Landsc Urban Plan 105(1):174–183

Poumadere M, Mays C, Le Mer S, Blong R (2005) The 2003 heat wave in France: dangerous climate change here and now. Risk Anal 25(6):1483–1494

Santé I, García AM, Miranda D, Crecente R (2010) Cellular automata models for the simulation of real-world urban processes: a review and analysis. Landsc Urban Plan 96(2):108–122

Silva EA, Clarke KC (2002) Calibration of the SLEUTH urban growth model for Lisbon and Porto, Portugal. Comput Environ Urban Syst 26(6):525–552

Skamarock WC, Klemp JB, Dudhia J, Gill DO, Barker DM, Duda MG et al (2008) A description of the advanced research WRF version 3, NCAR technical note. Mesoscale and Microscale Meteorology Division, National Center for Atmospheric Research, Boulder, CO, USA

Tobler WR (1970) A computer movie simulating urban growth in the Detroit region. Econ Geogr 46(sup1):234–240

Vermeiren L, Dequidt A, Afroun M, Guerra TM (2012) Motion control of planar parallel robot using the fuzzy descriptor system approach. ISA Trans 51(5):596–608

Wang J, Huang B, Fu D, Atkinson PM, Zhang X (2016) Response of urban heat island to future urban expansion over the Beijing–Tianjin–Hebei metropolitan area. Appl Geogr 70:26–36

Weng Q, Larson RC (2005) 10 Satellite Remote Sensing of Urban Heat Islands: current practice and prospects. Geo-Spatial technologies in urban environments: with 45 Figures and 14 Tables, 91

Yang X, Lo CP (2003) Modelling urban growth and landscape changes in the Atlanta metropolitan area. Int J Geogr Inf Sci 17(5):463–488

Yin H, Kong F, Luo Z, Yan W, Sun C, Xu F (2013) Suitability assessment of construction land in the central and southern parts of Hebei Province, China based on potential-limitation model. Chin J Appl Ecol 24(8):2274–2280 (in Chinese)

Yin H, Kong F, Hu Y, James P, Xu F, Yu L (2015) Assessing growth scenarios for their landscape ecological security impact using the SLEUTH urban growth model. J Urban Plann Dev 142(2):05015006

Yin H, Kong F, Middel A, Dronova I, Xu H, James P (2017) Cooling effect of direct green façades during hot summer days: an observational study in Nanjing, China using TIR and 3DPC data. Build Environ 116:195–206

Yin H, Kong F, Yang X, James P, Dronova I (2018) Exploring zoning scenario impacts upon urban growth simulations using a dynamic spatial model. Cities
Yow DM, Carbone GJ (2006) The urban heat island and local temperature variations in Orlando, FL. Southeast Geogr 46:297–321

Dr. Fanhua Kong is a Full Professor at the International Institute for Earth System Science (ESSI), Nanjing University, China. Her recent work focuses on the green space function and ecosystem services. She has conducted research to address multi-functions of green space and developed new methods to quantify the green space ecosystem services. She has published over 60 peer-reviewed papers.

Dr. Haiwei Yin is an Associate Professor in the School of Architecture and Urban Planning at the Nanjing University, China. His recent work focuses on the development of planning methods to examine urban ecological resilience and urban green infrastructure. Yin has authored or co-authored multiple peer-reviewed papers.

Dr. Fei Jiang is an Associate Professor at the International Institute for Earth System Science (ESSI), Nanjing University, China. His research focuses on regional air quality modeling and prediction, inverse modeling of air pollutants and carbon fluxes. He has published over 50 peer reviewed papers.

Jiayu Chen is a postgraduate student at the Nanjing University, China. His research mainly focuses on GIS and dynamic spatial modeling.

Linking Land Use with Water Pollution in Coastal Watersheds of China

Jinliang Huang, Ayu Ervinia and Yaling Huang

Abstract Land-water relationship studies are important for watershed and coastal management. Changes in watershed-wide land use may exacerbate water pollution due to non-point source (NPS) and point-source (PS) pollutants, two major contaminants to the watershed-coast continuum. We take two largest coastal watersheds in Fujian Province, namely, Jiulong River watershed (JRW) and Minjiang River watershed (MRW), as the study site to examine the land use/cover and water pollution linkage, thereby suggesting land-water management implications. Land transformations have been accelerating across the JRW during the past 30 years. Urban watersheds were significantly more polluted than agricultural and natural watersheds. The spatial and temporal variations of water quality in the JRW are largely governed by the combination of natural factors (e.g. hydrology and biochemical processes) and anthropogenic inputs from industrial, agricultural, and urban activities. Climatic variability also affected the land use and water quality relationship and the performance of empirical regression models. Land use can affect river pollution, and such relationships vary spatially. Land use-water quality correlation is weakening under simultaneous point-source pollution. Furthermore, we developed three watershed modeling tools, namely, GIS-based empirical models, AnnAGNPS model, and SPARROW, to estimate non-point source nitrogen (N) and phosphorus (P) pollution, determine the sources of contaminants, and develop nutrient management strategies, respectively. Finally, we discuss the implications for water resource management.

Keywords Land-use · Water pollution · Linkage · Watershed modeling Watershed management

J. Huang (✉) · A. Ervinia · Y. Huang
College of the Environment & Ecology, Xiamen University, Xiamen 361102, Fujian, People's Republic of China
e-mail: jlhuang@xmu.edu.cn

X. Yang and S. Jiang (eds.), *Challenges Towards Ecological Sustainability in China*, https://doi.org/10.1007/978-3-030-03484-9_11

1 Introduction

Water pollution is among the most important environmental issues worldwide because people depend heavily on river water for residential, agricultural, and industrial uses (Mouri et al. 2011; Huang et al. 2013a, b). Rivers act as receiving water bodies that carry pollutant from a wide range of different sources, including agricultural runoff, domestic wastewater, industrial discharge, urban runoff, and ultimately discharging them into the estuarine and coastal water (Ouyang et al. 2006; Somura et al. 2012). It is not surprising that many rivers across the world suffer from severe deterioration in water quality due to rapidly increasing human population and economic development (UN WWAP 2009). China's Ministry of Environmental Protection estimated that nearly forty-three percent of major Chinese rivers are moderately to seriously polluted (MEP 2011). Therefore, evaluation of river water quality in the context of the escalating environmental pressure is critical for water resource management at various scales particularly in coastal watersheds due to their important role in regional ecosystem services.

The combination of natural and human-dimensional factors, along with basic hydrological and biogeochemical processes, determines the spatiotemporal variations of river water quality (Baker 2003; Li et al. 2013; Smith et al. 2013; Huang et al. 2014). Watershed characteristics, such as topography, soil type, and bedrock are recognized as the important natural factors governing a spatial variation in river water quality. For example, the upper catchment with steeper slope has higher erosion rates. The fertilizer application, domestic wastewater, and industrial effluents are the major pollution sources from anthropogenic activities across various land use categories. On the other hand, hydrological and biogeochemical processes are important and should be considered in explaining the seasonal variation of watershed water quality. Normally, nutrients are firstly accumulated within the ground surface, vadose zone or the groundwater and are then displaced into streams from groundwater during dry seasons or surface runoffs during wet seasons. Biogeochemical processes and hydro-meteorological settings control the temporal variability of stream water quality. Understanding how human-dimensional and natural factors affect water quality and how such relationships vary spatiotemporally can help water resource managers to determine appropriate scales and most critical issues that can help improve the management success.

Watershed-wide land use affects stream quality mainly by way of non-point source pollution, the major contaminant contributor to the watershed-coast continuum (Swaney et al. 2012). Land-based pollution was estimated to contribute over eighty percent of the contaminants discharging into seas (UN General Assembly 2004). Land use changes often lead to widespread nutrient enrichment in surface waters to a point where it is difficult for water resource managers to mitigate eutrophic conditions including algal blooms, low dissolved oxygen, and resulting fish kills (White et al. 2008). This issue is quite relevant in China because of two major reasons. Firstly, China has being undergone rampant land transformations caused by population and urban spatial growth during the past thirty years. Second, stream water quality in China is generally poor and degrading because of point-source and non-point source pollutions. The former mainly includes industrial and municipal sewage discharges, and the latter comes from fertilizers, pesticides and manure contained in agricultural and aquacultural runoffs (Liu and Diamond 2005). Accordingly, the 2011 Chinese State Government Work Report vowed to strengthen governance and control of stream water quality, particularly in large river systems. Therefore, investigating the linkages between land use and water quality is of great significance to watershed management practices for predicting pollution potential and developing strategy to improve water quality.

The Jiulong River Watershed (JRW) is a good example, and offers a rare opportunity to examine the land use/cover (LULC) and water quality linkage. Assessing land use-water quality relationship is essential for understanding the underlying mechanisms of water pollution variations so that the impacts of changing land use and other human activities on water quality can be predicted and modeled. JRW is a coastal watershed with medium size that is located in southeastern China. While the Jiulong River watershed is important for the region's economic and ecological health, it has undergone extensive land changes and resultant water quality deterioration during past three decades. This chapter is organized into several sections. After an introduction in Sects. 1, 2 describes the study areas. Section 3 discusses the land use/cover change dynamics in the watershed and the long term water quality trend, and the spatial and temporal variation of water quality. Section 4 discusses the linkage between LULC and water quality through several statistical methods. Section 5 presents the result of watershed modeling of non-point sources (NPS) pollution that we constructed for the study site, with the main objective to estimate non-point source nitrogen (N) and phosphorus (P) pollution, to determine the sources of contaminants in the Jiulong River estuary, and to develop nutrient management strategies by conducting pollution control scenario analysis. Finally, we discuss the implications for watershed management with appropriate strategies and actions to improve water quality in the Jiulong River watershed (Sect. 6).

2 Overview of the Jiulong River and Minjiang River Watersheds

The Jiulong River Watershed (JRW) and the Minjiang River Watershed (MRW) lies on the southeastern coastal area of China. These two watersheds ranked the first and second largest watershed in Fujian Province. The watershed area of the MRW is 60,992 km^2 and that of the JRW is 14,700 km^2. The JRW encompasses eight counties/districts, extending from 116°46′55″E to 118°02′17″E and from 24°23′53″ to 25°53′38″N (Fig. 1), while the MRW includes 31 counties/districts, extending from 116°23′ to 119°43′E and from 25°23′ to 28°19′N (Fig. 2). As located in a subtropical zone, the watersheds are characterized by monsoon climate with average annual rainfall ranges between 1400 and 1800 mm and average annual temperatures range

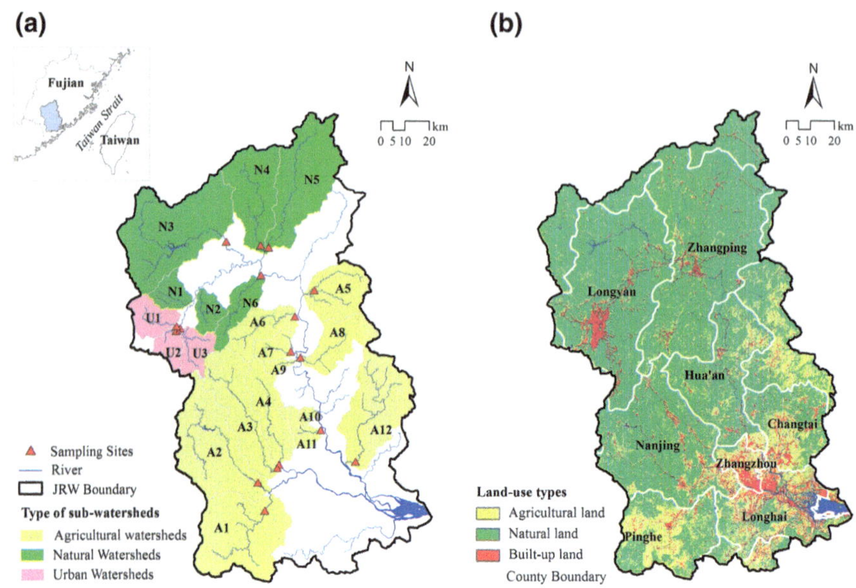

Fig. 1 The Jiulong River Watershed **a** water sampling sites and **b** land use and counties

between 19 and 21 °C. Rainfall intensity distributes unevenly throughout the year of which 70% occurs between April and September. Both watersheds play a pivotal role in the regional ecological health and economic development. Approximately 10,000,000 people utilize the water from the Jiulong River for a variety of purposes, while nearly 12,000,000 residents rely upon the water from the Minjiang River. The gross domestic product (GDP) of the Jiulong River and Minjiang River watersheds accounts for 25% and 30% of Fujian Province's economic output, respectively.

The JRW includes the North and West River reaches, with the total annual flow of approximately 1,200 cubic kilometers being transported downstream into estuarine and coastal waters. The topography of the JRW is predominantly mountainous region in the upstream (almost 70% of watershed has a slope steeper than 18%) (Huang et al. 2004). The major geological rock types of the JRW are granite, volcanic tuff, and sandstone (Cao et al. 2006). Zhangzhou plain surrounding the JRW downstream end belongs to China's most developed regions, particularly in agricultural production. The main agricultural products cultivated in Zhangzhou plain include flowers and various types of subtropical and tropical fruits. The predominant N fertilizers applied for most cropland including urea, ammonium hydrocarbonate and NPK compound fertilizers, with higher application rates occurred in spring (over 200 kg N/hm^2). Excessive fertilizer use has been identified as critical non-point source (NPS) pollution in agricultural watersheds. On the other hand, industrial wastewater and sewage are ineligible point sources (PS) pollution in the JRW due to low municipal wastewater treatment rate.

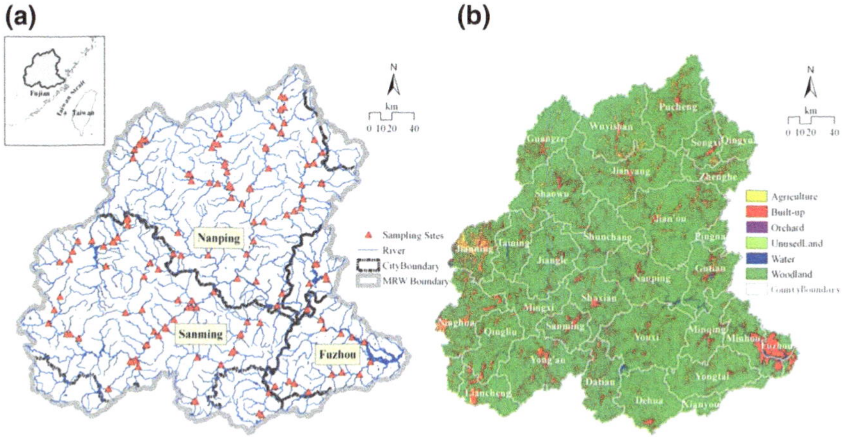

Fig. 2 The Minjiang River Watershed **a** water sampling sites and **b** land use and counties

The MRW produces an annual average discharge of 1760 m^3/s into Minjiang River Estuary and the river discharge is the seventh highest annual runoff in China. The upstream of the Minjiang River is predominated by forest and is characterized by steep slopes and high stream densities. The major soil type is yellow-red soil originated from various parent materials (see Hu et al. 2014). One of the major forests in the south of China lies in Nanping City covering the central part of the watershed. The most important industrial base for raw material as well as the main agricultural production base for Fujian Province is located in Sanming City in the western part of the MRW. Fuzhou City in the downstream end of the study area contributes largely to non-point sources pollution from residential land and poultry production. Similar to the JRW, the point-source pollution in the MRW is still severe because of inadequate sewage treatment and the discharge of highly concentrated pollutants from sewage treatment facilities.

3 Landscape and Water Quality Dynamics

3.1 Land Use/Cover Change

LUCC is driven by complicated interactions between the human and physical environments, which led to regional to global environmental changes, and China is no exception (Manandhar et al. 2010; Sarma et al. 2008). As one of the world's fastest-growing economy, China experiences rapid anthropogenic environmental changes, most notably in coastal watersheds, resulting from urbanization, agricultural intensification, and deforestation. Due to the intensified environmental pressures and their crucial role in regional ecosystem services, coastal watersheds becoming a consid-

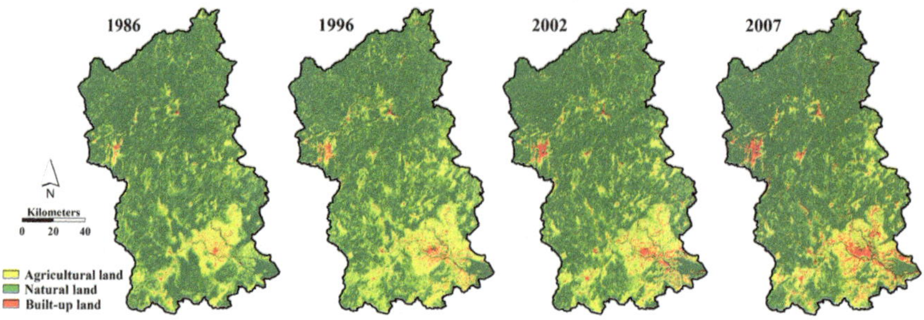

Fig. 3 Land use/cover maps of the Jiulong River Watershed in 1986, 1996, 2002, and 2007 (*Source* Huang et al. 2012)

erable concern for land change scientists (Hobbie 2000). Land use/cover changes in coastal basins impose threats to ecological health and have brought about socio-economic consequences (Johnson et al. 1997; Li et al. 2001; Fukushima et al. 2007). Therefore, quantifying and understanding land changes in coastal watersheds of China are of fundamental importance to develop effective land management strategies.

Comprehensive analysis of the LUCC patterns and processes for the Jiulong River Watershed has been conducted using remote sensing and geographic information systems. A novel quantitative analytic method called intensity analysis has been performed on the three land categories, namely, Natural, Agriculture, and Built during the period of 1986 to 2007 (Huang et al. 2012). Intensity analysis allows quantitatively analyzing the annual change intensity at three different levels, namely, time interval, category, and transition. Figure 3 presents maps of the three land categories between 1986 and 2007. The dominant land cover type in JRW is Natural land, accounting for 70–79% of the total area. Agricultural land covers 21–29% of the landscape, while Built-up land only occupies 2–4%. Figure 4 shows land-use patterns during 1986–2015. The watershed land cover has been extensively changed in the Jiulong River Watershed. Agricultural land showed a net increase during 1986-1996 but began to shrink after 1996 in the total area. Natural land significantly decreased during 1986–1996, indicating the deforestation occurred in this period. Most notably, the area of Built-up land increased steadily due to conversion of croplands and forests (Huang et al. 2012; Zhou et al. 2014).

The transition from agricultural to built-up land is considered to be systematic because the decrease of agricultural land occurred since built-up land continually increased during 1996–2007 (Fig. 4). Most of the targeting transition from agriculture to built-up land was attributed to urbanization process where most of the area surrounding the built-up land locations is used for agricultural activities. Human settlements are normally developed in the area of level ground, and so do the agricultural activities. Therefore, built-up land expansion is likely at the cost of agricultural land. Urbanization brings many benefits to economic growth in developing regions, but it may also lead to problems like water quality degradation, air pollution, soil erosion,

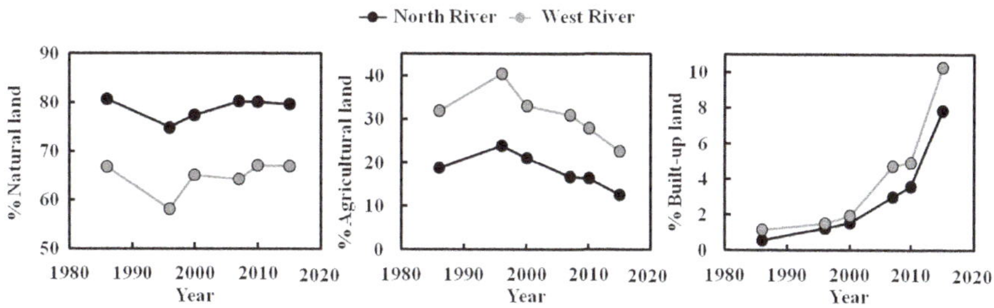

Fig. 4 Land-use patterns in the Jiulong River Watershed during 1986–2015. Land use/cover data were based on Huang et al. (2012) and Zhou et al. (2014)

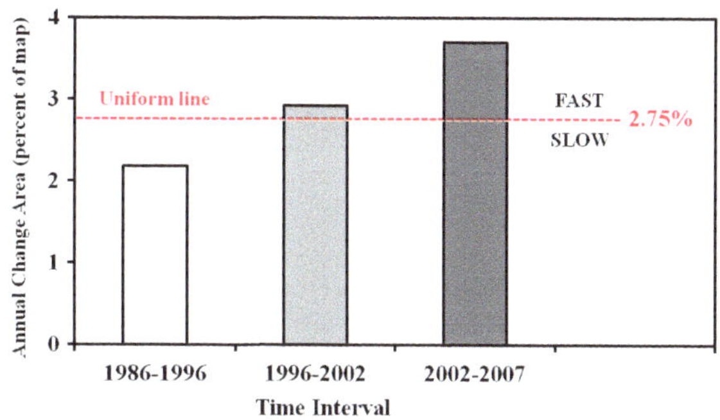

Fig. 5 Time interval intensities for 1986–1996, 1996–2002, and 2002–2007 (*Source* Huang et al. 2012)

etc., leading to negative impacts on environmental and social health (Withers and Jarvie 2008).

Land change rate is generally accelerating through time (Fig. 5). It was rather slow during 1986–1996, but became faster after that, especially during the 2002–2007 interval. A similar accelerating land transformation was also observed in other coastal areas. Urban settlements expanded dramatically at the expense of cropland in many coastal regions such as the Bohai Sea (Zuo et al. 2011), Zhejiang province (Ding and Wang 2006) and Shanghai (Han et al. 2009).

Anthropogenic land changes are frequently driven by economic forces and opportunities, especially the influx of foreign investments and large-scale development have resulted in urban expansion in the coastal regions (Lambin et al. 2001; Wang et al. 2008; Seto and Kaufmann 2003). The information from Fig. 6 supports this hypothesis as both socio-economic development and agricultural practices have intensified in the watershed. The statistical data indicated that watershed's GDP as well as total N fertilizer application has increased significantly in the Jiulong River Watershed, resulting in water quality degradation over the past 20 years, threatening the local drinking water security and the ecological health of adjacent coastal waters.

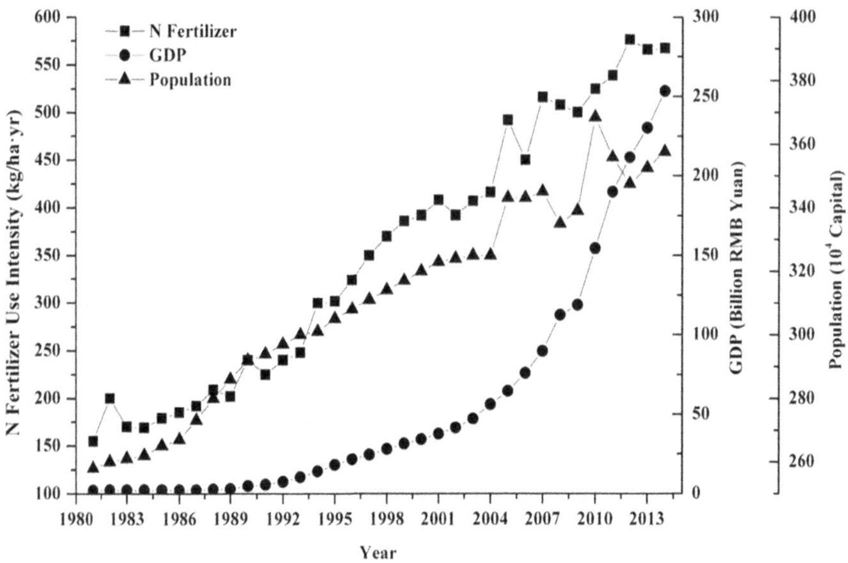

Fig. 6 Nitrogen (N) fertilizer use intensity, population, and Gross Domestic Product (GDP) for the Jiulong River Watershed which includes eight counties/districts: Zhangzhou, Xinlou, Zhangping, Hua'an, Changtai, Pinghe, Longhai, and Nanjing (*Source* yearbooks of Zhangzhou, Longyan city during 1981–2014)

3.2 Long Term Water Pollution Trends

Agricultural activities and urbanization are two most major underlying causes driving water pollution, which is projected to increase in the coming years because of the rapid population growth, the increasing demand for food production, and socio-economic development (Mouri et al. 2011; Duan et al. 2012; Carpenter et al. 1998). For this reason, evaluation of long-term water quality data is necessary to further understand how water pollution is worsened by anthropogenic changes. Data of monthly DIN concentrations, i.e. nitrate–N (NO_3^-–N) and ammonia–N (NH_4^+–N), from 1961 to 2014 at the downstream of the two river reaches (i.e. North and West Rivers) of the Jiulong River watershed were obtained from Fujian Province's Hydrological Bureau and Environmental Protection Bureau. Based on the monitored DIN concentrations and discharges at Punan and Zhendian, the annual flux was estimated by applying the global mean (GM) and flow-weighted (FW) methods (Huang et al. 2018).

Results showed a significantly increase in riverine DIN exports from Jiulong River Watershed from 1961 to 2014 (Fig. 7), which is consistent with urban expansion, intensive agricultural activities, and climate variability. In comparison, we had similar observation on total phosphorus (TP) concentrations in the Jiulong River, especially in Longyan County, also increased significantly during the years of 2003 and 2008

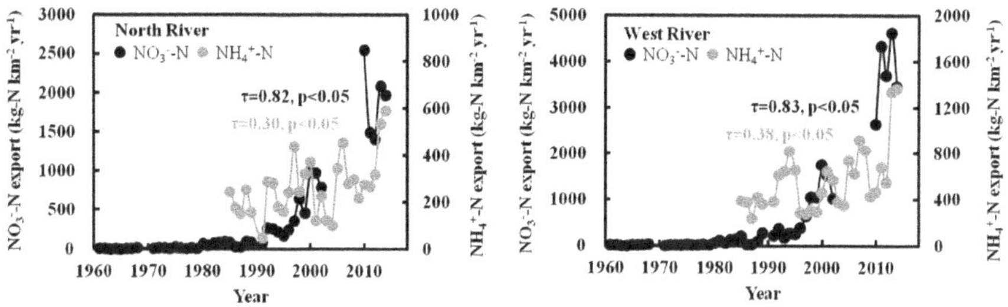

Fig. 7 Long-term trend in nitrate–N and ammonia–N exports from the two largest reaches of the Jiulong River Watershed during 1961–2014 (*Source* Huang et al. 2018)

(Huang et al. 2013a, b). Moreover, the TP concentration substantially exceeds China's National Standard for Surface Water Environmental Quality (GB3838-2002) and it is even higher than in the extremely polluted Taihu Lake and most rivers in Fujian Province (EMB 2008).

Land changes caused by rapid population and urban physical growth in Asia contributed to the long-term N increase and the related coastal eutrophication (Dumont et al. 2005). Our studies indicate that different land change types can impose various potentials to increase watershed-wide N export that are eventually transported into coastal waters in China. Based on the response of N export to runoff variability, we identified three periods of nitrogen exports with different dominant controls, namely, nitrogen deficit period (1961–1980), nitrogen rapid increase period (1981–2002), and nitrogen enriched period (2010–2014) (Fig. 8). The first period named nitrogen deficit period occurred in JRW during 1961–1980 as river exported lower nitrate–N exports, which were correlated with runoff. The result indicated runoff exerted control over nitrate export. One possible reason for lower nitrate–N exports is a limited application of chemical fertilizers in agricultural practices under communal households when Chinese farmers were encouraged to use recycled N fertilizers. Nevertheless, anthropogenic disturbances increase in watershed does not always translate into elevated DIN concentrations or exports (Morse and Wollheim 2014). Nitrogen deficit in rivers and streams can also be attributed to watershed retention (Duan et al. 2007). Many studies suggest that streamflow can be an important variable affecting N export in minimally disturbed watersheds (e.g., Lewis 2002).

The use of fertilizers largely determined the annual nitrate–N export dynamics in the JRW during 1981–2002, which is known as nitrogen rapid increase period. During this time period, although the agricultural land did not substantially change, watershed has undergone rapidly increasing fertilizer use, the intensity of which might exceed the watershed retention capacity, resulting in amplified DIN concentrations and exports. A weak correlation between N export and runoff variability during this period suggested that the fast fertilizer use (not streamflow) increase controlled the watershed-wide N exports. Similar results were also observed in many other rivers in agricultural regions, such as Mississippi and Changjiang (Turner and Rabalais 1991; Duan et al. 2000).

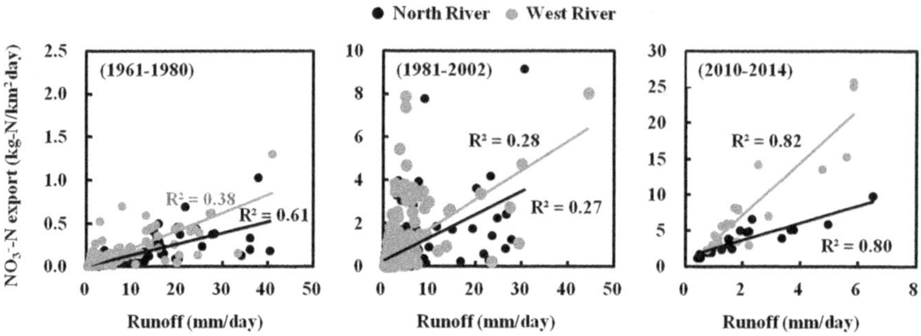

Fig. 8 Nitrate–N exports plotted against runoffs at two largest reaches of the Jiulong River Watershed: North River and West River, during 1961–1980, 1981–2002 and 2010–2014 (*Source* Huang et al. 2018)

During the recent years, the interplay between land use and climate variability has greatly affected watershed-wide N exports (Huang et al. 2013a, b; Zhang et al. 2015). Meanwhile, the increase rate of fertilizer use, population and GDP ceased or slowed down, and so did the N concentrations in streams and N exports. Again, climate variability or streamflow dominated with more N exports during wet years but not during dry years. The climate variability impact upon the N exports was also observed in some N-concentrated rivers during the post-increase period, e.g., Mississippi River (Turner and Rabalais 1991).

3.3 Spatial and Temporal Variations of Water Quality

Human activities and watershed characteristics can affect the spatial variability of the watershed-wide stream water quality (Huang et al. 2014). Meanwhile, the temporal variability of stream water quality is commonly influenced by internal factors, including the biogeochemical processes, in-stream and terrestrial N cycle process, bedrock weathering, and hydro-climatic regime. The river health depends heavily upon the health of the surrounding watershed, thus alterations of watershed land cover can affect the surface water quality, posing threats to drinking water and food production security. Natural land cover types (i.e. forest, wetland) have higher retention capacity that allows excess nutrient and water to be retained within the soil and groundwater. Unlike the natural landscape, more developed watersheds receive too much pressure from human activities. Agricultural watersheds for instance, fertilizer, animal manure, and pesticides are intensively added to cropland and subsequently discharged into the river system via baseflow or storm events. Meanwhile, an increase of impervious surfaces (e.g. roads, parking loads, drainage, and human settlements) in urban watersheds accelerates the rainfall-runoff process, instead of allowing the water to percolate through soil (Ji 2008). Urban runoffs also carry pollutants from stormwater, vehicular use, and untreated domestic wastewater.

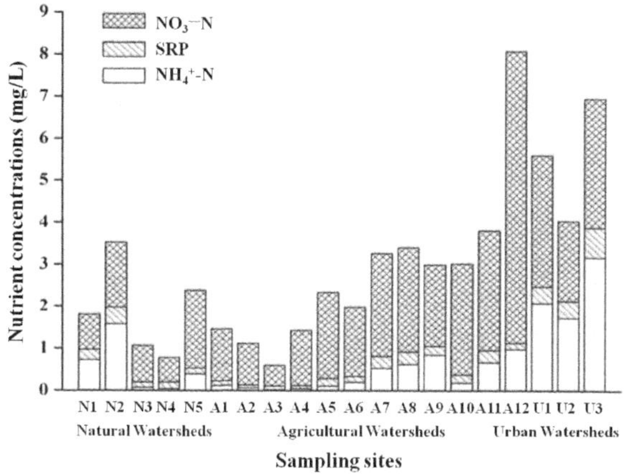

Fig. 9 Comparison between nutrient concentrations among the three types of watersheds (*Source* Huang et al. 2014)

Table 1 LSD Post Hoc comparisons of water quality variables among the three types of watersheds

Parameter	Urban watersheds-natural watersheds		Urban watersheds-agricultural watersheds		Agricultural watersheds-natural watershed	
	Mean difference	Sig.	Mean difference	Sig.	Mean difference	Sig.
NH_4^+–N	1.757[a]	0.000	1.958[a]	0.000	−0.201	0.450
NO_3^-–N	1.566	0.136	0.450	0.617	1.116	0.144
SRP	0.291	0.001	0.329[a]	0.000	−0.037	0.510

Source Huang et al. (2014)
[a]Indicates significant at $p < 0.05$

To examine land use/cover and water quality linkage, we collected surface water samples of baseflow from twenty headwater sub-watersheds in the Jiulong River watershed in the flood, dry and transition seasons during 2010–2012 (Huang et al. 2014). Sub-watersheds were divided into three major land-use categories, namely agricultural, natural, and urban watersheds.

Water quality varied greatly across agricultural, natural, and urban watersheds (Fig. 9). The NH_4^+–N and SRP concentrations were higher in urban watersheds than in natural and agricultural ones. The NO_3^-–N concentration in agricultural watersheds was generally higher than in natural and urban ones. Moreover, the LSD Post Hoc comparisons revealed urban watersheds exhibited more significance in NH_4^+–N, than natural and agricultural watersheds (Table 1).

Land use and land cover regulates the fate and transport of pollutants in catchment, thus changing natural land cover can exacerbate surface water pollution. Compared to natural and agricultural watersheds, urban watersheds were more seriously polluted given the higher urbanization level at the selected sub-watersheds. Other factors that contribute to urban water pollution are mining activities, steel plants, and livestock farming that mostly located at the upstream of JRW. In other words, the water quality

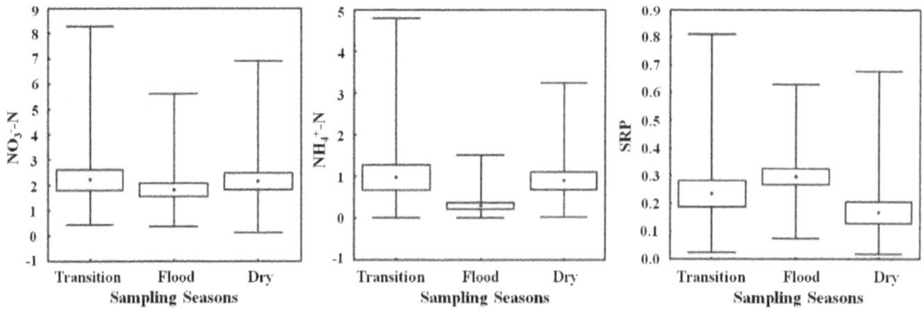

Fig. 10 Comparison between nutrient concentrations among the three sampling seasons (*Source* Huang et al. 2014)

in the urban watersheds was affected by both PS and NPS pollution from industrial, domestic, and livestock wastes. Moreover, natural watersheds had the lowest N and P concentrations than agricultural watersheds except for the N2 sub-watershed with a high proportion of forest but being heavily polluted due to mining activities.

Water quality shows a distinct temporal variation (Fig. 10). The highest mean concentrations of NH_4^+–N and NO_3^-–N were observed in the dry and transition seasons, whereas the lowest concentrations detected in the flood season. A distinct seasonal variation in DIN concentrations suggest that water quality dynamics in the JRW may be controlled by several factors including anthropogenic N emissions associated with vadose zone leaching and groundwater seepage, in-stream N immobilization rate, and hydrological regime. In the dry and transition seasons, decreased streamflow as well as in-stream N immobilization rate may increase in nutrient concentrations. Conversely, increased streamflow in the flood season may lead to the lowest N concentrations in response to effective dilution effect associated hydrologic regime.

Interestingly, the temporal variation of nitrogen exports was different from that of nitrogen concentrations. The mean daily nitrate and ammonia exports in the flood season were higher than in the transition and dry seasons, suggesting that high daily runoffs would result in high daily nitrogen export (Fig. 11). As we expected, the nitrate export was higher in urban and agricultural watersheds than in natural ones within the three sampling seasons, while the ammonia export was the highest in urban watersheds than in agricultural and natural ones. The nitrate export in the flood season had the greatest variation in agricultural watersheds than in other sub-watersheds types, indicating the large temporal variation of precipitation and thereafter non-point source pollution.

In contrast with nitrogen concentrations, the highest soluble reactive phosphorus (SRP) concentration was detected in the flood season (Fig. 10), which could be linked with the releasing sediment-bound P input being related to hydrological processes. During the flood season, the SRP load was exported from watershed non-point sources via storm water runoff and was released from in-stream internal sources as a result of increasing streamflow (May et al. 2001; Huang et al. 2013b). Besides hydrological process, other factors that can affect P exports include biogeochemi-

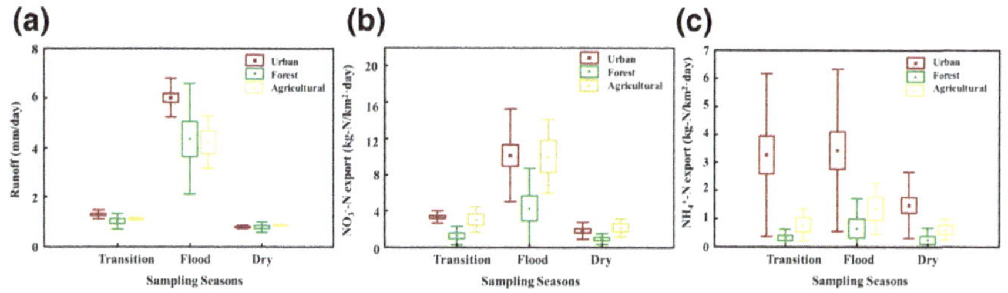

Fig. 11 Average daily runoff, nitrate–N, and ammonia–N exports for each sampling season from three types of headwater streams of Jiulong River Watershed during 2010–2014 (*Source* Huang et al. 2018)

cal process and anthropogenic input. For example, Duan et al. (2012) observed the high SRP concentration in the low-flow condition due to the increased biochemical reaction with rising temperatures during summer. Similar result was also revealed by Bowes et al. (2005), however wastewater effluent was more important than biochemical reaction for explaining high P concentrations during summer low flow. Our analysis in the JRW suggests it is the hydrological process rather than biogeochemical process that determines P export's seasonal variation, which may be due to the fact that the inter-annual precipitation had a greater variability than of the inter-annual temperature in the watershed.

The seasonal river water chemistry variations are mostly related to climatic and biotic conditions. Thus, they are mostly controlled by watershed processes especially natural or anthropogenic induced disturbances (Sliva and William 2001; Duan et al. 2012). Nitrogen (N) and phosphorus (P) exported from watershed through different transport pathway. Inorganic N defined as water soluble nutrients may be leached into the groundwater and transported into streams through subsurface water (Petry et al. 2002). Conversely, insoluble phosphorus often exists in particulate forms and therefore phosphorus is transported along with sediments through the surface water pathways, especially during the rainy season (Willett et al. 2004; Chang and Kwon 2007). Therefore, bio-reactive elements can cycle differently, releasing from or remaining in sediments responding to shifting redox conditions and warming. Hence, the nutrient source can vary. For instance, during the flood season, the stream nitrate concentration can be higher because of the contribution from shallow groundwater in nitrate-rich hill slopes, whereas during the dry season, stream water largely originates from denitrified bottomland groundwater (Altman and Parizek 1995; Hill 1996) and deep fractured aquifers (Pauwels et al. 2001).

4 The Land Use and Water Quality Linkage

Land use/cover (LULC) plays a significant role in water quality variation at the watershed scale (Huang et al. 2013a). To some extent, land use types influence pollution sources, transport of pollutants, and watershed retention. Land use-induced non-point source (NPS) pollution has become a major cause of water quality degradation across the catchment-coast continuum (Griffith 2002). For this reason, investigating the linkages between land use and water quality is of great significance to watershed management practices for predicting pollution potential and developing strategy to improve water quality (Xiao and Ji 2007; Uuemaa et al. 2007).

Many studies have assessed the landscape and water quality relationship (e.g., Johnson et al. 1997; Rhodes et al. 2001; Griffith et al. 2002; Fedorko et al. 2005; Tu et al. 2007; Rothenberger et al. 2009; Wilson and Weng 2010). Nevertheless, these studies revealed inconsistent results in land use-water quality relationships since such relationship is scale-dependent and varies over time and space. A watershed-scale approach has been widely used to analyze the linkage between landscape and water quality. In doing so, watershed land use/cover types are characterized, which are then associated with water quality indicators. Increasing availability of multi-temporal remote-sensor data is very useful in the landscape-water quality studies at different scales. Field sampling and monitoring are important because land use/cover-water quality relationship can be affected by the sampling strategy (Baker 2003).

4.1 Ordinary Least Square (OLS) Regression

Ordinary least square (OLS) regression can be used to model the linear relationship between one or more independent variables and a dependent variable. OLS regression has been widely applied to study the global relationship between landscape characteristics and water pollutant at the watershed scale. For this reason, we applied two OLS regression methods, namely Pearson correlation analysis and multiple linear regression analysis, to examine the land-water relationship in the Jiulong River Watershed. The Pearson correlation was estimated by plotting pollutant concentration (dependent variable) against percentage of land use type (independent variable) in the sub-watersheds for the 21 sampling sites (Huang et al. 2015). According to Pearson correlation results, watershed land use types can cast multiple influences on water quality (Table 2). The cropland proportion was significantly correlated with nitrate–N (NO_3^-–N). The percentage of built-up land had strong positive relationships with ammonium–N (NH_4^+–N) and soluble reactive phosphorus (SRP). Unlike built-up and cropland, the percentage of woodlands was negatively correlated with N and P concentrations.

A positive correlation between nitrate concentrations and cropland was associated with agricultural non-point source pollutions. Nitrate–N sources mainly came from fertilizer application to crops and fields. On the other hand, strong correlations between built-up land and ammonia–N as well as SRP might have affected urban watersheds to be easily exposed to pollutants from domestic wastewater, livestock waste, and industrial effluents. Among three land use types, only forest land presented a negative association with all pollutants, especially with NO_3^-–N, which means higher forest proportion in agricultural sub-watersheds can help reduce nitrate–N

Table 2 Pearson correlations between nutrient parameters and land use classes

Parameter	Cropland	Built-up	Forest
NH_4^+–N	−0.20	0.77[a]	−0.22
NO_3^-–N	0.45[a]	0.28	−0.72[b]
SRP	−0.06	0.63[a]	−0.19

Source Huang et al. (2015)
[a]Indicates significant at $p < 0.05$
[b]Indicates significant at $p < 0.01$

Table 3 Multiple linear regression results for the calibration procedure

Landscape variables	1996 (dry)		2002 (dry)		2007 (wet)	
	NH_4^+–N	COD_{Mn}	NH_4^+–N	COD_{Mn}	NH_4^+–N	COD_{Mn}
PD	−0.580[b]					
SHDI				0.695[a]		
%AGR					−0.905[b]	
%NA						
%BL	0.834[b]	0.661[a]	0.675[a]		0.940[b]	0.741[b]
R^2	0.712	0.437	0.455	0.438	0.892	0.549

Source Huang et al. (2013a)
PD patch densities, *SHDI* Shannon's diversity index, *%AGR* proportion of agricultural land, *%NA* proportion of natural land, %BL proportion of built-up land
[a]Indicates significant at $p<0.05$
[b]Indicates significant at $p<0.01$
Number of observations for the multiple regression analysis for each of the three years is 11

concentration. In other words, reforestation could reduce the influence from fertilizers application in agricultural sub-watersheds considering the ability of forest to absorb extra nutrients.

The influence of landscape characteristics on water quality was further evaluated during dry and wet years by the multiple linear regression analysis (Huang et al. 2013a). For this purpose, we divided the JRW into 11 sub-watersheds using the 11 gauge stations established by Fujian province's Environmental Protection Bureau (EPB) for regular monitoring of water quality. In the multiple linear regression models, ammonium–N (NH_4^+–N) and chemical oxygen demand (COD_{Mn}) were considered as dependent variables at three points in 1996, 2002, and 2007 with varying hydrological conditions. Results showed the landscape and water quality linkage varied at different time periods (Table 3). Landscape variables had stronger relationship with ammonia–N during the wet year of 2007 with R^2 of 0.892 than during the dry years of 1996 and 2002 with R^2 of 0.712 and 0.455, respectively. The R^2 order follows the same order as the annual flow did in the three years, suggesting that climatic variability may affect the landscape characteristics and water quality linkage (Huang et al. 2013a).

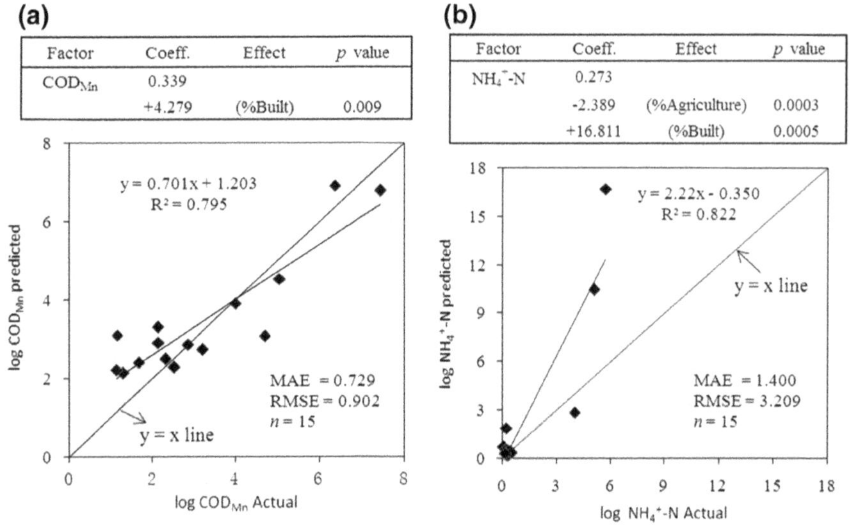

Fig. 12 Two plots of predicted values versus observed values. The equation above each plot is the source of prediction. The equation inside the plot gives the least squares line and the R^2 of the least squares line. Cases **a** and **b** predict across space by using some sub-watersheds in 2007 to predict other sub-watersheds in 2007 (*Source* Huang et al. 2013a)

Landscape variables exerted more influences on water quality variation in wet years. This is understandable because rainfall-runoff process facilitates transport of non-point sources pollutions from watershed into streams, leading to water quality degradation. The empirical regression models for COD_{Mn} and NH_4^+–N in wet year of 2007 were constructed (Fig. 2). The COD_{Mn} model consider a single landscape metric of %BL as independent variable, while the NH_4^+–N model take into account two metrics of %Agriculture and %BL. NH_4^+–N model had a higher R^2 value of 0.89 in the calibration process than COD_{Mn} model with R^2 value of 0.55. Despite the low R^2 value, the COD_{Mn} model presented a higher predictive capability spatially than the NH_4^+–N model did. Therefore, it is important to note that greater R^2 may not necessarily ensure accurate predictions based on the validation data (Fig. 12).

A significant relationship between %BL and water quality suggests some problems with wastewater treatment management. Effective wastewater treatment is rare in China as only about 20% of domestic wastewater has been treated in the country. Therefore, the drainage and the rainfall runoff with high COD_{Mn} and NH_4^+–N concentration are usually discharged into streams directly (Huang et al. 2013a, b). The empirical models established here should be quite suitable for water quality predictions in the watersheds without wastewater treatment plants (WWTPs). For the watersheds with WWTPs, using the urban areal coverage as a proxy for non-point source nutrients can generate false conclusions because urban land cover is mostly impervious and the drainage is usually routed to WWTPs first and then discharged into local streams.

When examining the relationship between land-use and water pollution using OLS regression (e.g., Pearson correlation analysis and multiple regression models), researchers tend to explain the potential influencing factors of such relationships, such as rainfall-runoff process, topography, agricultural activities, or a spatial scale effect. But exceptions do exist. For instance, Mehaffey et al. (2005) noted that less than 10% cropland could still contribute to serious water pollution. We concluded that OLS regression has its advantages on exploring the linkage between land use and water quality, but it also has some limitations in examining such relationships from the perspective of local spatial variability.

4.2 *Geographically Weighted Regression (GWR)*

Land use can affect water pollution, and such a relationship may or may not vary spatiotemporally (Huang et al. 2015). The global statistics being frequently employed to examine the overall association for the whole study site, including Pearson correlation analysis and multiple regression, may not reveal some local relationships, especially among the watersheds being dominated with various land use types, such as agriculture, forest, or urban (Tu and Xia 2008; Tu 2011). Given the limitation of ordinary least square (OLS) regression, we attempted to investigate the spatial variations in the land use and water pollution relationships using geographically weighted regression (GWR). GWR has been widely applied to evaluate the spatial variation of the land-water relationship (e.g., Brunsdon et al. 1998; Pratt and Chang 2012). We collected surface water samples of baseflow from headwater sub-watersheds during 2010–2013. In our GWR models, the percentage of one land use type was considered as an independent variable to examine its association with a specific pollutant. Three land use types were considered, namely, cropland, forest land, and built-up land. All sampling sites were included to build the GWR models for each land use type, and the results were then grouped into three categories according to the three watershed types for further comparison and interpretation (Huang et al. 2015).

4.2.1 Effect of Cropland on Water Pollution

The GWR model allows examining the spatial variation of the cropland and water pollution relationship. Cropland was weakly associated with water pollution for all three sub-watershed groups (Fig. 13). All the R^2 values with regards to cropland models are less than 0.4 and most are less than 0.2. However, the relationships between the cropland and water pollutants showed a distinct spatial variation.

The proportion of cropland was strongly correlated with nitrate–N and SRP, especially in agricultural sub-watersheds. The stronger correlation between cropland and water pollution in the agricultural sub-watersheds suggested that cropland may contribute largely to water pollution. This is understandable because of the large amount of fertilizers being applied in the agricultural sub-watersheds with over 10% crop-

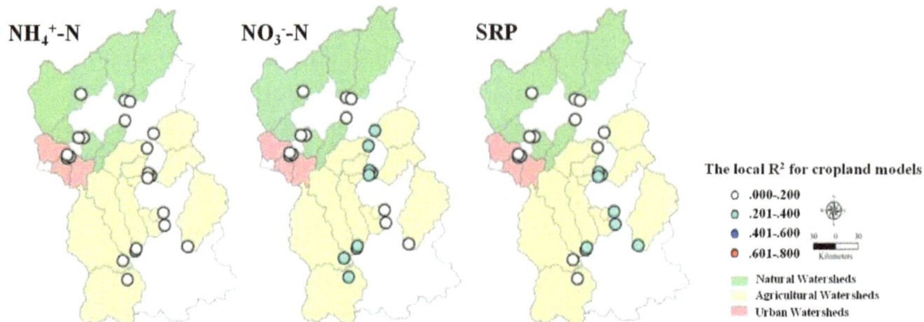

Fig. 13 Local R^2 values for the GWR cropland models among three types of sub-watersheds. The area in yellow, red, and green represents sub-watersheds dominated by agricultural, built-up, and forest land, respectively (*Source* Huang et al. 2015)

land. On the contrary, cropland was less correlated with pollutants in natural and urban sub-watersheds. For natural sub-watersheds, it might be related to the nutrient retention in the forests area that prevents nutrient and sediment leaching, which possibly contribute to the lower concentrations of pollution. Meanwhile, agricultural activities become negligible pollution sources in urban watersheds especially when the proportion of built-up is much higher than that of cropland.

4.2.2 Effect of Built-up Land on Water Pollution

The built-up land was positively associated with water pollution that mainly comes from residential wastewater, municipal discharge, and, industrial effluent. Therefore, increases in proportion of built-up area may exacerbate water pollution. Interestingly, the local built-up land and water pollution relationship was stronger for the agricultural and natural sub-watersheds (Fig. 14). The local R^2 values were greater than 40% for all natural sub-watersheds, which implied the built-up addition greatly increases pollution where both the built-up proportion and pollution are low.

Our analysis suggested that urbanization can greatly affect water pollution in rural areas than in urban areas in the watershed, which may be related to the lower wastewater treatment capacity in rural areas. Besides, the urban watersheds have been already polluted that current water pollutions might be less affected by an additional percent of built-up area. Similar results were revealed by Tu (2011) that the same built-up land increase magnitude can have a stronger influence on water pollution in non-urban sub-watersheds. Finally, we can consider an emergent urbanizing sub-watershed as a critical source area (CSA), often generating a disproportionate amount of pollution (Dickinson et al. 1990; Tripathi et al. 2003). Therefore, implementing the best management practices should target CSAs to ensure effective pollution reduction.

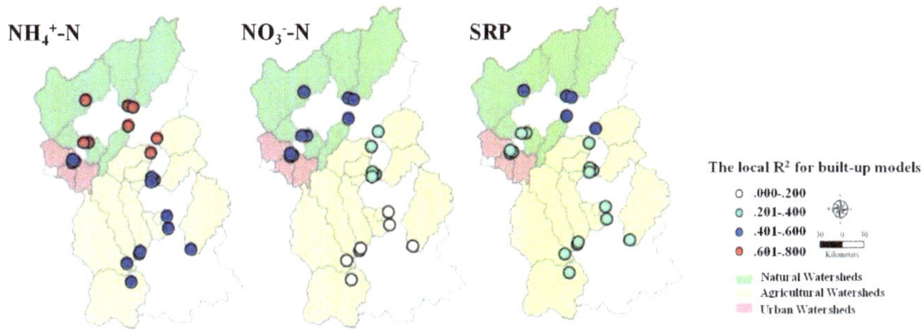

Fig. 14 Local R^2 values for the GWR built-up models among three types of sub-watersheds. The area in yellow, red, and green represents sub-watersheds dominated by agricultural, built-up, and forest land, respectively (*Source* Huang et al. 2015)

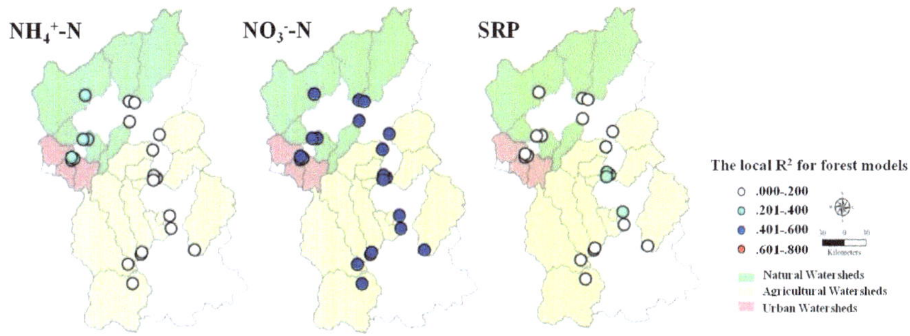

Fig. 15 Local R^2 values for the GWR forest models among three types of sub-watersheds. The area in yellow, red, and green represents sub-watersheds dominated by agricultural, built-up, and forest land, respectively (*Source* Huang et al. 2015)

4.2.3 Forest Land and Water Pollution

Forest is normally linked with low water pollution. The application of GWR in our study showed the forest and water pollution relationship varied spatially (Fig. 15). The forest land proportion had a stronger negative relationship with ammonium–N (NH_4^+–N) and SRP in the urban sub-watersheds. Moreover, it had a stronger negative relationship with nitrate–N (NO_3^-–N) in some agricultural sub-watersheds, indicating a higher forest proportion can lower nitrate concentration and suppress the fertilizers application influence in JRW. This could be a warning for deforestation practices in watershed since it may negatively impact the ecosystem function of the forest on controlling water pollution.

4.3 Self-Organizing Map (SOM) Based Approach

Understanding the land-water association can help better manage land-sourced pollution in coastal watersheds (Zhou et al. 2016). Watershed management strategies to control nutrient pollution were likely to target both point and non-point source pollutions. In a typical agricultural watershed, diffuse N and P pollutions are mainly from agricultural sources, such as fertilizer application and livestock breeding (Smith et al. 2013; UNEP 2014). Meanwhile, domestic and industrial activities were considered as major point-source pollution leading to water quality deterioration in urban sub-watersheds (Wu and Chen 2013). However, how land use could influence water quality when considering point source (PS) pollution simultaneously is not quite clear. To this end, we selected Minjiang River Watershed (MRW), the largest watershed in Fujian Province, to explore the land use and water quality relationship using self-organizing map (SOM) based approach (Zhou et al. 2016). We collected water samples from 139 headwater sub-watersheds including six major land use/cover types, i.e., Built-up, Agriculture, Orchard, Woodland, Unused land, and Water (see Fig. 2).

SOM is an unsupervised learning approach to mine multi-dimensional data in a two-dimensional lattice (Kohonen 2001). The procedure allows us to present the specific land use types and water quality correlations in an intuitive way by adjusting mask value to zero for all the land use types. Furthermore, a total of 139 sampling sites were grouped into six clusters according to six water quality parameters: total nitrogen (TN), nitrate–N (NO_3^-–N), ammonium–N (NH_4^+–N), total phosphorus (TP), soluble reactive phosphorus (SRP), and chemical oxygen demand (COD_{MN}). We further identified the local land-water quality relationship among four clusters with sufficient sample sizes.

The SOM result suggests some patterns of similarity among water quality parameters (Fig. 16). For example, we detected the highest value of N and P pollutants in the low left part of the SOM plane, which were consistent with the pattern of agricultural and built-up land uses. In opposite, the percentage of woodland was lowest at the low left part of the plane, indicating that sampling sites with high forest coverage has relatively low concentration of water quality parameter.

Water quality parameters showed significantly different patterns among the six clusters (Fig. 17). Cluster 1 represented the sampling sites which have lowest concentrations for all the water quality parameters. The sub-watersheds in the cluster 1 cover Dehua, Pucheng, and Taining counties, which of forest coverages rank among the highest in the Minjiang River Watershed. In contrast, the highest nutrient concentrations were observed in sub-watersheds belongs to clusters 4, 5, and 6. According to China's Environmental Quality Standards for Surface Waters (GB 3838-2002), water in clusters 4–6 it is not suitable for drinking water source. The nutrient concentrations in clusters 2 and 3 were less than in clusters 4–6. This could be due to woodland's purification ability and a low-level pollution as a result of moderately intensified agricultural activities.

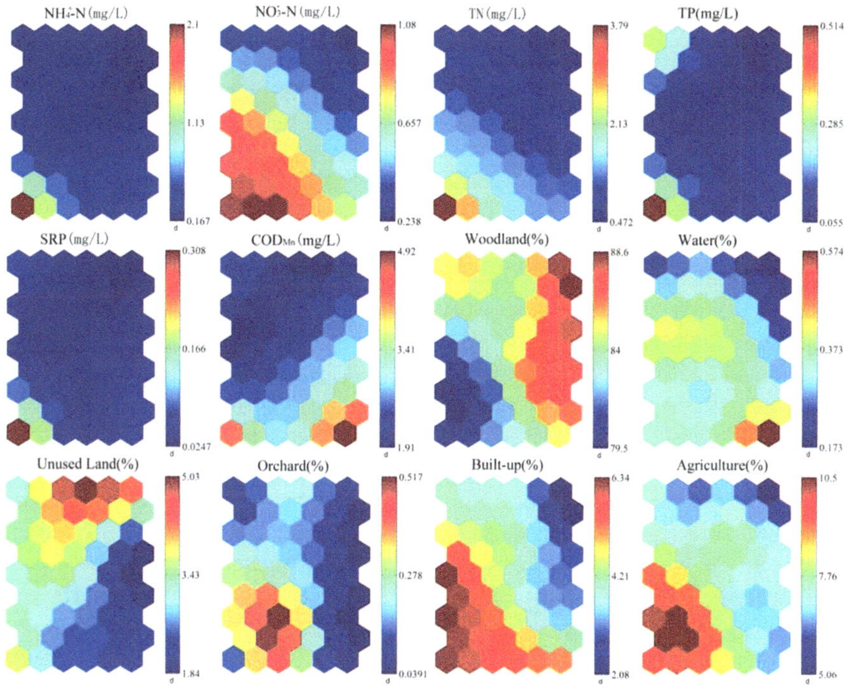

Fig. 16 Visualization of each variable in the trained SOM (self-organizing map). *Note* "d" under legend means denormalization of the SOM results (*Source* Zhou et al. 2016)

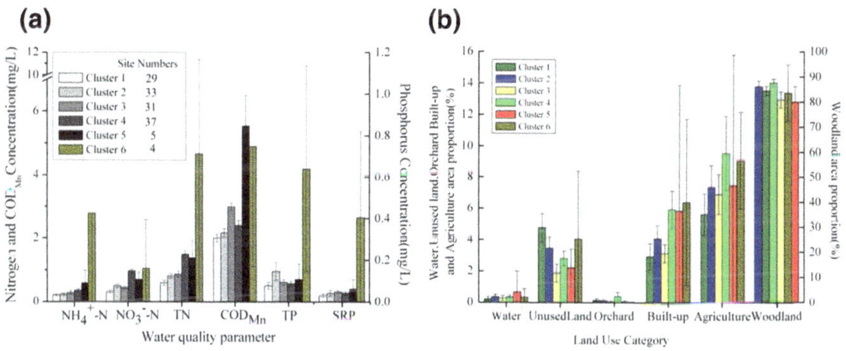

Fig. 17 Average values and 95% confidence intervals for **a** the Water quality parameters and **b** land use categories for each cluster. The Kruskal–Wallis test indicates the differences between water quality parameters among clusters are significant ($p < 0.01$). "Site Numbers" in the legend for panel A indicates the number of sites in each cluster (*Source* Zhou et al. 2016)

The sub-watersheds in clusters 4–6 mainly occupied Yong'an, Sanming, Shaxian, Nanping, and Jianning counties. These regions were characterized by the high built-up land and agricultural land proportions and the low woodland proportion.

The industrial discharges, domestic wastewater, and agricultural pollutants could be the major sources of contaminants in clusters 4–6. Although waste water treatment facilities are available in many cities (e.g. Sanming and Yong'an), however the final products of WWTPs still contaminate the riverine water because of very high pollutant concentrations before treatment (Zhou et al. 2016).

The land use and water quality correlation magnitude varies spatially throughout the watershed (Table 4). In general, forest can absorb nutrients but urban built-up land and farmland are largely related to point and non-point source pollutions. This can help explain the negative correlation between NH_4^+–N, SRP and the woodland proportion in the first cluster and the positive correlation between several pairs including NH_4^+–N and built-up in the first two clusters, TP and built-up in Cluster 4, and agricultural land and NH_4^+–N, SRP in Cluster 1. Comparatively, COD_{Mn} was positively correlated with the woodland proportion and negatively correlated with the built-up, orchard, and unused land proportions in the first cluster (Zhou et al. 2016). This may be due to the facts that the organic matter from decaying plants is refractory while the one from industrial sources decomposes easily after being transported into streams through surface runoffs (Nguyen and Hur 2011; Lee et al. 2014). The woodland increase in the first cluster can lead to the refractory organic matter accumulation, while the orchard, built-up, and unused land increase can help boost the labile organic matter being decomposed at the tributary outlet. Another likely cause is that agricultural activities in the first cluster may help substantially cut down the organic matter loss into the runoff from vegetated lands as compared with those from forest (Yang et al. 2013). Therefore, the orchard increase may help cut down the organic matter loss into the runoff.

We attempted to identify whether point source (PS) or non-point source (NPS) pollution affects water pollution more in the six clusters by excluding the data in each cluster individually and performing backward stepwise regression (Fig. 18). An increased R^2 in the regression models will be an indicator of the point-source pollution affects the water quality in the excluded cluster. Results showed that excluding cluster 1 will increase R^2, but only for TP parameter. Excluding cluster 2 will increase R^2 for all the water quality models while taking out the third cluster will increase R^2 for TN, NO_3^-–N, NH_4^+–N, and SRP. When taking out the fourth cluster 4, R^2 increases for NH_4^+–N, TP, SRP and $COD_{Mn.}$ Taking out the fifth cluster will increase R^2 for TN and NO_3^-–N. Leaving out the sixth cluster will increase R^2 for TN, NO_3^-–N, NH_4^+–N, and COD_{Mn}.

In the first three clusters, both the NPS and PS pollutions are not severe and most of water quality parameters satisfy the cleanliness standard. The sub-watersheds in the first three clusters are mainly located at the counties with less intensive industrial activities and thus low PS-pollution contribution. Moreover, the high woodland percentage and the low built-up, agriculture, and orchard percentages may help explain the low NPS pollution in these clusters.

In the fourth cluster, both NPS and PS pollutions influence river water quality. The point-source pollution affects all parameters, particularly TP, SRP, NH_4^+–N, and COD_{Mn}, which may be caused by intensive industrial activities and domestic waste water discharge in Jianning, Jian'ou, Sanming, Shaxian, and Yong'an counties.

Table 4 Pearson's correlation coefficients between land use classes and water quality parameters

	Woodland	Water	Unused land	Orchard	Built-up	Agriculture
Cluster 1						
NH_4^+–N	−0.345	0.167	−0.098	0.051	**0.474**[b]	0.312
NO_3^-–N	−0.227	0.036	−0.128	−0.047	0.316	0.261
TN	−0.195	0.191	−0.335	−0.156	0.331	0.331
TP	−0.013	0.089	−0.035	0.132	0.178	−0.129
SRP	−0.128	−0.093	**0.431**[a]	−0.034	−0.065	−0.031
COD_{Mn}	**0.436**[a]	0.079	**−0.476**[b]	**−0.505**[b]	**−0.458**[a]	−0.070
Cluster 2						
NH_4^+–N	**−0.415**[a]	−0.151	−0.263	0.074	**0.478**[b]	**0.413**[a]
NO_3^-–N	−0.052	0.288	0.313	0.260	0.037	−0.162
TN	−0.190	0.074	0.018	0.203	0.250	0.074
TP	0.081	−0.206	−0.115	−0.233	−0.019	0.000
SRP	**−0.457**[b]	0.011	−0.126	0.245	0.338	**0.454**[b]
COD_{Mn}	−0.045	0.026	−0.194	−0.141	0.030	0.147
Cluster 3						
NH_4^+–N	0.095	0.224	0.118	0.042	0.114	−0.252
NO_3^-–N	0.018	0.011	0.143	0.107	0.028	−0.105
TN	0.114	0.042	−0.042	−0.020	0.092	−0.169
TP	−0.260	0.191	0.129	−0.169	**0.393**[a]	0.070
SRP	−0.100	**0.516**[b]	0.169	−0.271	0.128	−0.062
COD_{Mn}	0.048	−0.302	0.071	−0.218	−0.118	0.013
Cluster 4						
NH_4^+–N	−0.148	0.148	−0.266	0.061	0.272	0.101
NO_3^-–N	−0.005	−0.200	−0.043	0.301	0.048	−0.033
TN	0.016	0.050	−0.153	0.188	0.072	−0.051
TP	−0.166	0.035	0.053	−0.054	0.085	0.179
SRP	0.032	−0.110	0.129	−0.225	−0.174	0.050
COD_{Mn}	0.070	0.094	−0.123	0.025	−0.189	0.026

Source Zhou et al. (2016)
[a]Indicates significant at $p < 0.05$
[b]Indicates significant at $p < 0.01$
We did not perform the correlation analysis for clusters 5 and 6 because the sample sizes are small

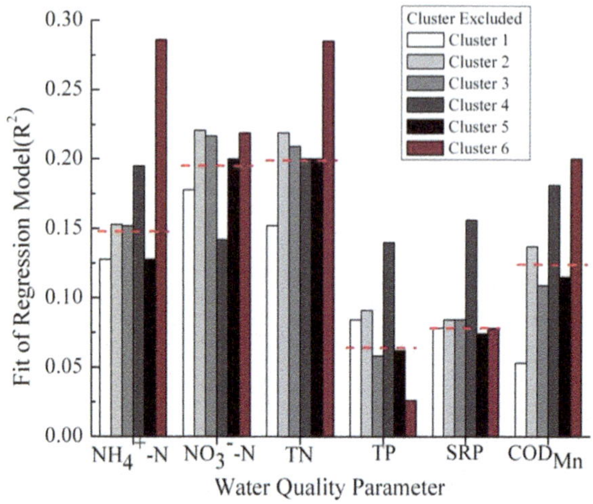

Fig. 18 Fits of the backward stepwise regression models. Legend indicates the excluded cluster when performing the regression. The dashed horizontal lines indicate the R^2 values of the models based on all the data (*Source* Zhou et al. 2016)

Fortunately, the NPS and PS pollution in cluster 4 are not serious enough to make all of the above parameters except TN to be substandard. The possible cause for this may be due to the combined PS pollution from urban and NPS pollution from agriculture, orchard, and built-up.

In the fifth cluster, point-source pollution seems to be significant in determining river water quality. The TN and NO_3^-–N models end with higher R^2 values, an indication of the point-source pollution influence. The sub-watersheds are within several highly urbanized counties, such as Jian'ou and Nanping, where the treated sewage percentage is quite low. We believe that the serious point-source pollution was from the industrial and domestic discharge. However, the non-point source pollution is not serious, largely because of agricultural use and orchards are relatively smaller in terms of their area. Since TN is the only parameter being substandard, it is believed that this has been controlled by the point-source pollution. However, the degree of point-source pollution may not be serious enough to make the NO_3^-–N substandard.

In the sixth cluster, it is the point-source pollution that determines TN, NH_4^+–N, and COD_{Mn}, and non-point source pollution that causes the high values of TP, NO_3^-–N, and SRP. The intensive industrial activities in Gutian and Shaxi counties are responsible for much of the point-source pollution. Meanwhile, the COD_{Mn} concentration was the lowest in this cluster. The COD_{Mn} concentration in the discharge from waste water treatment facilities is standard, but the TN, NH_4^+–N, and TP concentrations surpass the Class V limit (FJEPB 2015). This suggests that large point-based pollution may not necessarily bring about high COD_{Mn} concentration. It seems that the high agricultural land proportion may be responsible for the high TP, NO_3^-–N, and SRP concentrations.

Our results showed that point-source pollution affects the land use and water quality correlation in each cluster. The PS pollutions were less severe in clusters 1–3 and more severe in the other three clusters. The more serious the point-source pollution is, the fewer the pairs of significant land-water correlation exist (Table 4). The most significant correlation pairs were noted in the first cluster but less visible in the fourth cluster. The point-source pollution may overshadow the land-water correlation in the fourth cluster and dilute the correlation in the first three clusters. Therefore, water management should pay more attention on controlling non-point source and point-source pollution in sub-watersheds of the fourth and sixth clusters. For the fifth cluster, attentions should be paid upon controlling the point-source pollution. Although the water quality in the first three clusters meets the cleanliness standard, the point-source pollution still exists.

5 Watershed Modeling and Management

A model is a (simplified) representative of a system allowing to investigate the properties of the system and sometimes predict the future outcome. Watershed modeling has been considered as important and effective tools in water sources management. Historically, water management efforts only focused on single source pollutions, for instance by controlling industrial discharges, sewage discharge, and waste water treatment plant effluents. Very often, addressing single source pollutions was not successful in meeting water quality standard because in some areas, stream pollution might be affected mainly by non-point source pollutions. For this reason, water managers began to raise concerns over the integrated watershed management (IWM) as an important measure to control stream pollution by considering all the values and perspectives among water users or managers. Heathcote (2009) listed six characteristics of a successful IWM: (a) a watershed management plan considering all uses of the water system and other activities affecting water quantity and quality; (b) information about the full hydrological regime for the watershed; (c) an analytical system or model being capable of revealing all the impacts of watershed-wide particular uses and development; (d) specific watershed management objectives with criteria to assess management alternatives objectively; (e) the participation of all relevant regulatory agencies; and (f) provisions for public participation in determining objectives and in management decisions.

The model selection should be based on the simulation goal, the study site scale, the data availability, the expected accuracy, and the temporal and financial costs (Grizzetti et al. 2005; Kovacs 2006; Kliment et al. 2008). Various watershed models have been developed over time, including hydrological modeling and modeling non-point source pollutions. Watershed models enable us to simulate the field-observed stream flow or water quality by incorporating factors and processes controlling river flow and nutrients dynamic, which will be helpful to estimate pollutant loadings, identify critical pollution source areas, and gain insight of natural processes in watershed scale (Pullar and Springer 2000; Borah and Bera 2003). For the Jiulong River watershed, we have constructed at least three watershed modeling tools, namely GIS-based empirical models, AnnAGNPS model, and SPARROW, which will be further discussed in this section. The main objectives of applying these models were to quantify

Table 5 Sources of diffuse total N and total P loadings in Jiulong River watershed

Unit (t yr^{-1})	Loading	Soil losses	Excess fertilizer use	Domestic wastewater	Livestock breeding
Total N	12460.9	1542.2	6649.3	1655.8	2613.6
Total P	1416.9	208.1	578.0	186.0	444.8

Source Huang and Hong (2010)

non-point N and P emissions, to determine the critical sources area of contaminants, and to develop nutrient management strategies (Huang and Hong 2010; Zhou et al. 2018).

5.1 GIS-Based Empirical Modeling

GIS-based empirical modeling can help combine grid-based GIS with three specific empirical models: universal soil loss equation (USLE), soil conservation service curve number (SCS-CN), and nutrient losses equations. Considering the simple model structure and data availability, empirical models were selected to model nutrient TN and TP losses processes in the Jiulong River watershed. Generally, the potential TN and TP loading values for the whole watershed were estimated by aggregating each cell into the unit of each land use/cover type. The data input, variables, and equations for GIS-based empirical are illustrated in more detail in Huang and Hong (2010).

The soil losses, fertilizer uses, rural domestic waste water, and livestock breeding are the major non-point sources pollutants associated with agricultural activities in the Jiulong River Watershed (Table 5). Our results indicate that the sources of TN loadings in descending order by percentage of their contributions were excessive fertilizer use, livestock breeding, rural domestic wastewater, and soil losses, which accounted for 53.4%, 21.0%, 13.3, and 12.4%, respectively. As for TP loadings, the contributions of the above items were estimated to be 40.8%, 31.4%, 14.7% and 13.1%, respectively. To summarize, fertilizer use and livestock breeding are important sources of TN and TP loadings in the Jiulong River Watershed (Table 5).

The GIS-based empirical model has the advantage to facilitate identification of critical source areas and source apportionment of diffuse TN and TP pollutions for the administrative units in the Jiulong River watershed. Our analysis indicated that sources and contributors for TN and TP loading vary across counties/districts (Fig. 19). For example, in Xinluo district, livestock breeding and soil losses were identified as the main TN and TP sources, accounting for 40.6% and 24.2% of TN, and 51.1% and 24.0% of TP, respectively. For Zhangping, fertilizer use and soil losses were the major TN sources, which contribute 38.8% and 29.4%, respectively. As for TP sources in Zhangping, soil losses and excessive fertilizer uses accounted for 34.1% and 27.8%, respectively. Fertilizer use and livestock breeding made a

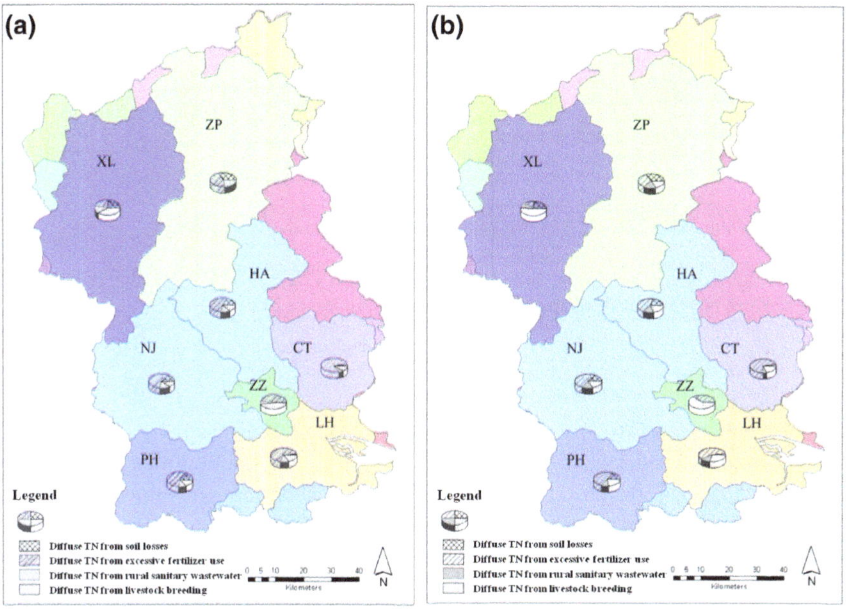

Fig. 19 Sources and contributors of N from NSP for major administrative units in Jiulong River Watershed **a** TN and **b** TP. *PH* Pinghe, *NJ* Nanjing, *LH* Longhai, *ZZ* Zhangzhou, *CT* Changtai, *HA* Hua'an, *ZP Zhangping*, *XL* Xinluo (*Source* Huang and Hong 2010)

major contribution to TN and TP loadings in Zhangzhou district, which contributed more than 50%. Similarly, excessive fertilizer use was the main TN sources in other counties including Changtai, Hua'an, Longhai, Nanjing, and Pinghe, accounting for over 50% of the TN loading.

5.2 *The AnnAGNPS Method*

The Annualized Agricultural Non-Point Source (AnnAGNPS) pollutant model is a physically based and continuous model for predicting surface runoff, suspended load, and nutrients in daily steps and it has been widely used to develop management alternatives of non-point source pollution in medium-sized river basins. The pollutant loading surface runoff sub-model within the AnnAGNPS simulates transport of soluble nitrogen and particulate forms of phosphorus using routines derived from the CREAMS model (Knisel 1980). We applied AnnAGNPS model to simulate TN and TP loadings at the downstream of the North River and West River reaches of the Jiulong River watershed, which were delineated into 2351 and 908 drainage areas, respectively. The predicted TN and TP loadings at the outlet of the watershed were

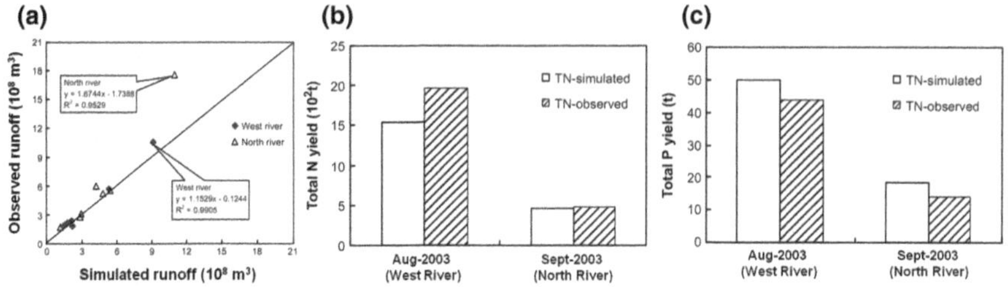

Fig. 20 Comparison of simulated and observed **a** runoff, **b** TN, and **c** TP for model validation for West river and North river watersheds (*Source* Huang and Hong 2010)

important indicators in the management alternative simulation for NPS controlling under changing land-use conditions (Huang and Hong 2010).

AnnANGPS model has the advantage in simulating the complex processes of non-point-source pollutions, including surface flow, sediment, and nutrient transport and retention, and therefore it provides detailed TN and TP results. Results showed that the surface runoff was simulated well by AnnANGPS model, given the deviation errors of $\pm 20\%$ and R^2 values of 0.95 and 0.99, for North River and West River, respectively. Meanwhile, the nitrogen and phosphorus export prediction was acceptable, with the average error of $\pm 30\%$ (see Fig. 20). Based on AnnAGNPS simulation results, the annual TN loadings were 10.28 kg hm^{-2} yr^{-1} and 24.76 kg hm^{-2} yr^{-1} in the North River and West River, respectively, and annual TP loadings were 0.4 kg hm^{-2} yr^{-1} and 0.67 kg hm^{-2} yr^{-1}, respectively.

Furthermore, AnnANGPS combined with scenario analysis were simulated to find suitable management alternatives for controlling non-point source pollutions in the Jiulong River watershed in response to land-use transformation. The changing land-use scenarios were simulated separately at the typical sub-watersheds, North River, and West River. First simulation at grid cell ID 92, namely Tianbao and Xiandu sub-watersheds showed that reforestation on the main slope might cause reduction in surface runoff, TN, and TP loadings by 21.6%, 96%, and 79.2%, respectively. Second simulation in the North River suggested that TN will decrease by 63.54% after removing all the hoggeries. Third simulation in the West River evaluated the impact of the cultivation plant change from banana to rice on TN and TP loadings, which were reduced by 23.83% and 9.08%, respectively.

5.3 SPARROW

SPARROW (SPAtially Referenced Regressions On Watershed attributes) is a hybrid statistical and process-based model (Schwarz et al. 2006). It has been widely used to identify the nutrient sources and transport processes at the watershed scale (e.g., Alexander et al. 2002; Moore et al. 2011; Saleh and Domagalski 2015; Kim et al. 2017). SPARROW's spatially detailed networks and travel time data can be used to

Table 6 Non-linear least square results of SPARROW TN and TP model in the Jiulong River watershed (JRW)

Parameter	R^2	R^2_{Adj}	R^2_{Yield}	MSE
TN	0.95	0.93	0.67	0.14
TP	0.94	0.92	0.65	0.15

Source Zhou et al. (2018)

estimate in stream-loss rates and also retain the detailed spatial information about all environmental factors used in the regression model. Generally, SPARROW modeling is used to estimate the TN and TP sources and transport within the JRW, which can be used to develop nutrient management strategies (Zhou et al. 2018). The model was calibrated with observation data at 30 locations across the watershed, including daily streamflow, monthly TN and TP concentrations in 2014. Results revealed that SPARROW TN and TP load models performed satisfactorily, with the R^2 values of 0.95 and 0.94, and MSE values of 0.14 and 0.15, respectively (Table 6).

The major nitrogen sources identified include sewage discharge, fertilizers, and livestock breeding (Table 7). Model coefficients were within the 95% confidence interval. Comparatively, the major phosphorus sources include industrial discharge, fertilizers application, and livestock breeding in the JRW (Table 8). Among the three sources, industrial discharge exhibited a p-value greater than 0.05, which mean that industrial discharge was not an important TP sources.

The TN model's land delivery factors consist of air temperature, terrain slope, and stream density, while water delivery factors include reach decay factor and reservoir decay factor (Table 7). Model coefficients lay within the 95% confidence interval. However, only air temperature was statistically significant ($p<0.05$).

As for the TP model, land-water delivery factors consist of precipitation, sand proportion, soil pH, reach decay, and reservoir decay. Model coefficients lay within the 95% confidence interval. Among all factors, sandy soil was the only factor with p-value of less than 0.05, showing that the sand proportion was the most important land delivery factor.

The total TN and TP fluxes were divided by watershed area to be expressed as the incremental TN and TP yields reflect the intensity of anthropogenic activities upon land, which can help identify the critical source areas with high pollution productivity. The areas with high TN yields included Punan, Huaan, Wanan, and Yanshi, while the areas with high TP yields included Punan, Longjin, Wanan-Yanshi, and Yanshi (Fig. 21). Generally, North River had relatively low TN and TP yields compared with West River, South River, and Longhai sections.

Based on SPARROW TN modeling, fertilizer application, livestock breeding, and sewage discharge were identified as the major TN pollution sources in the entire watershed, accounting for 43%, 39%, and 18%, respectively. In comparison, the major TP sources came from livestock breeding, fertilizer application, and industrial discharge with contribution equal to 46%, 46%, and 8%, respectively (Fig. 22). The major TN pollution sources varied among the sections in North River, e.g. sewage discharge in Punan and Longyan, livestock breeding in Wan'an, Yanshi,

Table 7 Evaluation of parameters of the SPARROW_TN model in the Jiulong River watershed (JRW)

	Model coefficient (NLLS)	Standard error of coefficient	p-value	Nonparametric bootstrap estimate of coefficient (Mean)	Lower 95% confidence interval for coefficient	Upper 95% confidence interval for coefficient
Sewage discharge	1.47	2.32	0.53	1.77	1.40	2.13
Fertilizer application	0.36	0.13	0.01	0.37	0.34	0.40
Livestock breeding	5.36	4.08	0.21	5.58	4.80	6.36
Slope of the terrain (°)	4.09	3.75	0.29	4.47	3.70	5.24
Air temperature (°C)	−7.61	2.81	0.02	−7.32	−7.90	−6.75
Drainage density (1/m)	2.75	2.87	0.35	3.55	3.00	4.10
Reach decay	0.0073	0.15	0.96	0.0046	−0.025	0.034
Reservoir decay	6.26	3.87	0.12	7.07	6.26	7.87

Source Zhou et al. (2018)

and Wan'an-Yanshi. Conversely, fertilizer application was identified as the main TN source in the West River. For the South River, the major TN sources consisted of fertilizer application and sewage discharge. Similar to TN pollution sources, the major TP sources also varied among the sections, e.g. livestock breeding and fertilizer application in Wa'nan, Yanshi, and Wan'an-Yanshi, fertilizer application in Hua'an, industrial discharge in Longjin and Punan, fertilizer application and livestock breeding in Yongfeng and Chuanchang, fertilizer application in Huashan and South River.

The main objective of SPARROW model development was to select appropriate management strategies for controlling nutrient pollutions in the JRW. For this reason, we designed the three pollution control scenarios with the main goal of making over 70% reaches to meet the surface water quality standard (class III) for TN and TP by 2020.

Table 8 Evaluation of parameters of the SPARROW_TP model in the Jiulong River watershed (JRW)

	Model coefficient (NLLS)	Standard error of coefficient	p-value	Nonparametric bootstrap estimate of coefficient (Mean)	Lower 95% confidence interval for coefficient	Upper 95% confidence interval for coefficient
Industrial discharge	0.057	0.12	0.64	0.098	0.064	0.13
Fertilizer application	0.036	0.011	0.007	0.032	0.028	0.037
Livestock breeding	0.14	0.004	0.005	0.15	0.13	0.17
Sand percentage	−6.54	2.96	0.04	−5.62	−6.74	−4.50
Precipitation	1.44	3.95	0.72	1.82	0.16	3.48
Soil pH	−0.32	0.23	0.18	−0.24	−0.33	−0.16
Reach decay	−0.061	0.18	0.74	−0.079	−0.14	−0.017
Reservoir decay	1.73	1.51	0.26	2.03	1.58	2.50

Source Zhou et al. (2018)

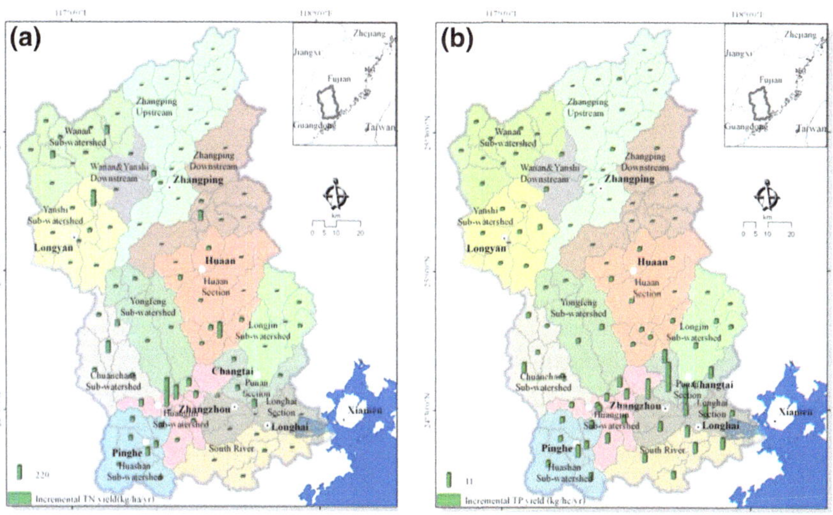

Fig. 21 Yield of **a** TN and **b** TP in each sub-watershed in the Jiulong River watershed (*Source* Zhou et al. 2018)

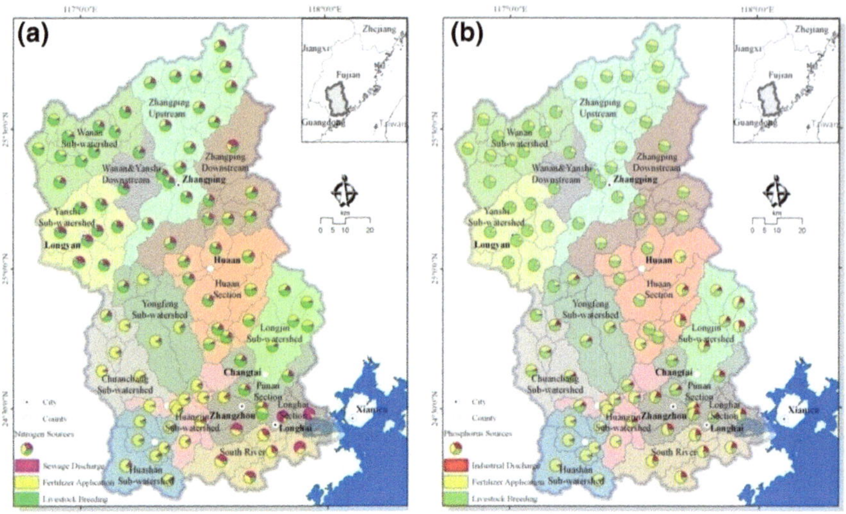

Fig. 22 Pollution sources of **a** nitrogen and **b** phosphorus at the sub-watersheds in the Jiulong River watershed (*Source* Zhou et al. 2018)

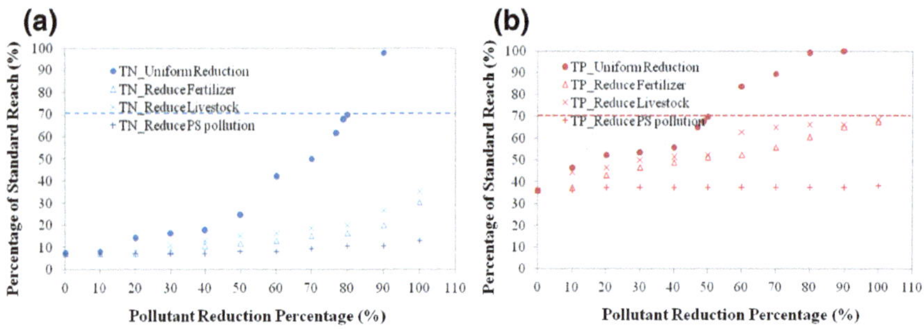

Fig. 23 Relationship between standard **a** TN and **b** TP proportion and pollution control (*Source* Zhou et al. 2018)

5.3.1 Scenario 1: Uniform Pollution Reduction

Here, the TN or TP pollution sources were uniformly cut down. Scenario analysis suggested that we need to reduce N and P pollutants in 2014 by 80% and 50%, respectively, in order to achieve the goal that 70% reaches could meet water quality standard (Fig. 23). That is to say that after 2014 we should cut down 13.3% of TN and 8.3% of TP at the yearly basis.

5.3.2 Scenario 2: Single Source Reduction

In Scenario 2, water quality improvement was expected by reducing each pollutant individually (Fig. 23). Results show the reduction of sewage discharge, fertilizer application, and livestock breeding by less than 10% each had a similar impact on increasing the standard TN proportion. When pollutant reduction was above 10%, the TN source reduction effectiveness declined in the order of livestock breeding, fertilizer application, and sewage discharge. If all TN pollutants from livestock breeding, fertilizer application, and sewage discharge were cut off, the reaches with standard TN were increased from 7% to 35%, 30%, and 13%, respectively (Fig. 23a).

Similarly, focusing on the reduction of the TP pollutant from livestock breeding became more effective in improving the standard TP reaches than from fertilizer application (Fig. 23b). If all TP pollutants from livestock breeding, fertilizer application, and industrial discharge were completely reduced, the reaches with standard TP were increased from 36% to 67%, 69% and 38%, respectively. It is clear that targeting a single pollution source reduction may hardly reach the goal with over 70% reaches being TN or TP standard. Thus, we need to target various sources in pollution reduction simultaneously.

5.3.3 Scenario 3: Simultaneous Major Source Reduction

Scenario 3 was designated to obtain an optimal strategy to improve water quality through simultaneous reductions of the major nutrient pollution sources, i.e. livestock breeding and fertilizer application. It is important to note that the industrial discharge and sewage discharge were assumed to be constant in this scenario. According to the scenario 3 results, it is impossible to accomplish a goal, of which 70% reaches meet standard TN, unless the reductions of fertilizer application and livestock breeding reduction are more than 90% and 65%, respectively. More specifically, the optimization for controlling TN pollution is to reduce fertilizer application by 90–95% and livestock breeding by 85%. Comparatively, the optimization for controlling TP pollution is to cut down the TP sources from fertilizer application by 55–50% and by approximately 50% from livestock breeding (Fig. 24).

6 Implications for Watershed Management

Environmental policies concerning water pollution problems, such as Clean Water Act, Water Framework Directive, Action Plan for Preventing and Treatment of Water Pollution have been established by many countries like United States, European Union, and China. These policies primarily aimed at managing water resources, controlling water pollution, and ensure a sustainable use of water. The Jiulong River watershed has encountered multiple anthropogenic disturbances resulting from economic development, urbanization, and population growth. Anthropogenic distur-

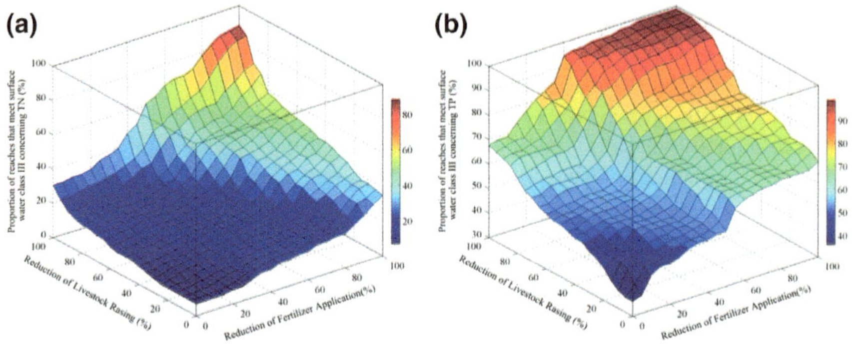

Fig. 24 Proportions of reaches that meet surface water class III concerning **a** TN and **b** TP when controlling livestock breeding and fertilizer application simultaneously (*Source* Zhou et al. 2018)

bances which include land changes, flow regulation, agricultural practices, and urban development, inevitably lead to water quality degradation and more frequent harmful algal blooms (HABs) in Jiulong River by various extents. According to Water Source Office of Fujian Province, nearly sixteen HABs events occurred in the Jiulong River Estuary and the Xiamen-Kinmen coastal waters between 1997 and 2003 (Office of Water Source of Fujian 2004). Six years later, a serious dinoflagellate bloom recurred in Jiulong River, because of excessive nutrient loads from livestock wastes cleanup. The algal bloom incidents pose a high risk to drinking water system safety of the Xiamen City, and therefore, management strategies for nutrient pollution control can be crucial to the environmental and human well-being in the study site. While current policies put much effort on reducing nutrient sources, either from non-point sources (e.g. N fertilizer use) or point sources (e.g. industrial discharge), we suggest that proper watershed spatial planning should be made to ensure spatially optimization of landscape elements so as to improve water quality in the developing watersheds. Our analysis revealed that landscape pattern and climate variability are the two major influencing factors for water quality variation. Cropland had strong positive correlations with nutrient concentrations, particularly in agricultural sub-watersheds, indicating that nutrient sources mainly came from agricultural activities. Built-up land also had strong positive correlations with most of the water pollutants, particularly in sub-watersheds with low proportion of built-up land, suggesting that management should ensure that new urbanization in rural places apply the best management practices for water pollution. Comparatively, the forest land percentage was negatively correlated with nutrient concentrations, indicating that higher forest coverage can reduce water pollution. For pollution control, the deforestation issue should be considered and more efforts should be made for forest restoration. Furthermore, the scenario analysis from SPARROW modeling offers a new perspective on watershed nutrient management by controlling the nutrient sources and reducing nutrient transportation. According to Action Plan for Preventing and Treatment of Water Pollution, about 85% of TN and 50% of TP loads in the JRW have exceeded standard set by the government. The optimized strategies to meet such a requirement were achieved by reducing TN by 90–95% from fertilizer application and 85% from livestock breeding and reducing TP by 55–60% from fertilizer application and 50% from livestock breeding. However, such massive reductions in fertilizer use and livestock breeding are not operationally feasible in the JRW, a typical agricultural watershed, because it would affect livelihood of farmers. Since transportation process also control the amount of nutrient at river system, reducing nutrient transportation should be encouraged. Some possible strategies should focus on the improvement of fertilization efficiency and the optimization of the land-water delivery processes. More specifically, livestock breeding should not be allowed near streams, more vegetation should be planned along river banks, and more waste water treatment plants should be constructed.

References

Alexander RB, Elliott AH, Shankar U et al (2002) Estimating the sources and transport of nutrients in the Waikato River Basin, New Zealand. Water Resour Res 38(12):1268, 4.1–4.23

Altman SJ, Parizek RR (1995) Dilution of nonpoint-source nitrate in groundwater. J Environ Qual 24(4):707–718

Baker A (2003) Land use and water quality. Hydrol Process 17(12):2499–2501

Borah DK, Bera M (2003) Watershed-scale hydrologic and nonpoint-source pollution models: Review of mathematical bases. Trans ASAE 46(6):1553–1566

Bowes MJ, Hilton J, Irons GP et al (2005) The relative contribution of sewage and diffuse phosphorus sources in the River Avon catchment, southern England: implications for nutrient management. Sci Total Environ 344(1–3):67–81

Brunsdon C, Fotheringham S, Charlton M (1998) Geographically weighted regression-modeling spatial non-stationarity. Statistician 47:431–443

Cao WZ, Hong HS, Zhang YZ et al (2006) Anthropogenic nitrogen sources and export in a village-scale catchment in Southeast China. Environ Geochem Health 28(1–2):45–51

Carpenter SR, Caraco NF, Correll DL et al (1998) Nonpoint pollution of surface waters with phosphorus and nitrogen. Ecol Appl 8(3):559–568

Chang H, Kwon WT (2007) Spatial variations of summer precipitation trends in South Korea, 1973–2005. Environ Res Lett 2(4):045012

Dickinson WT, Rudra RP, Wall GJ (1990) Targeting remedial measures to control non-point source pollution. Water Resour Bull 26(3):499–507

Ding H, Wang R (2006) Cropland change analysis in Zhejiang coastal region. J Zhejiang Univ (Agriculture & Life Science) 32(5):585–590 (in Chinese)

Duan SW, Zhang S, Huang HY (2000) Transport of dissolved inorganic nitrogen from the major rivers to estuaries in China. Nutr Cycl Agroecosys 57:13–22

Duan SW, Xu F, Wang LJ (2007) Long-term changes in nutrient concentrations of the Changjiang River and principal tributaries. Biogeochem 85:215–234

Duan SW, Kaushal SS, Groffman PM et al (2012) Phosphorus export across an urban to rural gradient in the Chesapeake Bay watershed. J Geophys Res 117:G01025

Dumont E, Harrison JA, Kroeze C et al (2005) Global distribution and sources of dissolved inorganic nitrogen export to the coastal zone: Results from a spatially explicit global model. Global Biogeochem Cycles 19:255–268

Environmental Monitoring Bureau of China (2008) National Environmental Quality. http://www.china.com.cn/environment/2009-01/16/content_17140320_2.htm. Accessed 16 Aug 2012

Fedorko E, Pontius RG, Aldrich SP et al (2005) Spatial distribution of land type in regression models of pollutant loading. J Spat Hydrol 5(20):60–80
Fukushima T, Takahashi M, Matsushita B et al (2007) Land use/cover change and its drives: a case in the watershed of Lake Kasumigaura, Japan. Landsc Ecol Eng 3:21–31
Griffith JA (2002) Geographic techniques and recent applications of remote sensing to landscape-water quality studies. Water Air Soil Poll 138:181–197
Griffith JA, Martinko EA, Whistler JL et al (2002) Preliminary comparison of landscape pattern-Normalized difference vegetation index (NDVI) relationships to central plains stream conditions. J Environ Qual 31:846–859
Grizzetti B, Bouraoui F, Marsily GD (2005) Modelling nitrogen pressure in river basins: A comparison between a statistical approach and the physically-based SWAT model. Phys Chem Earth 30:508–517
Han J, Hayashi Y, Cao X et al (2009) Evaluating land-use change in rapidly urbanizing China: case study of Shanghai. J Urban Plan D-ASCE 135(4):166–171
Heathcote IW (2009) Integrated Watershed Management: principle and practice. Wiley, Hoboken, New Jersey
Hill AR (1996) Nitrate removal in stream riparian zones. J Environ Qual 25(4):743–755
Hobbie JE (2000) Estuarine science: The key to progress in coastal ecological research, executive summary. Island Press, Washington DC
Hu XS, Wu CZ, Hong W et al (2014) Forest cover change and its drivers in the upstream area of the Minjiang river, China. Ecol Indic 46:121–128
Huang JL, Hong HS, Zhang LP (2004) Study on predicting soil erosion in Jiulong River watershed based on GIS and USLE. J Soil Water Conserv 18(5):75–79 (in Chinese)
Huang JL, Hong HS (2010) Comparative study of two models to simulate diffuse nitrogen and phosphorus pollution in a medium-sized watershed, southeast China. Estuar Coast Shelf Sci 86:387–394
Huang JL, Pontius RG, Li QS et al (2012) Use of intensity analysis to link patterns with processes of land change from 1986 to 2007 in a coastal watershed of southeast China. Appl Geogr 34:371–384
Huang JL, Li QS, Pontius RG et al (2013a) Detecting the dynamic linkage between landscape characteristics and water quality in a subtropical coastal watershed, southeast China. Environ Manage 51:32–44
Huang JL, Lin J, Zhang YZ et al (2013b) Analysis of phosphorus concentration in a subtropical river basin in China: implications for River Basin Management. Ocean Coast Manag 81:29–37
Huang JL, Huang YL et al (2014) Coupled effects of natural and anthropogenic controls on seasonal and spatial variations of river water quality during baseflow in a coastal watershed of southeast. China. PLOS ONE 3e91528:1–19
Huang JL, Huang YL, Pontius RG Jr et al (2015) Geographically weighted regression to measure spatial variations in correlations between water pollution versus land use in a coastal watershed. Ocean Coast Manag 103:14–24
Huang YL, Huang JL, Ervinia A et al (2018) Land use and climate variability amplifies watershed nitrogen exports in coastal China. Ocean Coast Manag (in press), https://doi.org/10.1016/j.ocecoaman.2018.02.024
Ji ZG (2008) Hydrodynamics and water quality: modeling rivers, lakes, and estuaries. Wiley, Hoboken, NJ
Johnson L, Richards C, Host G et al (1997) Landscape influences on water chemistry in Midwestern stream ecosystems. Freshw Biol 3:193–208
Kim DK, Kaluskar S, Mugalingam S et al (2017) A Bayesian approach for estimating phosphorus export and delivery rates with the SPAtially referenced regression on watershed attributes (SPARROW) model. Eco Inform 37:77–91
Kliment Z, Kadlec J, Langhammer J (2008) Evaluation of suspended load changes using AnnAGNPS and SWAT semi-empirical erosion models. CATENA 73:286–299
Knisel (1980) CREAMS: A field scale model for chemical, runoff, and erosion from agricultural management systems. USDA, conservation research report 26

Kohonen T (2001) Self-organizing maps. Springer, Berlin
Kovacs A (2006) Comparative study of two watershed scales to calculate diffuse phosphorus pollution. Water Sci Technol 53(2):281–288
Lambin EF, Turner BL, Geist HJ et al (2001) The cause of land-use and land-cover change: moving beyond the myths. Global Environ Chang 11:261–269
Lee Y, Hur J, Shin KH (2014) Characterization and source identification of organic matter in view of land uses and heavy rainfall in the Lake Shihwa, Korea. Mar Pollut Bull 84:322–329
Lewis WM (2002) Yield of nitrogen from minimally disturbed watersheds of the United States. Biogeochem 57–58:375–385
Li X, Peterson J, Liu G et al (2001) Assessing regional sustainability: the case of land use and land cover change in the middle Yiluo catchment of the Yellow River Basin, China. Appl Geogr 21:87–106
Li SY, Xia XL, Tan X, Zhang QF (2013) Effects of catchment and riparian landscape setting on water quality chemistry and seasonal evolution of water quality in the upper Han River Basin, China. PLoS ONE e53163
Liu J, Diamond J (2005) China's environment in a globalizing world. Nature 435:1179–1186
Manandhar R, Odeh IOA, Pontius RG (2010) Analysis of twenty years of categorical land transitions in the Lower Hunter of New South Wales, Australia. Agr Ecosyst Environ 135:336–346
May L, House WA, Bowes M et al (2001) Seasonal export of phosphorus from a lowland catchment: upper River Cherwell in Oxfordshire, England. Sci Total Environ 269(1–3):117–130
Mehaffey MH, Nash MS, Wade TG et al (2005) Linking land cover and water quality in New York City's water supply watersheds. Environ Monit Assess 107(1–3):29–44
Ministry of Environmental Protection of the People's Republic of China (2002) Environmental quality standards for surface water (GB 3838–2002)
Ministry of Environmental Protection of the People's Republic of China (2011) Report on the state of the environment In China 2009. http://english.mep.gov.cn/standards_reports/soe/soe2009/201104/t20110411208976.htm. Accessed 6 Sept 2011
Moore RB, Johnston CM, Smith RA et al (2011) Source and delivery of nutrients to receiving waters in the Northeastern and Mid-Altantic regions of the United States. J Am Water Resour Assoc 47:965–990
Morse NB, Wollheim WM (2014) Climate variability masks the impacts of land use change on nutrient export in a suburbanizing watershed. Biogeochem 121:45–59
Mouri M, Takizawa S, Oki T (2011) Spatial and temporal variation in nutrient parameters in stream water in a rural-urban catchment, Shikoku, Japan: effects of land cover and human impact. J Environ Manage 92:1837–1848
Nguyen HVM, Hur J (2011) Tracing the sources of refractory dissolved organic matter in a large artificial lake using multiple analytical tools. Chemosphere 85:782–789
Office of Water Source of Fujian 2004. Water resource report
Ouyang Y, Nkedi-Kizza P, Wu QT et al (2006) Assessment of seasonal variations in surface water quality. Water Res 40(20):3800–3810
Pauwels H, Lachassagne P, Bordenave P et al (2001) Temporal variability of nitrate concentration in a schist aquifer and transfer to surface waters. Appl Geochem 16(6):583–596
Petry J, Malcolm IA, Youngson AF (2002) Hydrological controls on nutrient concentrations and fluxes in agricultural catchments. Sci Total Environ 294(1–3):95–110
Pratt B, Chang H (2012) Effects of land cover, topography, and built structure on seasonal water quality at multiple spatial scales. J Hazard Mater 209–210:48–58
Pullar D, Springer D (2000) Toward integrating GIS and catchments models. Environ Model Softw 15:451–459
Rhodes AL, Newton RM, Pufall AA (2001) Influence of land use on water quality of a diverse New England watershed. Environ Sci Technol 35:3640–3645
Rothenberger MB, Burkholder JM, Brownie C (2009) Long-term effects of changing land use practices on surface water quality in a coastal river and lagoonal estuary. Environ Manage 44:505–523

Saleh D, Domagalski J (2015) SPARROW modeling of nitrogen sources and transport in rivers and streams of California and adjacent states, U.S. American Water Resources Association. J Am Water Resour Assoc 51:1487–1507

Sarma PK, Lahkar BP, Ghost S et al (2008) Land use and land cover change and future implication analysis in Manas National Park, India using multi-temporal satellite data. Curr Sci 95(2):223–227

Schwarz GE, Hoos AB, Alexander RB et al (2006) The SPARROW surface water-quality model: theory, application and user documentation. U.S. Geological survey, techniques and methods 6-B3

Seto KC, Kaufmann RK (2003) Modeling the drivers of urban land use change in the Pearl River Delta, China: integrating remote sensing with socioeconomic data. Land Economics 79(1):106–121

Sliva L, Williams DD (2001) Buffer zone versus whole catchment approaches to studying land use impact on river water quality. Water Res 35(14):3462–3472

Smith AJ, Thomas RL, Nolan JK et al (2013) Regional nutrient thresholds in wadeable streams of New York State protective of aquatic life. Ecol Indic 29:455–467

Somura H, Takeda I, Amold JG, Mon Y, Jeong J et al (2012) Impact of suspended sediment and nutrient loading from land uses against water quality in the Hii River basin, Japan. J Hydrol 450:25–35

Swaney DP, Humborg C, Emeis K et al (2012) Five critical questions of scale for the coastal zone. Estuar Coast Shelf Sci 96:9–21

Tripathi MP, Panda RK, Raghuwanshi NS (2003) Identification and prioritization of sub-watersheds for soil conservation management using SWAT model. Biosyst Eng 85(3):365–379

Tu J, Xia Z, Clarke KC et al (2007) Impact of urban sprawl on water quality in Eastern Massachusetts, USA. Environ Manage 40:183–200

Tu J (2011) Spatially varying relationships between land use and water quality across an urbanization gradient explored by geographically weighted regression. Appl Geogr 31(1):376–392

Tu J, Xia ZG (2008) Examining spatially varying relationships between land use and water quality using geographically weighted regression I: model design and evaluation. Sci Total Environ 407(1):358–378

Turner RE, Rabalais NN (1991) Changes in Mississippi River Water Quality This Century. Bio Sci 41:140–147

UN General Assembly (2004) Oceans and the Law of the Sea, Report of the Secretary-General of 18 August 2004. (A/59/62/Add.1, 29, para. 9. http://www.un.org/Depts/los/general_assembly/general_assembly_reports.htm. Accessed 17 Oct 2015

UN WWAP World Water Assessment Programme (2009) Water in a changing world. The United Nations World Water Development Report 3

UNEP (2014) In: Bringezu S, Schütz H, Pengue W, O'Brien M, Garcia F, Sims R, Howarth R, Kauppi L, Swilling M, Herrick J (eds), Assessing global land use: balancing consumption with sustainable supply. A report of the working group on land and soils of the international resource panel

Uuemaa E, Roosaare J, Mander U (2007) Landscape metrics as indicators of river water quality at catchment scale. Nordic Hydrol 38(2):125–138

Wang X, Zheng D, Shen Y (2008) Land use change and its driving forces on the Tibetan Plateau during 1990–2000. CATENA 72:56–66

White JR, Hendrickson J, Conkle JL (2008) Effects of changing land use on nutrient loads and water quality in a Southeastern US Blackwater River Estuary. In: Fares A, El-Kadi AI (eds) Coastal watershed management. WIT Press, Southampton, Boston, pp 199–218

Willett VB, Reynolds BA, Stevens PA et al (2004) Dissolved organic nitrogen regulation in freshwaters. J Environ Qual 33(1):201–209

Wilson C, Weng Q (2010) Assessing surface water quality and its relation with urban land cover changes in the Lake Calumet area, Greater Chicago. Environ Manage 45:1096–1111

Withers PJA, Jarvie HP (2008) Delivery and cycling of phosphorus in rivers: a review. Sci Total Environ 400:379–395

Wu Y, Chen J (2013) Investigating the effects of point source and nonpoint source pollution on the water quality of the East River (Dongjiang) in South China. Ecol Indic 32:294–304
Xiao H, Ji W (2007) Relating landscape characteristics to non-point source pollution in mine waste-located watersheds using geospatial techniques. J Environ Manage 88:529–551
Yang Y, He Z, Wang Y et al (2013) Dissolved organic matter in relation to nutrients (N and P) and heavy metals in surface runoff water as affected by temporal variation and land uses-a case study from Indian River Area, south Florida, USA. Agric Water Manag 118:38–49
Zhang ZY, Huang JL, Huang YL et al (2015) Streamflow variability response to climate change and cascade dams development in a coastal China watershed. Estuary Coast Shelf 166:209–2017
Zhou P, Huang JL, Pontius RG et al (2014) Land classification and change intensity analysis in a coastal watershed of Southeast China. Sensors 14:11640–11658
Zhou P, Huang JL, Pontius RG et al (2016) New insight into the correlations between land use and water quality in a coastal watershed of China: does point source pollution weaken it? Sci Total Environ 543:591–600
Zhou P, Huang JL, Hong HS (2018) Modeling nutrient sources, transport and management strategies in a coastal watershed, Southeast China. Sci Total Environ 610–611:1298–1309
Zuo LJ, Xu JY, Zhang ZX et al (2011) Spatial temporal land use change and landscape response in Bohai Sea coastal zone area. J Remote Sens 15(3):604–620 (in Chinese)

Dr. Jinliang Huang is a Full Professor of Environmental Geography at the Xiamen University, China. His research interests include watershed monitoring, modeling and management. Huang has authored or co-authored two books and over 70 articles in these areas.

Ayu Ervinia is currently a doctoral student majoring in Environmental Management at the Xiamen University, China. Her research focuses on watershed hydrology and watershed modeling. Eryinia has authored or co-authored two articles in these areas.

Yaling Huang is currently a doctoral student majoring in Environmental Management at the Xiamen University, China. Her research explores the mechanisms on the interactive effects of climate variability and land use on nitrate export at the watershed scale. Huang has authored or co-authored 10 articles in these areas.

Water and Ecological Security at the Heart of China's Silk Road Economic Belt

Yaning Chen, Zhi Li, Gonghuan Fang and Wei Bian

Abstract Many countries along the Silk Road economic belt are in the arid climate zone, which are typically rich in natural resources but very fragile in their ecological environment. This chapter presents the recent research progress on regional water security and ecosystem stability in Central Asia, the heart of the Silk Road economic belt. Due to climate changes, the solid-state water resources (i.e., glacier and snow) have retreated, leading to an overall decrease of solid water storage and a higher uncertainty of water system. Additionally, because of climate changes and anthropogenic activities (e.g. unplanned over-expansion of agricultural area), this region is facing various eco-environmental problems such as intensified drought and deterioration of ecological systems. Future water resources and possible ecological responses deserve further studies in order to provide a healthy ecological environment for building the Silk Road.

Keywords Climate changes · Glacier retreats · Terrestrial water storage Ecological problems · Management strategies · Central Asia · Silk road

1 Introduction

China's "Belt and Road initiatives" consist of the land-based Silk Road economic belt (the belt) and the ocean-based maritime Silk Road (the road). The Silk Road economic belt is considered as a critical bridge connecting developed countries in Europe with the energetic Asia-Pacific economic circle (Liu 2015). More than 60 counties are involved along the Silk Road. Ecological sustainability should be pursued to provide a better connection between Asia and Europe.

In this chapter, we will systematically analyze the water and ecological security of Central Asia in the heart of Silk Road economic belt. It includes northwestern

Y. Chen (✉) · Z. Li · G. Fang · W. Bian
State Key Laboratory of Desert and Oasis Ecology, Xinjiang Institute of Ecology and Geography, Chinese Academy of Sciences, Urumqi, Xinjiang 830011, China
e-mail: chenyn@ms.xjb.ac.cn

X. Yang and S. Jiang (eds.), *Challenges Towards Ecological Sustainability in China*, https://doi.org/10.1007/978-3-030-03484-9_12

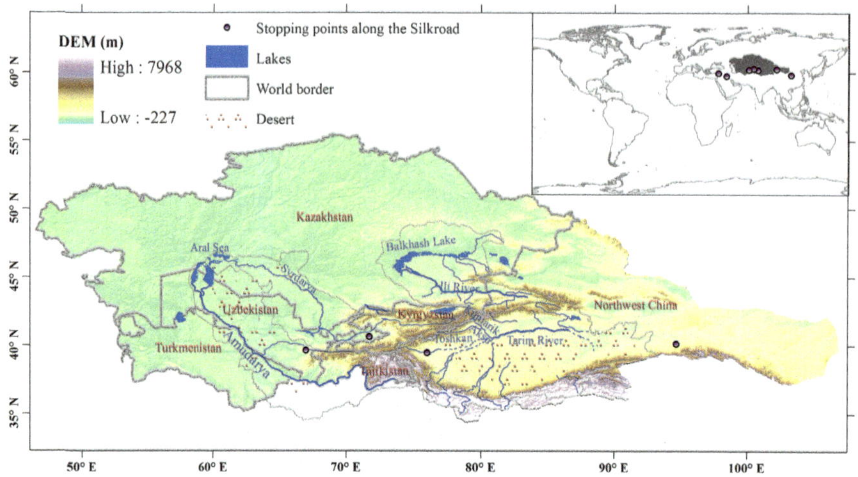

Fig. 1 Location (top right), topography and river systems of the Central Asia

China, Kazakhstan, Kyrgyzstan, Tajikistan, Uzbekistan and Turkmenistan (Fig. 1). With rich natural resources and an extremely vulnerable ecological environment, water security and ecological stability have become the most prominent problem in this region. We will discuss changes in water resources and the evolution in ecological systems, along with some policy recommendations for sustainable developments.

2 Water Security for the Water-Limited Areas Along the Silk Road

In recent years, changes in hydrological processes of inland rivers and water resources caused by climate changes and anthropogenic activities have drawn attention from scientific communities, the public and governments. The heart area of the Silk Road is extremely fragile with a continental climate with scarce precipitation. The oasis area only accounts for 9% of the total land area, and approximately 5% are artificial oases. The oases are surrounded by deserts with sparse vegetation and a simple community structure with low regeneration ability. This region is limited in water availability which is highly dependent on mountainous waters. There would be no oasis without water. The amount of water coming from mountainous areas directly affects the oasis stability and the development of society and economy.

Water is normally generated in the mountainous region of the arid Central Asia. Most rivers are originated from the alpine Tienshan Mountains, normally deemed as "water tower" of Central Asia, providing water resources to oasis and desert ecosystems in the lower reaches. Global warming leads to temporal and spatial changes of precipitation magnitude as well as precipitation patterns, accelerating the ablation and shrinkage of glaciers, and changing the runoff pattern and composition. Therefore, the water resources and the hydrological components in the Tienshan Mountains will be discussed with details.

2.1 Glacier and Snow Cover

In the past fifty years, the global warming rate was 0.175 °C per decade (°C/10a) since the 1960s, while that rate reached 0.36–0.42 °C/10a in Central Asia based on CRU TS3.23 data (University of East Anglia 2017). Obvious warming posed great impacts on water resources, especially for glaciers and snow cover. Glaciers and snow cover are stored in mountains as "solid reservoirs", and are important components of water resources in arid alpine areas. Climatic warming has accelerated the retreat of glaciers, intensified the ablation of snow and frozen earth, and led to the shrinkage of "water tower" at the heart of the Silk Road.

2.1.1 Shift of Snowfall to Rainfall

Under the context of global climate warming, the temporal and spatial distribution of mountainous precipitation, precipitation pattern (such as snowfall, rain-to-snow ratio) will change correspondingly, altering the watershed convergence processes. For example, delay of the first-snow-day and advance of last-snow-day, decrease of snow duration days, etc., should be normally expected (Chen et al. 2016a).

Based on the meteorological data from the China Meteorological Data Service Center (https://data.cma.cn/) and the Global Historical Climatology Network-Daily (http://ncdc.noaa.gov/ghcnd/), we analyzed the changes of snowfall fraction in the past four decades in four different parts of the Tienshan Mountains. The snowfall fraction showed an overall decreasing trend with an increased temperature and increased precipitation. The snowfall fraction in the Tienshan Mountains decreased from 11–24% during 1960–1998 to 9–21% during 1990–2015. This is consistent with the findings of Guo and Li (2015) in the Chinese Tienshan Mountains. Among them, the snowfall fraction in the mid-altitude area (1500–2500 m) decreased most significantly, while the variation was not obvious in high-altitude areas (>3500 m). In addition, a decreasing trend of snowfall duration has also been detected (Fig. 2), which is consistent with the findings in the Kunlun Mountains, the Swiss Alps and the continental United States (Feng and Hu 2007; Knowles et al. 2006; Serquet et al. 2011). For example, in the continental United States, the snowfall rate was generally decreasing, especially in the areas with a daily average temperature of −5 °C and a temperature increment of 0–3 °C (Knowles et al. 2006). Future projections of snowfall fraction also indicated decrease in snowfall fraction (Yang et al. 2017). Decreases in snowfall fraction will cause a shift of the Budyko line towards a more arid and fragile condition (Berghuijs et al. 2014).

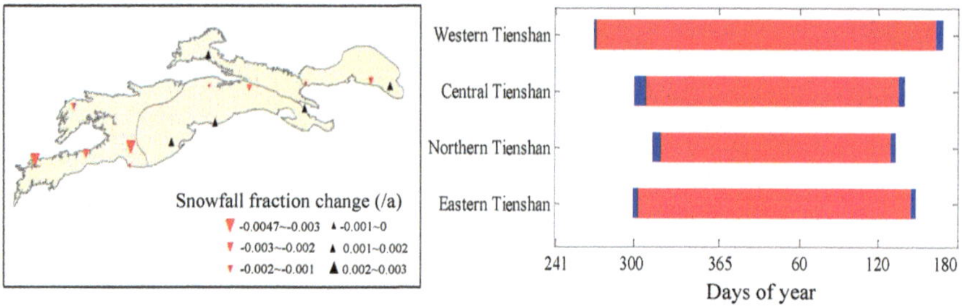

Fig. 2 Snowfall fraction (ratio of snowfall to precipitation) change and the first-snow-day and last-snow-day over the Tienshan mountains

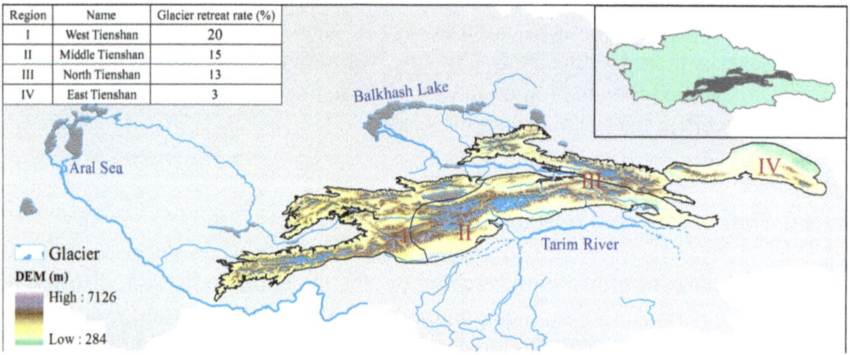

Fig. 3 Spatial distribution of glaciers and their retreat rates for the West Tienshan (I), the Middle Tienshan (II), the North Tienshan (III) and the East Tienshan (VI)

2.1.2 Glacier Shrinkage

Glaciers as solid reservoirs are a critical component of the Central Asia's water towers. According to the Randolph Glacier Inventory (https://www.glims.org/RGI/rgi50_dl.html), 10,778 glaciers are found in the Tienshan Mountain region, which cover a total area of approximately 13,566.6 km^2, feeding as high as 50% of river runoff (Zhou 1999).

Increased temperature has accelerated glacial melt in the Tienshan Mountains. Most glaciers have been shrinking over the past half century. About 97.52% of the glaciers retreated, while only less than 3% of glaciers showed increase or no obvious change during 1960s to 2010s. The largest glacier retreat rate (−20%) was observed in the west Tienshan Mountains, while the middle Tianshan Mountains has a slightly lower glacier retreat rate (15%). In contrast, the retreat rates were much slower in the northern and eastern Tienshan Mountains (Fig. 3). The glacier area of the Aksu River basin, the largest headwater of the Tarim River, has decreased by 22.23% (equivalent to 1071.72 km^2) during 1975–2015 interpreted from Landsat images.

Table 1 Snow cover changes of the Tienshan Mountains in Central Asia during 2002–2013 based on MOD10A1 data. Maximum (minimum) snow cover fraction is the ratio of maximum (minimum) snow covered area to region area. Adapted from Chen et al. (2017a, b)

	Maximum snow cover			Minimum snow cover		
	Snow cover fraction (%)	Absolute area change rate (km^2/a)	Relative area change rate (%/a)	Snow cover fraction (%)	Absolute area change rate (km^2/a)	Relative area change rate (%/a)
Western Tienshan	87	2.3	0.001	1.4%	16	0.085
Central Tienshan	79	−672	−0.32	4.7	−60	−0.029
North Tienshan	90	−78	−0.28	0.51	−20	0.02
East Tienshan	54	−240	−0.09	2.1	17.6	0.08

Accelerated glacier retreats were observed during recent decades compared to the middle twentieth century for the North Tienshan Mountains and Bogda Peak in the east Tienshan Mountains with retreat rates of 13.8% and 7.45%, respectively (Chen et al. 2016a). Interestingly, since 2000, the glacier retreat in West Tienshan Mountains slowed down slightly, with retreat rates of 8.1%.

2.1.3 Snow Cover Variations

Changes in snowfall will eventually translate into snow cover changes. Mountainous snow cover changes can affect water resources in the arid inland area. Using the 2002–2013 snow cover data from MODIS (Moderate Resolution Imaging Spectroradiometer; MOD10A1, https://modis.gsfc.nasa.gov/data/), the snow-cover area change in the Tienshan Mountains was analyzed. The results indicate that the snow-cover area had decreased in most of the Tienshan Mountains. Geographically, the maximum snow cover area and minimum snow cover area in the middle Tienshan Mountains shrinked significantly, with 672 km^2 per year (abbreviated as "/a" in the following parts) and 60 km^2/a, respectively. The maximum and minimum snow cover areas shrinked at much smaller rates in the North Tienshan Mountains, which were 78 km^2/a and 20 km^2/a, respectively. At the same time, the maximum and minimum snow cover areas were found to increase in the West Tienshan Mountains, with increasing rates being 2.3 km^2/a and 16 km^2/a, respectively (Table 1).

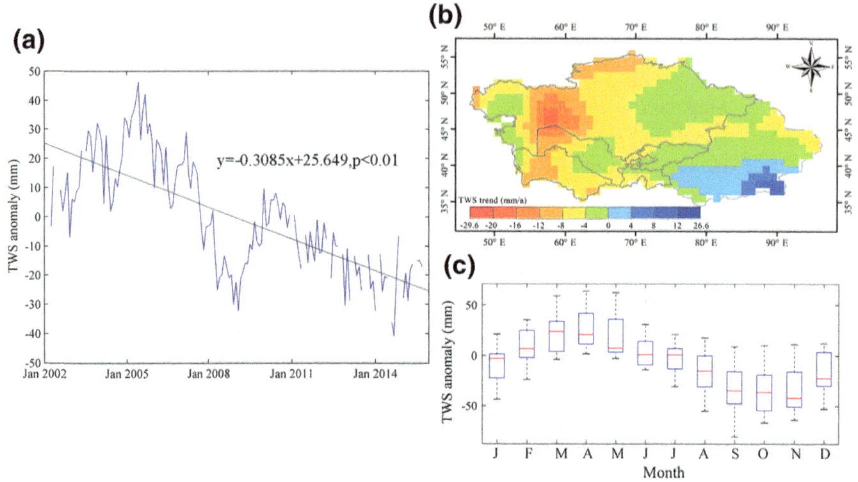

Fig. 4 Spatial (**a**), temporal (**b**) and intra-annual (**c**) changes of terrestrial water storage in Central Asia during 2002–2015

2.2 *Terrestrial Water Storage Variations*

Glaciers and seasonal snow cover changes may further alter the water storage in the Tienshan Mountains, which can significantly affect the oasis water supply (Berghuijs et al. 2014; Sorg et al. 2012). Based on the GRACE data (https://www2.csr.utexas.edu/grace/), the terrestrial total water storage (TWS) in the Tienshan Mountains has decreased at 3.1 mm/a during the period of 2002–2015 (Fig. 4). This indicates that Central Asia has lost a great amount of water at about 22.3×10^8 m^3/a. Future TWS were projected to continue decreasing based on the GCM models (Chen et al. 2016a). Seasonal differences were expected in the TWS variations with a positive anomaly in spring (MAM) (30.3 mm) and a negative anomaly in autumn (SON) (48.7 mm) (Deng and Chen 2017).

TWS spatial variations suggest a major spatial differentiation throughout Central Asia (Fig. 4). Most regions experienced a decreasing trend (0 to −8 mm/a), while the southeastern area (i.e., the north slope of the Kunlun Mountains) had an increasing rate of 0–12 mm/a. The western Central Asia had the largest decreasing rate (16–20 mm/a), which was mainly caused by the shrinking Aral Sea (Singh et al. 2012). Decreasing TWS were also detected in the Tarim River Basin. The agricultural water consumption increased by nearly 30% due to the increasing irrigation areas (Xu et al. 2013), leading to a rapid decline in groundwater levels during 2004–2010 (0.3–1.5 m/a).

2.3 Runoff Changes

The hydrological processes are affected by the land surface pattern and climate changes significantly. With the increase of temperature and changes in snow and glacier melt, the elements of water cycle have changed accordingly. The impact of climate change on basin-scale hydrological processes has become more complicated, leading to a more vulnerable water system.

2.3.1 Increased Melt Runoff Due to Warming

The water resources in the Tienshan Mountains are mainly composed of alpine glacier and snowmelt water in the alpine areas, rainfall in mid-latitude forest belt and groundwater in lower mountainous area. The mechanism of water production and convergence is complex and is typical for other arid areas of the world. Climate warming affects the distribution of glaciers and snow cover and also leads to changes in water supply and runoff components. This poses new challenges for the future water resources in a changing environment (Chen et al. 2017a, b).

Due to continuous warming and the resulted glacier shrinkage, the solid-state water resources have been transformed into liquid water, leading to increased runoff. The rivers in Central Asia, mainly recharged by glacier melt, generally showed a tendency of increasing runoff. Furthermore, a watershed with a higher level of glacierization is likely to have a higher level of runoff increment (Kong and Pang 2012; Zhang et al. 2016a, b). For example, a decreased rainfall recharge proportion (from 9.8% in 1966–1975 to 7.8% in 2000–2008) has been observed in the Turks River, headwater of the Ili River originated from the western Tienshan Mountains (Xu et al. 2015). Runoffs of the Kumalike and Toxkan rivers (glacier area ratios are 16.2% and 4.2% respectively), two headwaters of the Aksu River originating in the middle Tienshan Mountains, increased by 26.2 and 14.9% since 1998, of which 35–48 and 9–24% of runoff increments can be accredited to the increased glacial meltwater (Duethmann et al. 2015). Runoff of the Urumqi River, originating from the northeast Tienshan Mountains, increased by 10% during 1950 to 2009, 69.7% of which was attributed to glacial meltwater with glacial meltwater proportion increased from 62.8 to 72.1% (Sun et al. 2013, 2015). For some basins in the eastern Tienshan Mountains (e.g. Toudaogou and Erdaogou watersheds), the increases of runoffs were not obvious, or even decreasing trends were detected due to the relatively low proportion of glaciers. In the Karatal River basin, the runoffs of sub-basins with glacierisation above 5% increased while runoffs increased insignificantly or even decreased for subbasins with glacierisation less than 2% (Kaldybayev et al. 2016).

The above examples prove that the higher the glacierization of a watershed is, the higher the runoff increment under global warming would be. The same rule also existed in other mountains. For example, the runoffs in the Swiss Alps increased when catchment glacierisations were above 10%, while the runoffs show significant decreasing trends when glacierisations were below 10% (Birsan et al. 2005).

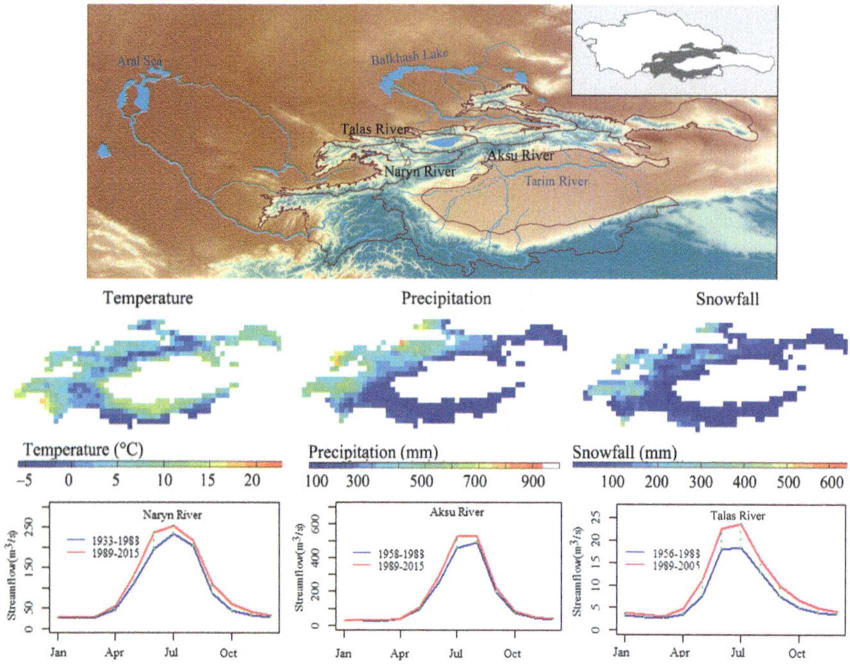

Fig. 5 Location of the mountainous region of Central Asia and its temperature, precipitation and snowfall variations. Intra-annual variations of runoffs for the Nargn River, the Aksu River and the Talas River before and after 1998 are included

2.3.2 Variation of Intra-annual Streamflow Peak

Climate changes also affect the intra-annual streamflow patterns. We have examined the spatial variation of temperature, precipitation (including snowfall) during the past decades in the high mountains of Central Asia, including the high Tienshan Mountains and the northern flank of the Kunlun Mountains where most water is generated (Fig. 5). In general temperatures were higher in low-altitude areas but lower in high-altitude areas. Precipitation was relatively higher in the west while lower in the east and south, with a spatially averaged value of 237 mm (ranging from 33.4 to 833 mm). The spatial pattern of snowfall was similar to that of precipitation (Yang et al. 2017).

Changes in streamflows were observed in the Tienshan Mountains, which showed a higher rate of glacier and snow melting with earlier peaks (Berghuijs et al. 2014). Snowmelt-dominated rivers experienced earlier maximum runoff peaks, whilst the summer runoff either remained unchanged or decreased (Chen et al. 2018). The glacier melt-recharged rivers generally demonstrated enhanced flood peaks during June to September (Shen et al. 2013). The streamflow's temporal shift is important for reservoir and agricultural water management because it could cause a water supply decline during the crop-growing season. For example, the streamflow peaks of the Naryn River Basin, a tributary of the Syrdarya, showed significant temporal changes, shifted from summer season to spring (Fig. 5). The same phenomenon has been observed in the Aksu River and the Talas River.

2.3.3 Climate Change Intensified Hydrological Variations

Under the context of global warming, hydrological variations have intensified and resulted in more vulnerable water resources. Complex responses of watershed hydrological processes to climate change were detected in the Tienshan Mountains. It seems that catchments with a larger share of glacierized area (e.g., Middle Tienshan Mountains) tend to increase in runoff, whilst catchments with less or no glacierization (e.g., East Tienshan Mountains) show a larger variation. With global warming, the hydrological processes have altered significantly for some rivers reaching their melting points. Runoff variations increased obviously for most rivers, which will likely aggravate the complexity of the hydrological processes. Higher uncertainties are expected in runoff prediction for rivers jointly recharged by glacier/snow melt and rainfall. With the shrinkage of glaciers, the contribution of glacier melt will gradually weaken in the 21st century (as was projected by GCM models in CMIP5). Future runoff variations will further increase as precipitation uncertainty getting enlarged (IPCC 2013).

3 Ecological Degradation in Central Asia

The Silk Road economic belt with a complex climate is a sensitive region experiencing climate changes. The temperature increase can speed up the global water cycle, intensify the extreme hydrological events, exacerbate desert vegetation degradation, and cause the water-dependent ecosystem to be much more fragile in Central Asia.

3.1 Dried-up Riverways and Shrinking Terminal Lakes

Central Asia including northwest China belongs to the area where water availability is very limited. Water resources originate from mountains, are recharged with ice-snow melting water from alpine mountains and precipitation from mid-altitude forest, and finally fissure water in the lower hilly areas.

The essential problem of water resources in Central Asia is caused by excessive exploitation. Taking the transboundary rivers of the Amudarya River and Syrdarya River for example, we find that the cultivated land increased rapidly in the river basins. The cultivated land area increased from 227.5×10^4 hm^2 in 1960 to 481.3×10^4 hm^2 in 2010 (Fig. 6). With the rapid expansion of irrigation area and agricultural water use, the water inflowing into the Aral Sea decreased significantly. The

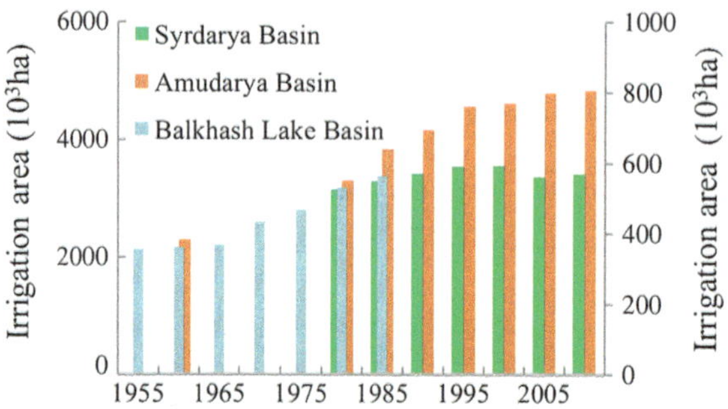

Fig. 6 Evolution of irrigation areas for the Syrdarya Basin, the Amudarya Basin (left axis) and the Balkhash Lake Basin (right axis)

level of the Aral Sea decreased significantly, and 6.6×10^4 km^2 lake area was dried up (http://www.cawater-info.net/index_e.htm).

For the Balkhash Lake Basin, the agricultural land has also expanded rapidly from 35.2×10^4 hm^2 in 1955 to 56.0×10^4 hm^2 in 1980s (Kezer and Matsuyama 2006). At the same time, the water discharging into the Balkhash Lake from several main streams has dropped from 41.2×10^8 m^3 to 22.3×10^8 m^3.

For Northwest China, the water consumption for socio-economical development far exceeds the planning goal. According to the Xinjiang Water Resources Bulletin, the total amount of water resources was 900×10^8 m^3, while the water consumption was 585×10^8 m^3, far beyond the planned water consumption in 2015 (515×10^8 m^3) and in 2020 (526×10^8 m^3).

What worth noting is that the proportion of agricultural water consumption with more than 90% and as high as 96% in the Tarim River Basin is much higher than that in Asia (81%) and the global average (69%) (http://www.fao.org/nr/water/aquastat/water_use/index.stm). The water use per unit GDP in the Tarim River Basin was 1130 m^3/10,000 RMB, much higher than the Xinjiang average (487 m^3) and the national average (89 m^3). This massive irrigation consumed much of the ecological water, leading to exacerbated water conflicts. As a result, the over-use of ecological water led to decreased groundwater table and decayed desert vegetation. The ecological crisis was getting increasingly prominent.

The leading reason behind these ecological problems is that there is no consensus on water resources carrying capacity in the heart areas of the Silk Road. Without considering limited water resources, water has been over-exploited, which seriously threatened the sustainable development.

3.2 *Vegetation Dynamics and Drought Risk*

Vegetation dynamics respond quickly to climate changes, especially in arid regions. The high volatility of temperature can break the original natural balance, which, when coupling with an obvious decrease in water storage and a decline of shallow groundwater levels, can accelerate plant evapotranspiration and soil moisture evaporation and cause the shallow-root-plants to damage and die, leading to the reduction of species diversity and decrease of land cover.

3.2.1 Negative Ecological Effects

The dramatic temperature increase can have negative ecological consequences. Recent studies indicate that the vegetation coverage and the Normalized Difference Vegetation Index in Central Asia tended to increase before 2000 but decrease after 2000 (Li et al. 2015). Vegetation dynamics can be affected by climate changes and human-induced land changes (Li et al. 2014, 2015; Wang et al. 2014). Vegetation dynamics in arid areas can be more sensitive to climate changes than in other regions. The temperature and pan evaporation increase and the precipitation decrease can jointly cause a severe soil moisture loss, which makes shallow desert plants to deteriorate and cease living, finally reducing species diversity and vegetation cover. The rapid growth of population and expansion of agricultural land were mainly achieved by reclaiming unutilized land and natural grassland (Chen et al. 2015a, b). Agricultural economy is dominant in Oases. Because of the limitation of natural conditions and regulations, the vegetation/crop distribution shows general patterns and also random variations (Wang et al. 2014). Some forestland patches were converted into grassland, cultivated land and bare land, while forestland was divided by cultivated land patches. These changes indicate that the economic development in this region is mainly dependent upon the agricultural water and soil development. Desert riparian forests in river basins decreased continuously.

For the Taklamakan Desert, natural desert vegetation plays a crucial role in maintaining a stable desert environment, preventing desert expansion, and safeguarding the oasis safety ecologically. Based on the NDVI data from Global Inventor Modeling and Mapping Studies (GIMMS) database, the NDVI of natural vegetation in the Tarim River Basin increased at 0.012 per decade before 1998, and then decreased at 0.005 per decade after 1998 (Fig. 7). In addition, bare surface area increased at 7.8% per decade during 1982–2013 (Chen et al. 2016b).

Artificial cultivated land in the Tarim Basin increased from 2.17×10^4 km^2 (in the 1980s) to 3.46×10^4 km^2 (in the 1990s), and then to 3.93×10^4 km^2 (in the 2000s). The natural desert vegetation area, at NDVI $<$ 0.3, increased from 9.28×10^4km^2 (in the 1980s) to 9.7×10^4 km^2 (in the 1990s). However, the vegetation area of natural desert decreased to 8.87×10^4 km^2 in the 2000s (Chen et al. 2016b).

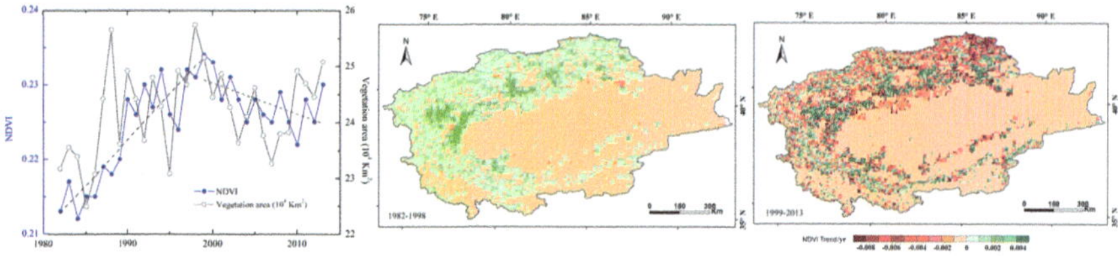

Fig. 7 Temporal changes of NDVI and vegetation area in the Tarim River Basin during 1982–2013 (Left) together with the spatial patterns of NDVI changes during 1982–1998 (middle) and 1999–2013 (right)

3.2.2 Expanded and Browned Desert

Generally, NDVI of 0.1 was taken to distinguish vegetated area from unvegetated areas (Iwasaki 2006). Assuming that bare land is extracted by NDVI below 0.1, its area within the desert has grown by 8.7% per decade since 1998 (Jeong et al. 2011). The desertification process of the Tarim Basin further intensified due to expansion of the bare land and browning of the Taklamakan Desert boundaries in the 2000s (Chen et al. 2016a, b). Recent studies indicate that dust activities over arid and semi-arid areas had increased by the end of the 1990s, especially in the Sahara Desert, East Africa, southern Argentina, western Australia, and Central Asia (Donohue et al. 2009; Jeong et al. 2010, 2011).

The irrigation area of the Tarim River Basin increased at a rate of 686.42 km^2/yr during 2001–2013 based on the remote sensing data (Chen 2015). Large shares of the increase were mainly distributed in the northern basin, including the lower reaches of the Aksu River Basin, the Ogan-Kuqa River Basin, and the Yanqi Basin and the surrounding area of the mainstream of the Tarim River. For example, the irrigation area along the mainstream increased by 80%. Further, the agriculture land was converted from woodlands, wetlands and other landscapes steadily and rapidly. With the warming climate and expanding irrigation areas, regional transpiration and evapotranspiration have intensified as crops consume 2–3 times more water than bare land does (Li and Zhao 2010). Therefore, the newly reclaimed cropland increased a water consumption of 32.8×10^8 m^3 in the Tarim River Basin over the past decade, that is to say, water use for ecological purposes has been squeezed out by irrigation (Chen et al. 2016a, b).

3.2.3 Shrub Encroachment into Grassland

Shrub encroachment generally leads to the formation of a tessellated landscape in which shrub patches are mixed with grass patches. Such a type of grassland was described as shrub-encroached grassland (Chen et al. 2015a, b). Shrub encroachment into grassland has been widely observed in arid and semiarid savanna regions (Knapp et al. 2008; Maestre et al. 2009). Compared to moist and cool areas, shrub-covered

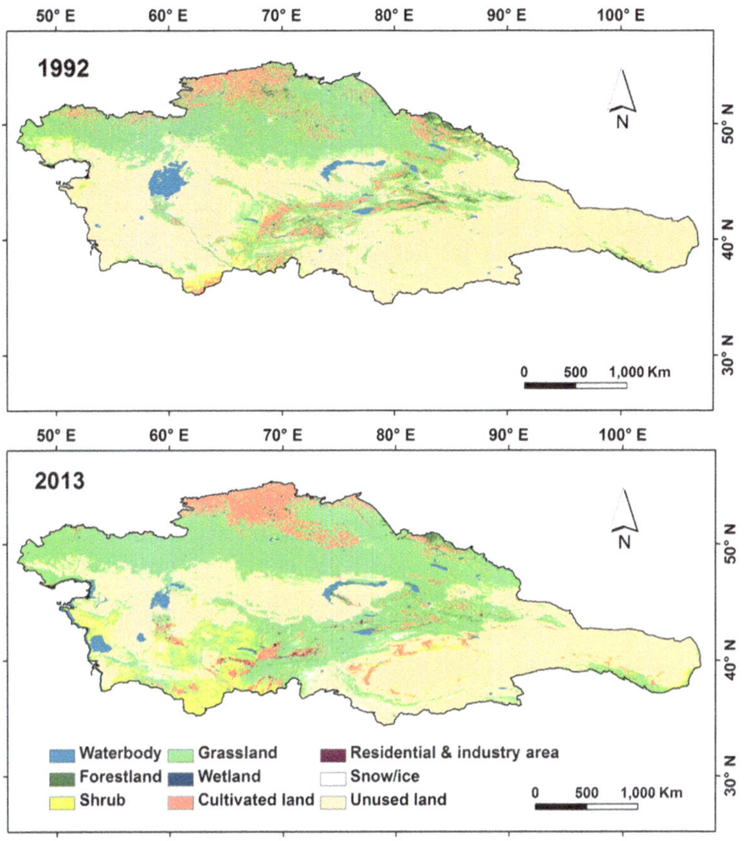

Fig. 8 Land use/cover maps for 1992 (upper) and 2013 (lower) in Central Asia

areas and patch sizes were greater in dry and warm habitats (Li et al. 2013; Chen et al. 2015a, b). Shrub encroachment can be driven by complex biogeological processes involving many interrelated factors at varying spatial and temporal scales.

For Central Asia region, the land use/cover has experienced extensive changes during the past 30 years. These changes include grassland degradation, forest loss, agricultural land expansion, and shrub encroachment. The grassland degradation was most prominent in northern Kazakhstan and the substantial increase of open shrublands was observed in Turkmenistan, Uzbekistan, and the Tarim River Basin. These changes indicated a strong shrub encroachment into grassland (Fig. 8). The underlying reasons of the thrub encroachment maybe related to an overwhelming warming, which in turn accelerates the evapotranspiration of the shallow soil layer, and eventually reduces the soil moisture (Li et al. 2015).

Based on the GIS-based landscape matrices method, the land use/cover conversion pattern was analyzed, which shows that grassland shrinked by approximately 25.9% and open shrubland increased by about 13.5%. If we examine the land conversion

Fig. 9 Shrub (left) encroachment into grassland (right)

closely, about 8% of the grassland was transformed into shrubland, and about 9.9% of the forest land was changed into shrubland from 2000 to 2013. Shrub encroachment into native grasslands alters species interaction, and may result in biodiversity loss and finally influence the ecosystem function (Li et al. 2015).

Generally, the vegetation in Central Asia is composed of the mesophyte plant and the xerophilous plant. The mesophyte plants rely on precipitation, while the xerophilous plants primarily take water source from groundwater. The evidence of shrub encroachment suggests that some xerophilous plants are replacing the mesophyte ones, causing the water-dependent ecosystem to be more vulnerable under the influence of climate changes in Central Asia (Fig. 9). Additionally, the vegetation pattern can be affected by water and heat conditions, especially the water condition (Li et al. 2015). Under the context of continuous warming, the vegetation growth will be further modulated by precipitation-related soil moisture in this arid region.

3.2.4 Assessment of Droughts

How will regional droughts evolve due to the ongoing climate change? Previous studies on drought indices indicate that the drought frequency and magnitude are likely to elevate due to accelerating evaporation together with the decreased precipitation (e.g., Dai et al. 2004; Fu and Feng 2014; Huang et al. 2016). However, the calculation procedures to derive drought indices (e.g., Palmer Drought Severity Index) were different from each other, resulting in controversial findings in drought assessment. Additionally, previous studies have pointed out some flaws in calculating the drought index (e.g., Van der Schrier et al. 2011; Hobbins et al. 2012; Sheffield et al. 2012; Zhang et al. 2016a, b).

PDSI-$_{PM}$ method in estimating the PDSI. To calculate the PDSI, potential evapotranspiration (PET) is one important input. PET estimates using the Penman-Monteith method are generally considered to be physically realistic although it needs additional types of input data.

PDSI can be calculated using the difference between the observed and required monthly "precipitation". The required precipitation is then estimated under climatically appropriate for existing conditions (CAFEC).

$$\widehat{p} = \widehat{E_T} + \widehat{R} + \widehat{R_O} - \widehat{L} \tag{1}$$

where $\widehat{E_T}$ indicates actual evapotranspiration, $\widehat{R}$ is soil moisture, $\widehat{R_O}$ is runoff, and $\widehat{L}$ represents moisture loss in soil system

$$\widehat{E_T} = \alpha PET, \quad \alpha = \frac{\overline{E_T}}{\overline{PET}} \tag{2}$$

$$\widehat{R} = \beta PR, \quad \beta = \frac{\bar{R}}{\overline{PR}} \tag{3}$$

$$\widehat{R_O} = \gamma PR_O, \quad \gamma = \frac{\bar{R}_O}{\overline{PR_O}} \tag{4}$$

$$\widehat{L} = \delta PL, \quad \delta = \frac{\bar{L}}{\overline{PL}} \tag{5}$$

where $\overline{PET}$ is the mean potential evapotranspiration, $\overline{PR}$ is average recharge, $\overline{PR_O}$ is average runoff, and $\overline{PL}$ is average moisture loss.

In this chapter, the daily potential evapotranspiration (mm d^{-1}) was estimated using the Penman-Monteith method.

$$PET = \frac{0.408\Delta(R_n - G) + \gamma \frac{900}{T_{mean} + 273} U_2 (vp_s - vp)}{\Delta + \gamma(1 + 0.34U_2)} \tag{6}$$

In Eq. 6, PET (mm) represents the potential evapotranspiration, R_n (MJ m^{-2} d^{-1}) is the net radiation at the crop surface, G (MJ m^{-2} d^{-1}) denotes the soil heat flux density. T_{mean} (°C) and U_2 (m s^{-1}) denote the air temperature and wind speed at 2 m height, respectively. vp_s (kPa) and vp (kPa) are the saturation vapor pressure and actual vapor pressure, respectively. Δ (kPa °C^{-1}) is the slope of vapor pressure curve, and γ (kPa °C^{-1}) is the psychrometric constant. We further transformed the observed wind speed data at 10 m above the ground to 2 m.

Available soil moisture (PR) and available runoff (PR_0) can be estimated based on Eqs. 7 and 8.

$$PR = AWC - (S_s + S_u) \tag{7}$$

$$PR_0 = AWC - PR = (S_s + S_u) \tag{8}$$

where S_s and S_u denote the available soil moisture stored in the surface layer and in the underlying levels at the start of the month, respectively. AWC is the available water capacity of the soil system.

Soil moisture loss (PL) could be estimated by the following equations:

$$PL = PL_s + PL_u \tag{9}$$

$$PLs = \min(PET + S_s) \tag{10}$$

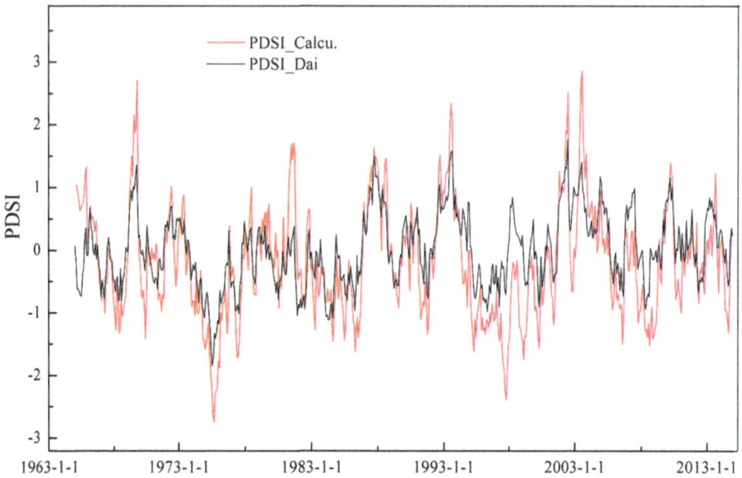

Fig. 10 Time series of our calculated PDSI (*PDSI_Calcu.*) and Dai's result (*PDSI_Dai*)

$$PL_u = \frac{(PET - PL_s)S_u}{AWC}, PL_u \leq S_u \tag{11}$$

where PL_s and PL_u are soil moisture loss from the surface layer and the underlying layer, respectively.

Finally, the monthly Palmer Drought Severity Index is defined as:

$$\mathrm{PDSI}_j = 0.897\,\mathrm{PDSI}_{j-1}Z_j/3 \tag{12}$$

$$Z = K_j d_j \tag{13}$$

$$d = P - \widehat{P} \tag{14}$$

where d denotes the difference between the actual and the CAFEC precipitation, Z is the monthly moisture anomaly index, j denotes month, K_j is related constant for month j.

Figure 10 plots the variations of the PDSI (*PDSI_Calcu*, displayed as mean ± error). For comparison, the PDSI calculated based on Dai's method (*PDSI_Dai*) was also plotted. *PDSI_Calcu.* agrees well with the *PDSI_Dai* (Li et al. 2017). In addition, *PDSI_Calcu* remained at a stable level, demonstrating a near-zero trend (0.0036 per year) during the study period. Besides the slight overall trend in PDSI, the PDSI showed a drying trend since the 2000s with a decreasing rate being −0.023 per year, which is consistent with the dramatic warming in Central Asia.

The PDSI shows a spatial pattern mixing drying (42% of the total land area) with wetting (58%) from 1965 to 2014, which jointly generated a slightly wetting tendency. The increase of PDSI (wetting) was observed in the northern Xinjiang, the Tienshan Mountains, and the western part of Kazakhstan. With a state of high temperature variability (over the past decade), the decrease of PDSI (drying) was dominant throughout most of Central Asia, about 65% of the land area, especially in the eastern and western Kazakhstan, Kyrgyzstan, Turkmenistan, and the Tarim River Basin (Li et al. 2017).

The above-mentioned accelerated drought, coupled with the rapidly expanding population, will lead to land degradation, even desertification, particularly over the dryland in some developing countries.

3.3 *Sustainable Water and Ecosystem Management*

3.3.1 Positive and Negative Feedbacks Between Water and Ecological Problems

The increased temperature has accelerated glacier melting (Barnett et al. 2005), reduced snow cover, induced precipitation shift from snow to rain in the mountains (Berghuijs et al. 2014), and intensified extreme hydrological events (Chen 2014). These ongoing events accompanied by global warming have led to changes in the hydrological processes and also the ecological processes of Central Asia (Fig. 11). Over the last few decades, Central Asia has experienced pronounced warming. Climate changes have already produced significant impacts on water resources and ecological system (e.g., accelerating glacier melt water and increasing runoff, which benefit water resources and ecology). More precipitation falling as rain rather than snow was related to lower runoff, causing negative impacts on water and ecological systems. Additionally, the rising temperature also resulted in decrease in soil moisture by enhancing the evapotranspiraion (ET). The overexpansion of irrigation area leads to approximately 95% of the water being used in agriculture, and also resulted in the over exploitation of groundwater. As groundwater levels dropped, the ecological water availability has reduce and eventually caused ecological security crisis. The outlook for water security in this arid region is not optimistic. The current increase in runoff has been threatened by the reductions in total water storage (i.e. glaciers and snow cover). Therefore, the sustainable development of oasis is not guaranteed. For the desert ecosystem, the decreasing soil moisture due to rising temperatures and therefore increased potential evaporation, caused the desert plants with shallow roots to die and further intensified the desert boundary browning. The water-dependent ecosystem will be more vulnerable, thus increasing the potential risk of oasis collapse. Synthetically, negative climatic responses including droughts and the subsequent ecological, social, economic, and political responses will likely become more critical if future warming outpaces precipitation increasing. Therefore, the impact of global warming on water resources and ecosystems is accelerating (Chen et al. 2016a, b).

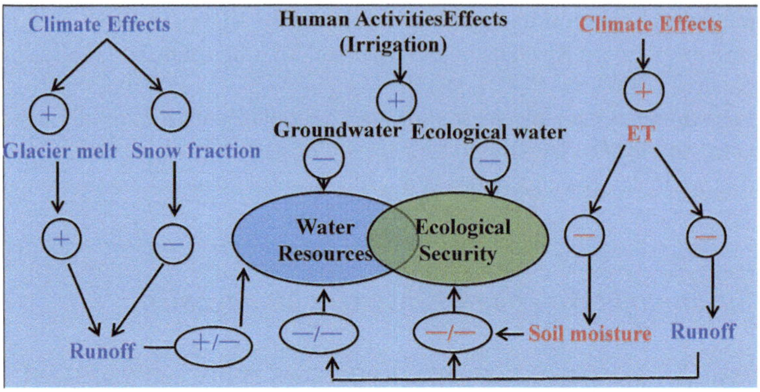

Fig. 11 A schematic diagram of the potential impacts of climate change and human activities on water resources and ecological security at the heart of China's Silk Road

3.3.2 Development and Disappearance of Ancient Civilizations Along the Ancient Silk Road

Understanding the causes and processes leading to the development and disappearance of ancient oasis civilizations along the Silk Road can provide some useful insight into human adaptation to current and future climate and environmental changes in the extremely arid area.

The consistent change between climate and ancient civilizations has drawn much international attention (Sandweiss et al. 1999). Many studies concluded that the disappearance of complex ancient societies was related to the deteriorated climate conditions (e.g., Weiss et al. 1993; Buckley et al. 2010). For instance, the persistent drought accompanied by several strong monsoons in the 14th–15th centuries resulted in the eventual collapse of Angkor, the capitol of the Khmer Empire in Cambodia (Buckley et al. 2010). The victories of Chinggis Khan's (Genghis Khan's) in the Mongol Empire were largely resulted from the regional warm climate and persistent wet conditions. The opposite is also true. The dry conditions agreed well with the extremely instable period of politics in Mongolia during the 1180s to the early 13th century (Pederson et al. 2014). In addition, the collapse of most dynasties in China corresponded well with low temperature during the last 2485 years (Liu et al. 2009). A recent study believed that temperature increase could trigger more violence (Hsiang et al. 2013) and the persistent hot summers could elevate human aggressiveness.

Established over 2100 years ago, the ancient Silk Road, was designed to promote commercial intercourse between China and Europe (Xu et al. 2010), and to stimulate cultural prosperity and economic growth along this road. In China, the Taklimakan Desert, the second largest sandy desert in the world, bordered these famous ancient civilizations. Under it extremely arid climate conditions, even a slight shift of climate change will cause dramatic changes in water availability, causing the ecosystems and agricultural systems exposed to water shortage risk (Li et al. 2016). The consequence

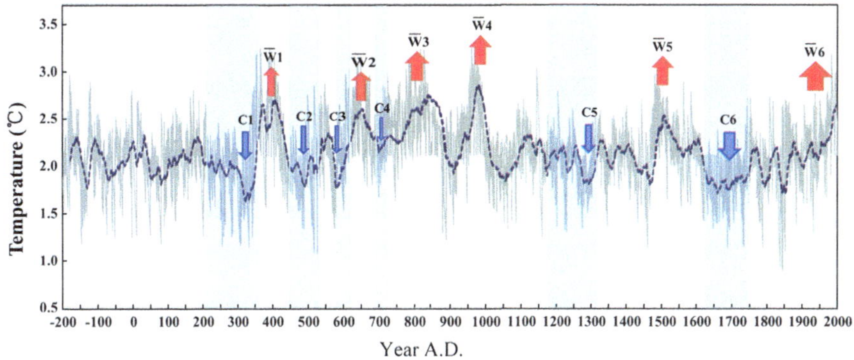

Fig. 12 Reconstruction of temperatures over the past 2485 years in the Qinghai-Tibet Plateau around the Tarim River Basin

of water disavailability was the old oases civilizations being engulfed by the surrounding desert. Therefore, distinguish the driving forces leading to the collapse of oasis civilizations along the ancient Silk Road will shed light on adaptation strategies to climate changes in the nowadays and future.

The climate variations (including temperature and precipitation) in the Northwestern China during the past over 2000 years were reconstructed based on the Guliya ice cores data (Shi et al. 1999), and the tree rings (Liu et al. 2009) collected in the Qinghai-Tibet Plateau near the Tarim River Basin (Fig. 12).

In the ancient time, human societies were quite weak in enduring extreme weather and unheralded climate changes, largely due to the limited water resources. As almost all the water resources are from the rivers originated from the alpine mountains, the productivity declined dramatically whenever cold and dry climate prevailing. In contrast to cold and dry conditions, abundant water resources, generated under the warm-wet conditions, could easily satisfy agricultural water needs. With the increased land productivity, population had increased and oasis had expanded, along with the increasing demand for food and energy. For the cold-dry phase, water generation decreased, leading to a deficiency in water resources used for irrigation. Other subsequent responses include dried–up river ways, intensified desertification, poor harvest, and unstable social environment. For the regions along the ancient Silk Road, the demise of the 21 ancient oases civilizations mostly occurred in the 4th, 5th, 7th and 8th centuries (Li et al. 2016). These periods were described as long lasting drought together with cold climate. Deficient precipitation and little glacier meltwater accelerate the desertification processes, which lead to the water-dependent oasis to become more vulnerable and ultimately to disappear. Therefore, it is believed that the cold-dry conditions contributed a lot to the demise of civilizations along the ancient Silk Road.

The temperature and precipitation in the Tarim River Basin have experienced abrupt change around 1986 (Chen 2014). Temperature has seen a strong increase (sharp increase) in 1997 and then it has stayed at a high value with high volatility

(Chen 2014). After 1986, the average annual PDSI is 0.38 higher than before based on our calculation along the ancient Silk Road in the Tarim River Basin. However, PDSI began to decline again within the recent fifteen years. If future climate warming overpaces precipitation increasing, heat droughts will occur similar to the ancient dry-periods, and the social, economic, and political responses will be probably not optimistic, and eventually the contradiction between water supply and water demand in the oasis and desert will be more challenging (Li et al. 2016).

3.3.3 Policy Recommendations Towards Sustainable Development

To create a water-safe and ecology-friendly environment, governments should work together to construct a feasible and stable ecological environment for Silk Road construction. More specific recommendations are discussed below.

Conducting Ecological Planning of the Silk Road Timely

Firstly, we should recognize that far more water than available has been consumed in Central Asia. Glacier and snowmelt, precipitation and streamflow from alpine mountains are very limited and can no longer meet the regional growing water demands. Moreover, pumping more groundwater is the main anthropogenic response to drought and expansion of irrigation area (Voss et al. 2013; Famiglietti 2014), and persistent groundwater depletion is likely to intensify the water shortage. As groundwater level drops, wells get dry, and in turn, much deeper wells are needed at an enormous financial cost. The above-mentioned problems will be aggravated and more challenging by both population growth and expansion of irrigation areas.

Ecological planning is to facilitate decision making to implement water and land development by taking environmental, social, political, economical and governmental factors into account. The aim is to provide a holistic framework and hence to achieve sustainable development. Therefore, more strength should be put on ecological and environmental planning, controlling the irrigation area under limited water resources. The planning, as a fundamental requirement for sustainable development, should be recognized by all relevant parties. Based on ecological planning (e.g., returning farmland plan to pasture and forestry), governments are urged to speed up the implementation of the plan (Chen et al. 2016a, b).

Evaluating the Carrying Capacity of Water Resources

To avert the potentially disastrous scenario, some measures should be taken to mandate high water use efficiency in agriculture. Agricultural use takes more than 90% of the total water utilization in the core area of the Silk Road economic belt, and approaching 96% in the Tarim River Basin, which is much higher than in the world average (69%). Even modest improvement in agricultural water use efficiency will lead to tremendous amount of water being saved. This amount of water could be used for environment or for other more economic-efficient uses such as economic growth, energy production and industry development (Famiglietti 2014). To this end, we should optimize the productivity layout, upgrade industrial structure, build effective water use industry, and foster efficient water-saving irrigation measures (e.g., drip irrigation, film mulching and microirrigation).

Based on the carrying capacity of water resources, jointly management of surface water and groundwater resources is an effective way to improve water use efficiency.

Establish Comprehensive Monitoring System for Water and Ecological Systems

Monitoring programs of water and ecological systems can vary significantly in scope, ranging from farmland-scale, to large-scale or even global scale by monitoring collaboratively worldwide (Tallis et al. 2012). The monitoring system in Central Asia has been stopped with the collapse of Soviet Union, and data are scarce in this region since 1989. Therefore, promoting the establishment of the Science Alliance for Sustainable Development of the Silk Road economic belt and International Joint Fund to promote cooperation research are urgently needed.

For example, groundwater is normally poorly monitored and barely managed along the agriculture intensive Silk Road. Excessive water use, including over pumping of groundwater, has exacerbated ecological degradation and the decline of groundwater levels. This led to more surface water needed to recharge groundwater. As the surface water management is getting more and more attention (especially for the Tarim River Basin and the Aral Sea Basin), there remains a severe problem of lacking of groundwater monitoring and management. Such work is urgently needed, given the continuing rapid expansion of the oasis area, which may result in the over exploitation of groundwater and further lead to the continuing decline of the groundwater level. In addition, the monitoring of irrigation area is also useful in controlling agricultural water use. Therefore, jointly management of both surface water and groundwater should be promoted urgently.

Coordination and Cooperation Towards a Sustainable Development

According to the "Belt and Road" Initiative, the jointly building of Silk Road economic belt is to "promote economic prosperity of countries along the routes and regional economic cooperation, strengthen exchange and mutual learning among different civilizations, and contribute to world peace and development" (https://eng.yidaiyilu.gov.cn/). As most important rivers run several countries in Central Asia, jointly management of ecological and water-related issues in the transboundary river basins is essential in the sustainable development. Higher standards and deeper levels of cooperation on ecohydrological protection and water use should be promoted to accomplish the sustainable development goals. Cooperation in water resources allocation, water use and irrigation, drought and flood, ecological protection should also be focused. Therefore, improvement of the comprehensive capacity in serving, supporting and ensuring ecological protection and improvement of water use efficiency are of great importance to construct the green, prosperous, and friendly Silk Road economic belt benefiting all countries along the route.

4 Summary

Significant climate change has been detected in the heart region of the Silk Road. Climate change led to temporal and spatial changes of temperature and precipitation as well as changes in precipitation patterns. Climate change accelerated the ablation and shrinkage of glaciers and snow cover, and changed the runoff composition and fluctuation patterns. Runoffs of alpine rivers with high glacierisation rates normally have increasing trends, while those dominated by rainfall manifested higher runoff variations under climate change.

Climate change and human activity induced changes in ecological systems. Unreasonable water use and warming accelerated droughts and expansion of desert browning area. The NDVI of natural vegetation in the Tarim River Basin tended to increase before 1998 but decrease after that. Negative ecological effects were getting more and more prominent.

Climate system and hydrological processes are closely related. If we take a look at what happened during the past centuries, we can learn that human livings were strongly affected by climate changes. Measures should be taken to cope with the already prominent water crisis and fragile ecological system. Several suggestions were presented, including ecological planning, water carrying capacity assessment, monitoring system and cooperation. Only with the jointly cooperation, water risk and ecological problems can be alleviated or coped with towards a sustainable and stable development of the Silk Road.

Acknowledgements The research is supported by the Strategic Priority Research Program of the Chinese Academy of Sciences (XDA19030204) and the National Natural Science Foundation of China (41630859).

References

Barnett TP, Adam JC, Lettenmaier DP (2005) Potential impacts of a warming climate on water availability in snow-dominated regions. Nature 438(7066):303–309. https://doi.org/10.1038/nature04141

Berghuijs W, Woods R, Hrachowitz M (2014) A precipitation shift from snow towards rain leads to a decrease in streamflow. Nat Clim Change 4(7):583–586. https://doi.org/10.1038/nclimate2246

Birsan MV, Molnar P, Burlando P et al (2005) Streamflow trends in Switzerland. J Hydrol 314(1–4):312–329. https://doi.org/10.1016/j.jhydrol.2005.06.008

Buckley BM, Anchukaitis KJ, Penny D et al (2010) Climate as a contributing factor in the demise of Angkor, Cambodia. PNAS 107(15):6748–6752. https://doi.org/10.1073/pnas.0910827107
Chen Y (2014) Water resources research in Northwest China. Springer
Chen Y (2015) Ecological protection and sustainable management of the Tarim River Basin. Science press, Beijing
Chen L, Li H, Zhang P et al (2015a) Climate and native grassland vegetation as drivers of the community structures of shrub-encroached grasslands in Inner Mongolia, China. Landscape Ecol 30(9):1627–1641. https://doi.org/10.1007/s10980-014-0044-9
Chen Y, Li Z, Fan Y, Wang H, Deng H (2015b) Progress and prospects of climate change impacts on hydrology in the arid region of northwest China. Environ Res 139:11–19. https://doi.org/10.1016/j.envres.2014.12.029
Chen Y, Li W, Deng H et al (2016a) Changes in Central Asia's water tower: past, present and future. Sci Rep 6:35458. https://doi.org/10.1038/srep35458
Chen Y, Li Z, Li W et al (2016b) Water and ecological security: dealing with hydroclimatic challenges at the heart of China's Silk Road. Environ Earth Res 75:881. https://doi.org/10.1007/s12665-016-5385-z
Chen Y, Li Z, Fang G et al (2017a) Impact of climate change on water resources in the Tianshan mountains, Central Asia. Acta Geogr Sin 72(1):18–26 [In Chinese with English Abstract]
Chen Y, Li W, Fang G et al (2017b) Review article: hydrological modeling in glacierized catchments of Central Asia–status and challenges. Hydrol Earth Syst Sci 21(2):669–684. https://doi.org/10.5194/hess-21-669-2017
Chen Y, Li Z, Fang G et al (2018) Large hydrological processes changes in the transboundary rivers of Central Asia. J Geophys Res 123:5059–5069. https://doi.org/10.1029/2017JD028184
Dai A, Trenberth KE, Qian T (2004) A global dataset of Palmer Drought Severity Index for 1870–2002: relationship with soil moisture and effects of surface warming. J Hydrometeorol 5(6):1117–1130
Deng H, Chen Y (2017) Influences of recent climate change and human activities on water storage variations in Central Asia. J Hydrol 544:46–57. https://doi.org/10.1016/j.jhydrol.2016.11.006
Donohue RJ, McVICAR T, Roderick ML (2009) Climate-related trends in Australian vegetation cover as inferred from satellite observations, 1981–2006. Glob Change Biol 15(4):1025–1039
Duethmann D, Bolch T, Farinotti D et al (2015) Attribution of streamflow trends in snow and glacier melt-dominated catchments of the Tarim River, Central Asia. Water Resour Res 51(6):4727–4750. https://doi.org/10.1002/2014wr016716
Famiglietti JS (2014) The global groundwater crisis. Nat Clim Change 4(11):945–948
Feng S, Hu Q (2007) Changes in winter snowfall/precipitation ratio in the contiguous United States. J Geophys Res 112(D15).D15109. https://doi.org/10.1029/2007JD008397
Fu Q, Feng S (2014) Responses of terrestrial aridity to global warming. J Geophys Res 119(13):7863–7875. https://doi.org/10.1002/2014JD021608
Guo L, Li L (2015) Variation of the proportion of precipitation occurring as snow in the Tian Shan mountains, China. Int J Climatol 35(7):1379–1393. https://doi.org/10.1002/joc.4063
Hobbins M, Wood A, Streubel D et al (2012) What drives the variability of evaporative demand across the conterminous United States? J Hydrometeorol 13(4):1195–1214. https://doi.org/10.1175/JHM-D-11-0101.1
Hsiang SM, Burke M, Miguel E (2013) Quantifying the influence of climate on human conflict. Science 341(6151):1235367. https://doi.org/10.1126/science.1235367
Huang J, Yu H, Guan X et al (2016) Accelerated dryland expansion under climate change. Nat Clim Change 6(2):166–171. https://doi.org/10.1038/nclimate2837
IPCC (2013) The physical science basis. Working group I contribution to the fifth assessment report of the intergovernmental panel on climate change. Cambridge, United Kingdom and New York
Iwasaki H (2006) Impact of interannual variability of meteorological parameters on vegetation activity over Mongolia. J Meteorol Soc Jpn Ser. II 84(4):745–762

Jeong S, Ho C, Kim K et al (2010) Potential impact of vegetation feedback on European heat waves in a 2xCO_2 climate. Clim Change 99(3–4):625–635. https://doi.org/10.1007/s10584-010-9808-7
Jeong S, Ho C, Brown ME et al (2011) Browning in desert boundaries in Asia in recent decades. J Geophys Res 116. https://doi.org/10.1029/2010jd014633
Kaldybayev A, Chen Y, Issanova G et al (2016) Runoff response to the glacier shrinkage in the Karatal river basin, Kazakhstan. Arab J Geosci 9(3):1–8. https://doi.org/10.1007/s12517-015-2106-y
Kezer K, Matsuyama H (2006) Decrease of river runoff in the Lake Balkhash basin in Central Asia. Hydrol Process 20(6):1407–1423. https://doi.org/10.1002/hyp.6097
Knapp AK, Briggs JM, Collins SL et al (2008) Shrub encroachment in North American grasslands: shifts in growth form dominance rapidly alters control of ecosystem carbon inputs. Global Change Biol 14(3):615–623. https://doi.org/10.1111/j.1365-2486.2007.01512.x
Knowles N, Dettinger MD, Cayan DR (2006) Trends in snowfall versus rainfall in the Western United States. J Clim 19(18):4545–4559. https://doi.org/10.1175/JCLI3850.1
Kong Y, Pang Z (2012) Evaluating the sensitivity of glacier rivers to climate change based on hydrograph separation of discharge. J Hydrol 434:121–129. https://doi.org/10.1016/j.jhydrol.2012.02.029
Li S, Zhao W (2010) Satellite-based actual evapotranspiration estimation in the middle reach of the Heihe River Basin using the SEBAL method. Hydrol Process 24(23):3337–3344. https://doi.org/10.1002/hyp.7748
Li X, Zhang S, Peng H et al (2013) Soil water and temperature dynamics in shrub-encroached grasslands and climatic implications: results from Inner Mongolia steppe ecosystem of north China. Agr Forest Meteorol 171:20–30. https://doi.org/10.1016/j.agrformet.2012.11.001
Li Z, Chen Y, Yang J et al (2014) Potential evapotranspiration and its attribution over the past 50 years in the arid region of northwest China. Hydrol Process 28(3):1025–1031. https://doi.org/10.1002/hyp.9643
Li Z, Chen Y, Li W et al (2015) Potential impacts of climate change on vegetation dynamics in Central Asia. J Geophys Res 120(24):12345–12356. https://doi.org/10.1002/2015JD023618
Li Z, Chen Y, Wang Y et al (2016) Drought promoted the disappearance of civilizations along the ancient Silk Road. Environ Earth Sci 75(14):1116. https://doi.org/10.1007/s12665-016-5925-6
Li Z, Chen Y, Fang G, Li Y (2017) Multivariate assessment and attribution of droughts in Central Asia. Sci Rep UK 7:1316. https://doi.org/10.1038/s41598-017-01473-1
Liu W (2015) Scientific understanding of the Belt and Road Initiative of China and related research themes. Prog Geogr 34(5):538–544 (in Chinese)
Liu Y, An Z, Linderholm HW et al (2009) Annual temperatures during the last 2485 years in the mid-eastern Tibetan Plateau inferred from tree rings. Sci China Ser D Earth Sci 52(3):348–359
Maestre FT, Bowker MA, Puche MD et al (2009) Shrub encroachment can reverse desertification in semi-arid Mediterranean grasslands. Ecol Lett 12(9):930–941. https://doi.org/10.1111/j.1461-0248.2009.01352.x
Pederson N, Hessl AE, Baatarbileg N et al (2014) Pluvials, droughts, the Mongol Empire, and modern Mongolia. PNAS 111(12):4375–4379. https://doi.org/10.1073/pnas.1318677111
Sandweiss DH, Maasch KA, Anderson DG (1999) Transitions in the mid-Holocene. Science 283(5401):499–500
Serquet G, Marty C, Dulex JP et al (2011) Seasonal trends and temperature dependence of the snowfall/precipitation-day ratio in Switzerland. Geophys Res Lett 38(7):L07703. https://doi.org/10.1029/2011GL046976
Sheffield J, Wood EF, Roderick ML (2012) Little change in global drought over the past 60 years. Nature 491(7424):435–438. https://doi.org/10.1038/nature11575
Shen Y, Su H, Wang G (2013) The responses of glaciers and snow cover to climate change in Xinjiang (I): hydrological effect. J Glaciol Geocryol 35(3):513–527 [In Chinese with English Abstract]

Shi Y, Yao T, Yang B (1999) Nearly 2000 a climate change of 10a dimension of Guliya ice cores and comparison with China Eastern literature record. Sci China Ser D 29(1):79–86
Singh A, Seitz F, Schwatke C (2012) Inter-annual water storage changes in the Aral Sea from multi-mission satellite altimetry, optical remote sensing, and GRACE satellite gravimetry. Remote Sens Environ 123(Supplement C):187–195. https://doi.org/10.1016/j.rse.2012.01.001
Sorg A, Bolch T, Stoffel M et al (2012) Climate change impacts on glaciers and runoff in Tien Shan (Central Asia). Nat Clim Change 2(10):725–731. https://doi.org/10.1038/NCLIMATE1592
Sun MP, Li ZQ, Yao XJ et al (2013) Rapid shrinkage and hydrological response of a typical continental glacier in the arid region of northwest China—taking Urumqi Glacier No. 1 as an example. Ecohydrology 6(6):909–916. https://doi.org/10.1002/eco.1272
Sun M, Li ZQ, Yao XJ et al (2015) Modeling the hydrological response to climate change in a glacierized high mountain region, northwest China. J Glaciol 61(225):127–136. https://doi.org/10.3189/2015JoG14J033
Tallis H, Mooney H, Andelman S et al (2012) A global system for monitoring ecosystem service change. Bioscience 62(11):977–986. https://doi.org/10.1525/bio.2012.62.11.7
Van der Schrier G, Jones P, Briffa K (2011) The sensitivity of the PDSI to the Thornthwaite and Penman–Monteith parameterizations for potential evapotranspiration. J Geophys Res 116(D3). https://doi.org/10.1029/2010jd015001
Voss KA, Famiglietti JS, Lo M et al (2013) Groundwater depletion in the Middle East from GRACE with implications for transboundary water management in the Tigris-Euphrates-Western Iran region. Water Resour Res 49(2):904–914. https://doi.org/10.1002/wrcr.20078
Wang Y, Roderick M, Shen Y et al (2014) Attribution of satellite-observed vegetation trends in a hyper-arid region of the Heihe River basin, Western China. Hydrol Earth Syst Sci 18(9):3499–3509. https://doi.org/10.5194/hess-18-3499-2014
Weiss H, Courty MA, Wetterstrom W et al (1993) The genesis and collapse of third millennium north Mesopotamian civilization. Science 261:995–995
Xu X, Yeats RS, Yu G (2010) Five short historical earthquake surface ruptures near the Silk Road, Gansu Province, China. B Seismol Soc Am 100(2):541–561. https://doi.org/10.1785/0120080282
Xu C, Chen Y, Chen Y et al (2013) Responses of surface runoff to climate change and human activities in the arid region of Central Asia: a case study in the Tarim River Basin, China. Environ Manag 51(4):926–938. https://doi.org/10.1007/s00267-013-0018-8
Xu BR, Lu ZX, Liu SY et al (2015) Glacier changes and their impacts on the discharge in the past half-century in Tekes watershed, Central Asia. Phys Chem Earth 89–90:96–103. https://doi.org/10.1016/j.pce.2015.03.003
Yang J, Fang G, Chen Y et al (2017) Climate change in the Tianshan and northern Kunlun Mountains based on GCM simulation ensemble with Bayesian model averaging. J Arid Land 9(4):622–634. https://doi.org/10.1007/s40333-017-0100-9
Zhang J, Sun F, Xu J et al (2016a) Dependence of trends in and sensitivity of drought over China (1961–2013) on potential evaporation model. Geophys Res Lett 43(1):206–213. https://doi.org/10.1002/2015GL067473
Zhang Y, Luo Y, Sun L et al (2016b) Using glacier area ratio to quantify effects of melt water on runoff. J Hydrol 538:269–277. https://doi.org/10.1016/j.jhydrol.2016.04.026
Zhou Y (1999) Xinjiang river hydrology and water resources. Xinjiang science and Technology Press, Urumqi

Dr. Yaning Chen is a Research Full Professor at the Xinjiang Institute of Ecology and Geography under the Chinese Academy of Sciences and Director of the State Key Laboratory of Desert and Oasis Ecology. His major research areas include hydrology and water resources, hydro-ecological processes, and ecological sustainability. Chen has authored or co-authored over 10 books and more than 200 articles in these areas.

Dr. Zhi Li is currently a Research Associate Professor at the Xinjiang Institute of Ecology and Geography under the Chinese Academy of Sciences. Her research interests include measurements and modelling of evapotranspiration and assessment of drought formation based on physical approaches linking to evapotranspiration in Central Asia. Li has authored or co-authored two books and 20 articles in these areas.

Dr. Gonghuan Fang is currently a Research Assistant Professor at the Xinjiang Institute of Ecology and Geography under the Chinese Academy of Sciences. Her research interests include distributed hydrological modelling in the arid regions of Central Asia. She has authored or co-authored more than 10 papers in these areas.

Wei Bian is a postgraduate student at the Xinjiang Institute of Ecology and Geography, Chinese Academy of Sciences. Her research focuses on the evaluation of data-limited mountainous precipitation.

The Spatial Network of Beibu Gulf Urban Agglomeration and Its Challenges for Coordinated Integral Development

Junneng Wang, Jianquan Cheng, Nianxiu Qin and Qingyi Chen

Abstract The Beibu Gulf Urban Agglomeration (BGUA), as one of the newly planned urban agglomerations in the southern China, has become a strategical focal point to support the future development of the region and its surrounding areas. From the perspective of city spatial network, this chapter aims to explore BGUA's spatial linkage and main characteristics by analyzing urban networks and their inter-city interactions. The results reveal BGUA's weak position in the network due to its insufficient competition with other cities. We argue that the BGUA, centered on Nanning, Zhanjiang, and Haikou, has been initially formed as a kind of "weak-nuclear" model and the central cities have not played their radial roles in promoting regional economy. The networking effect between cities is still dominated by inter-provincial linkages due to the negative influence from geographical and administrative divisions. There has been limited coordination between the cities within the urban agglomeration due to their varied stage of urban development. We conclude by suggesting the development strategy towards enhancing the networking effect and optimizing the inter-city coordination from four different perspectives, namely, transport infrastructure, economic development, regional governance, and urban/town development.

Keywords Beibu Gulf urban agglomeration · Networking · Spatial linkage
City flow · Inter-city interactions

J. Wang · N. Qin · Q. Chen
School of Geography and Planning, Guangxi Teachers Education University, Nanning 530001, Guangxi, China

J. Wang · J. Cheng (✉) · N. Qin
Key Laboratory of Environment Change and Resources Use in Beibu Gulf, Ministry of Education, Guangxi Teachers Education University, Nanning 530001, China
e-mail: Jianquancheng_China@hotmail.com

X. Yang and S. Jiang (eds.), *Challenges Towards Ecological Sustainability in China*, https://doi.org/10.1007/978-3-030-03484-9_13

1 Introduction

The economic globalization and rapid urbanization have shifted the competition between cities to the competition between urban agglomerations or city clusters. Urban agglomeration, as an emerging spatial unit participating in the global competition and international division of labor forces, will become a strategical focal point to support the future development of the region (Fang 2014, 2017; Ning 2011). The spatial linkage between cities within an urban agglomeration will massively affect the integral development at regional and urban levels. This has been a hot topic in economic geography, regional planning, and spatial development.

The published international literature in this area has primarily covered the spatial structural evolution of inter-city network and the spatial organization of urban agglomeration (Ullman 1957; Meywe 1980; Friedman 1973; Djankov and Freund 2002; Li and Shum 2001; Cabus and Vanhaverbeke 2006; Wheeler and Beebe 2011; Frideman 1986; Pyrgiotis 1991; Kunzmann and Wegener 1991). Meanwhile, a wealth of Chinese literature has extended to cover the inter-city spatial relationships and patterns, theoretically and empirically (Zhou 1998; Zhu and Yu 2002; Miao and Wang 2006; Gu et al. 2006; Fang et al. 2008; Xue 2008; Chen and Song 2011; Zheng et al. 2011; Meng and Zhao 2011; Wen et al. 2013; Zhang et al. 2014; Liu et al. 2014; Guan and Zhou 2014; Wang and Zeng 2014; Niu et al. 2015; Zhou et al. 2016; Sun and Xu 2011; Hu et al. 2016; Lu and Wang 2016; Jiao et al. 2016). The common research methods mainly include, but are not limited to, urban flow intensity analysis, gravity and potential models, spatial syntax, geo-economic theory, accessibility analysis (e.g. Cheng and Bertolini 2013) and social network analysis (Leng et al. 2014). The recent trend is dominated by the application of big data for urban flow space analysis (Castells 1996; Zhen et al. 2015, 2017; Wang and Zhen 2016; Xi et al. 2015). Particularly, there have been many successful case studies using urban gravity and flow intensity models to explore urban systems and agglomeration (e.g. Guan and Liu 2014) in particular in those well-developed regions, e.g. Yangtze River Delta agglomerations, Pearl River Delta agglomerations, and Beijing-Tianjin-Hebei agglomerations. These studies have been gradually extended to the central and southern parts of Liaoning province, the middle reaches of the Yangtze River (e.g. the central China, Chengdu and Chongqing) and other emerging urban agglomerations. With the further development of urbanization, along with the strategies of Western China Development and the Belt and Road Initiative, it can be predicted that the urban agglomeration in the Western China will be approved and promoted in the near future. For example, China's State Council has recently updated "Urban Agglomeration Development Plan in the Beibu Gulf" on 2 March, 2017. This plan at the national level aimed to not only extend the original Beibu Gulf Economic Zone in Guangxi province to three provinces, but also require the Beibu Gulf urban agglomeration to be responsible for promoting regional poverty alleviation and to implement the national strategy of Opening to the World.

Consequently, it is imperative to examine and explore the existing spatial linkages between all these cities at the early development stage of BGUA. There have been rare studies on this agglomeration as most studies are concentrated in Guangxi province as a whole (Guo 2015; Zhu et al. 2013; Zhang and Zhou 2012; Zhu 2013; Xu and Wu 2016). In this chapter, the spatial network among all the cities within the BGUA will be analyzed by using the urban flow intensity and interaction models, and the solutions to promoting these spatial linkages will be discussed to support for the new planning of this urban agglomeration. The analytical results will provide quantitative evidences to implement the Urban Agglomeration Development Plan in the Beibu Gulf.

2 Beibu Gulf Urban Agglomeration

The Beibu Gulf urban agglomeration is composed of three provincial-level administrative regions, which include specifically 11 prefecture-level cities, a county-level cities and 3 counties (e.g. Nanning, Beihai, Qinzhou, Fangchenggang, Yulin and Chongzuo in Guangxi Zhuang Autonomous Region; Zhanjiang, Maoming and Yangjiang in Guangdong province; and Haikou, Danzhou, Dongfang, Chengmai County, Lingao County, Changjiang County in Hainan province). The whole agglomeration has a total area of 116,000 square kilometres (Fig. 1), and a total coastline of 4,234 km. The total population was 41.41 million, and its total GDP 0.16295 trillion Yuan RMB in 2015, accounting for 3.01% and 2.25% of that of China respectively.

This region that is ASEAN-oriented and serves “three Souths” (southwest, central south, southern China), has been planned to form a city network to provide liveable and active environment with the blue bay for both living and doing business. The development pattern of “one bay and two central cities, one core city and two poles” has been formulated by positioning Nanning as a core city and Zhanjiang and Haikou as two central cities. It is proposed to jointly promote the construction of inter-city infrastructure network, improve the access to the “three Souths”, develop road network linked to ASEAN countries, and particularly enhance the motorway connection between cities and counties within the urban agglomeration. All these developments aim to make the Beibu Gulf Urban Agglomeration to have a modern system of road, sea and air transportation.

Though BGUA is regarded as one of the most promising urban agglomerations in southern China, most scholars argue that it is still at the initial stage of development according to the basic requirements for and spatial structure standard of urban agglomeration (Fang 2014; Pan and Liu 2014). Moreover, some nationally well-known scholars (Ning 2011; Yao and Zhou 2016) even have not listed it into the national network of urban agglomerations across China. Currently, for the policy-driven purpose, the spatial structures of and inter-city linkages within any urban agglomeration should be further investigated to provide more evidences for the plan-making at national and regional levels.

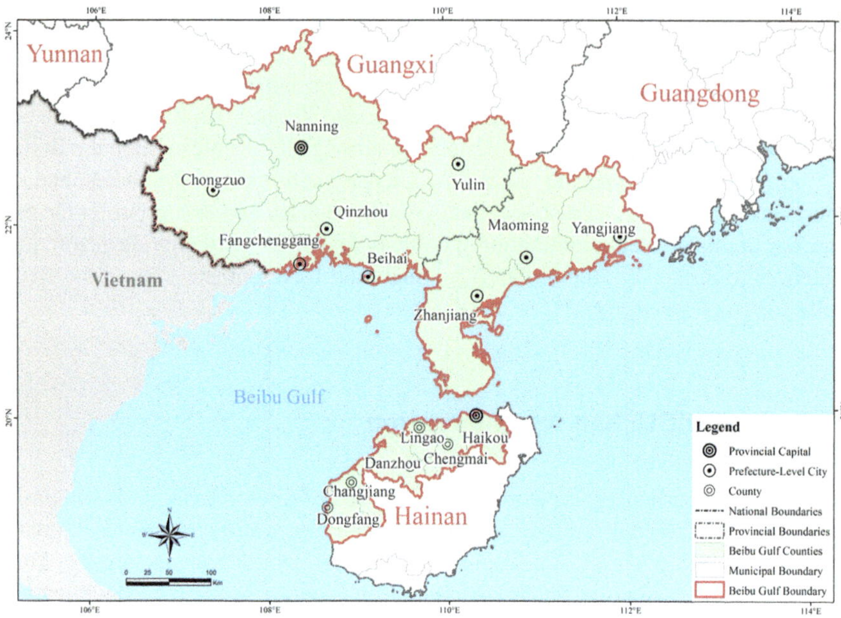

Fig. 1 The geographical location of the urban agglomeration in the Beibu Gulf

3 Data and Methods

3.1 Data Sets

In order to measure the energy of urban extroversion and the intensity of urban flows, 13 export-oriented industries, such as mining, have been selected as basic economic elements for quantitative analysis. Fourteen indexes including per capita GDP were chosen to evaluate the position or level of urban development. The original data of all these indicators are collected from national and provincial statistical yearbooks including "Chinese city statistics yearbook (2015)", "Hainan statistical yearbook (2015)", Guangxi statistical yearbook (2015)", and "Guangdong statistical yearbook" (2015). All the required spatial data sets were extracted from the Baidu web service (https://map.baidu.com/).

3.2 Urban Flow Intensity Analysis

The intensity of urban flow is generally used to analyze the external influence of the urban agglomeration and its radiation across a region, which can reflect the strength of the spatial linkage between the cities (Eq. 1).

$$F = N * E \tag{1}$$

where F is urban flow intensity; N is the functional efficiency of a city, which indicates the actual influence of extroversion functional capacity produced by each city. E is the city's external function capacity, which reflects the strength of city's extra-version function.

The level of extroversion functional capacity E mainly depends on the location quotient of the employees in a certain industry Lq. The location quotient (Lq_{ij}) for employee of j industry in city i, is calculated as follows (Eq. 2):

$$Lq_{ij} = \frac{G_{ij}/G_j}{G_j/G}(\mathrm{i} = 1, 2\ldots \mathrm{n}; \mathrm{j} = 1, 2\ldots, \mathrm{m}) \tag{2}$$

where G_{ij} is the number of employees in j industry of the city i; G_j means the total number of employees in industry j in the entire country; G_i the total number of employees in city i, and G the total number of employees in the country.

When the location quotient value is greater than 1, it indicates that this industry, as a specialized sector, has certain advantages over other cities across the region. By contrast, when the location quotient value is less than 1, it implies that this industry only serves the cities within the region without any advantage over others. The larger the location quotient value is, the higher the specialization degree of the industry is in the region, and vice versa. When $Lq > 1$ for an industry in a city, it indicates that the employment in this industry is better than the national average and the industry has high extroversion functional capacity. If $Lq < 1$ for an industry in the city, it indicates that the industry has low extroversion functional capacity, that is, $E = 0$.

The extroversion functional capacity of industry j in city i is calculated as E_{ij} (Eq. 3):

$$E_{ij} = G_{ij} - G_i(G_j/G) \tag{3}$$

The total extraversion functional capacity for M industries in city i is calculated as follows (Eq. 4):

$$F_i = \sum_{j=1}^{m} E_{ij} \tag{4}$$

The functional efficiency N_i of city i, based on the per capita GDP, is calculated as follows (Eq. 5):

$$N_i = GDP_i/G_i \tag{5}$$

At last, the urban flow intensity for city i—F_i is calculated as follows (Eq. 6):

$$F_i = N_i * E_i = E_i(GDP_i/G_i) = GDP_i(E_i/G_i) = GDP_i * K_i \tag{6}$$

where, K_i is the slope of urban flow, meaning the proportion of urban extroverted functional capacity over its GDP. To some extent, it reflects extroversion strength of city functional capacity and its relative position in the region.

3.3 Urban Gravity Model

The urban gravity model, which has existed for decades, is used to measure the gravitation of urban space. Its formula is listed as follows (Eq. 7):

$$R = \frac{\partial(M_a * M_b)}{d^2} \tag{7}$$

where R is the intensity of inter-city interaction or gravity; M_a and M_b, are the mass of cities a and b respectively, indicating their comprehensive development level. ∂ is a gravity coefficient. In simplicity, it is assumed $\partial = 1 \cdot d$ is the economic distance between the two cities a and b.

The assessment of city development level was conducted by using principal component analysis (Cheng 2000) based on the selected 14 indicators such as regional GDP. The statistical data were standardized and the principal component analysis was implemented by using SPSS 24.0. The economic distance d is measured as the shortest time between each pair of cities based on the highway network over the Baidu map web service (https://map.baidu.com/).

4 Results

4.1 The Location Quotient of Each Industry in the City

The location quotient was calculated for each of 13 industries within the Beibu Gulf Urban Agglomeration according to Eq. 2. These industries are specified as mining, manufacturing, construction, wholesale and retail trade, transportation, warehousing and postal service, accommodation and catering industry, scientific research and technical services, information transmission software and information technology services, public health and social work, finance, leasing and business services, culture, sports and entertainment, public management, social security and social organizations. This classification is based on the national standard of socio-economic statistics. Their results of location quotient are listed in Table 1.

Not a single city has achieved that the location quotient values of all industries are greater than 1. However, 10 industries in Haikou city have achieved the greater value of location quotient than 1, comparatively only 8 in both Nanning and Zhanjiang cities, 5 or less in the remaining 12 cities (or counties). All these values and variations

Table 1 The location quotient of the main export-oriented industries within the BGUA in 2014

City or County	Industries												
	Mining	Manufacturing	Construction	Wholesale and retail trades	Transport, storage and pose	Hotels and catering services	Information transmission, software and information technology	Financial intermediation	Leasing and business services	Scientific research and technical services	Health and social service	Culture, sports and entertainment	Public management, social security and social organization
Nanning	0.01	0.52	1.29	0.99	1.06	1.25	0.90	1.15	1.89	1.71	1.30	1.59	0.93
Beihai	0.19	0.94	0.65	0.43	0.69	1.10	0.64	1.48	0.85	0.93	1.57	1.05	1.28
Qinzhou	0.24	0.55	1.57	0.55	0.61	0.44	0.35	0.61	0.40	0.58	2.15	0.50	1.29
Fangchenggang	0.25	0.21	0.78	0.39	2.24	0.64	0.84	0.66	0.73	0.74	1.76	0.82	1.83
Yulin	0.02	0.91	0.86	0.55	0.64	0.37	0.54	0.77	0.72	0.68	1.83	0.62	1.12
Chongzuo	0.95	0.52	0.27	0.56	0.48	0.56	0.60	0.96	0.83	1.35	1.98	0.75	2.13
Zhanjiang	0.38	0.51	1.31	1.05	1.12	1.09	0.74	1.06	1.06	0.48	1.40	0.63	1.05
Yangjiang	0.02	0.90	1.29	0.80	0.72	0.98	0.61	0.84	0.33	0.30	1.23	0.56	1.53
Maoming	0.12	0.57	1.50	1.25	0.58	0.63	0.50	0.89	0.51	0.37	1.36	0.61	1.09
Haikou	0.01	0.36	0.60	1.63	1.55	2.23	1.43	1.49	1.27	1.36	1.12	1.82	1.01
Danzhou	0.03	0.28	0.40	2.23	1.12	2.49	0.27	0.04	0.64	0.31	0.22	0.40	0.19
Dongfang	0.07	0.13	0.17	1.80	0.89	2.59	0.20	0.02	0.20	0.08	0.19	0.21	0.23
Chengmai county	0.03	0.34	0.35	2.19	0.69	4.60	0.75	0.03	0.29	0.12	0.14	0.27	0.24
Lingao county	0.00	0.18	0.36	1.54	0.64	2.74	0.51	0.03	0.27	0.06	0.18	0.18	0.24
Changjiang county	1.33	0.18	0.13	1.32	0.55	1.97	0.14	0.03	0.10	0.17	0.26	0.27	0.38

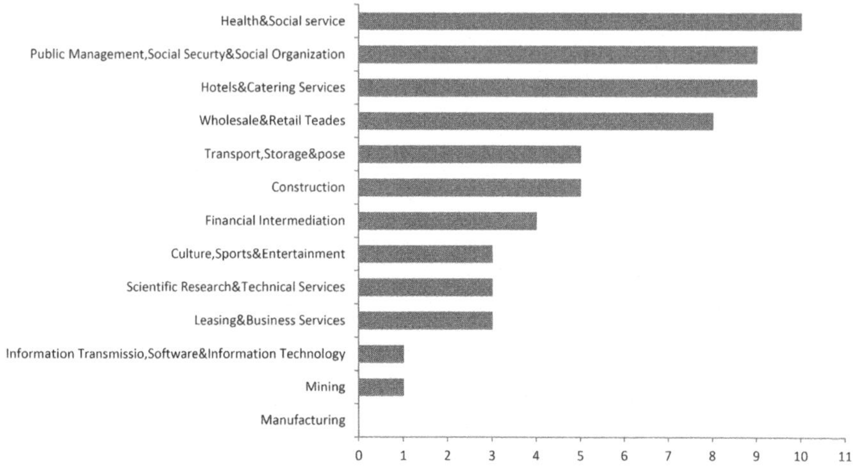

Fig. 2 The number of cities with the value of location quotient greater than 1 for export-oriented industry in the BGUA

between cities and counties indicate the weak specialization of the industries within the BGUA and relatively weak competition between these industries as well.

By contrast, six relatively strong industries within the BGUA are the wholesale and retail, lodging and catering, health and social work, public administration, social security and social organization. These four industries: manufacturing, information transmission, software and information technology services and other industries need to be further developed (Fig. 2). The results have also demonstrated homogeneous competitions in main industries between these cities and counties, or rather, there have been lacking good industrial cooperation within the BGUA.

The three central cities (Nanning, Haokou and Zhanjiang) have not played sufficiently leading roles in driving the industrial development within the urban agglomeration. Nanning city has the highest values of location quotient for four industries: leasing industry, business service industry, scientific research and technical service industry. Other four industries in Nanning: culture, sports and entertainment industries, information technology services and other industries, show some relative strength. However, the entire set of industries as a whole have not demonstrated strong and clear advantages over Haikou and Zhanjiang. Haikou city has the most industries with the value of location quotient greater than 1, particularly the industry of accommodation and catering has achieved a value up to 2.23, followed by the industries of transportation, culture and sports, which have been influenced by the tourism development in Haikou city. Zhanjiang city has 8 industries with the value of location quotient greater than 1. But the relatively smaller values of these industries indicate the weak position within the urban agglomeration.

Table 2 Total external functional capacity, flow slope and city flow intensity in the BGUA

City or County	GDP_i per person employed (100 million yuan RMB)	Total external functional capacity E	GDP_i/G_i	Flow slope K	City flow intensity F_i
Nanning	32.85	10.88	0.34	0.11	3.73
Beihai	59.44	0.96	4.12	0.07	3.96
Qinzhou	42.24	3.39	2.09	0.17	7.07
Fangchenggang	61.00	1.59	6.32	0.16	10.03
Yulin	38.64	1.65	1.11	0.05	1.84
Chongzuo	47.76	2.04	3.51	0.15	7.15
Zhanjiang	43.71	4.36	0.85	0.08	3.69
Yangjiang	47.77	2.49	1.95	0.10	4.86
Maoming	52.35	5.20	1.17	0.12	6.07
Haikou	21.29	6.45	0.42	0.13	2.68
Danzhou	9.98	3.92	0.23	0.09	0.89
Dongfang	6.09	1.42	0.28	0.06	0.39
Chengmai county	7.07	3.68	0.22	0.11	0.81
Lingao county	7.23	1.01	0.39	0.05	0.39
Changjiang county	7.79	0.51	0.64	0.04	0.32

4.2 *Urban Flow Intensity*

Based on the number of employees in 13 major export-oriented industries in the city, the urban flow intensity F_i (Eq. 1) and the urban flow slope K (Eq. 6) were calculated respectively (Table 2) and summarized as follows. Generally speaking, the overall strength of BGUA is relatively weak as the total urban flow intensity within the BGUA was 5.388 billion Yuan RMB in 2014 with an average urban flow intensity of 359 million Yuan RMB, which is less than 10% of the total urban flow intensity (57.136 billion yuan RMB) within the Yangtze River Delta Urban Agglomeration in 2008, and also lower than that of Wuhan metropolitan area (6.839 billion yuan RMB) and Zhongyuan Urban Agglomeration in 2008 (6.446 billion yuan RMB).

It demonstrates a "weak-nuclear" urban agglomeration without leading roles of central cities. The urban flow intensity of Nanning city is 373 million Yuan RMB, which is slightly larger than the average value of the region, but lower than that of other six cities (e.g. Fangchenggang city and Qinzhou city). The urban flow intensity values of the Yangtze River Delta, Wuhan metropolitan area and Zhongyuan urban agglomeration in 2008 have been about 53, 10, and 10 times more than that of BGUA in 2014 respectively. This gap has shown that the economic radiation capacity of

Nanning city is not strong, same as Haikou and Zhanjiang cities. The highest level of per capita GDP in the region is 61.00 billion Yuan RMB for Fangchenggang city, followed by Beihai city and Maoming city. However, as the central cities within the agglomeration, the per capita GDP values of three cities are not strong by contrast, e.g. 32.85 billion yuan RMB for Nanning, 43.71 billion yuan RMB for Zhanjiang and 21.29 billion yuan RMB for Haikou city.

Urban "short board" is evidential in this urban agglomeration. The total external functional capacity of Nanning is 10.88, which is the highest in the urban agglomeration. This has shown that the export-oriented industries account for a large proportion of the entire industries in this city. However, its GDP per person employed is relatively low, as well as its urban flow intensity of only 3.73. In contrast, GDP per person employed in Fangchenggang City is the highest and the urban flow intensity is the strongest in the BGUA, but its total external functional capacity is relatively low. Thereby, the overall strength of Fangchenggang city is low, same situation as Beihai, Qinzhou, Yulin, Yangjiang and Maoming cities.

Urban extroversion demonstrates a large variation between cities across the BGUA. Qinzhou, Fangchenggang, Chongzuo, Haikou, Maoming, Chengmai, and Nanning, have a larger urban flow slope and their overall extroversion of urban industry have certain advantages over others in the region. The urban flow intensity of Chongzuo, Qinzhou and Maoming cities are strongly influenced by the external factors such as trade ports and harbours. Yangjiang, Zhanjiang, Changjiang, Danzhou, Lingao, Yulin, Dongfang, and Beihai cities have a relatively low flow slope and low proportion of external functional capacity over the total functional capacity. This has reflected that the economic activities mainly serve the city itself. In addition to Haikou city, the other five cities (counties) within Hainan province have very low urban flow intensity so it means its external functional capacity of industries is not strong.

4.3 *Urban Interactions Within BGUA*

4.3.1 Principal Component Analysis of Urban Mass

The first step is to evaluate the comprehensive strength or quality of each city within the urban agglomeration region by using principal component analysis (PCA). The criteria for the assessment are dominated by the selected 14 indicators (Table 3) including economic, social and potential competitiveness, and radiation intensity. The selection of these indicators is based on the current development stage of the Beibu Gulf urban agglomeration and a literature review (e.g. Zhou et al. 2016).

The PCA analysis starts with KMO and Bartley sphericity tests, which are carried out after the normalization of all factors by using SPSS 24.0. The value of KMO test has reached 85.248%. The PCA analysis reveals that four principal components have contributed 80% in the cumulative variances, which were selected for further analysis by using the maximum variance rotation method (Table 4). Four principal component

Table 3 Criteria for evaluating the urban mass within the BGUA

Criteria	Indicators
Economic competitiveness	Gross domestic product (GDP—billion yuan RMB)
	Proportion of tertiary industry to GDP (%)
	Fixed investment in the whole society (100 million yuan RMB)
	Actual amount of foreign investment (million dollars)
	Per capita GDP (yuan/person)
Social competitiveness	Built-up area (square kilometer)
	Total retail sales of social consumption goods (billion yuan RMB)
	The number of doctors per hundred people
	Per capita living water consumption (per person million tons)
Radiation competitiveness	Total passenger volume (ten thousand persons)
	Total freight volume (ten thousand tons)
	Total export value of foreign trade (ten thousand dollars)
Potential competitiveness	Book collections of public library per 100 people (books)
	Number of students per hundred senior high school students

factors (FAC1-FAC4) are calculated for each city in Beibu Gulf urban agglomeration. The comprehensive strength or urban quality of each city is aggregated by using the variance contribution from the square sum of rotation load (Table 4) as weights (Sun and Xu 2011), that is, the mass of a city, indicating its comprehensive strength, is calculated as follows (Eq. 8):

$$M = FAC1 * 40.33\% + FAC2 * 24.67\% + FAC3 * 11.48\% + FAC4 * 8.97\% \quad (8)$$

The results in the table above indicate that there is still a large spatial variation in urban mass or quality between the cities in the BGUA. For example, only Nanning, Haikou, Zhanjiang and Qinzhou demonstrate a positive M value. According to the amended value Mq, all the cities and counties are ranked, among which Nanning takes the first place, contributed from its special location in the urban agglomeration, In general, the ranking is basically consistent with the central city of the planning space system, but it also reveals a weak nuclear pattern in the urban agglomeration. At the provincial scale, the three cities in Guangdong province, above the average level in ranking, show more balanced development. By contrast, the six cities in Guangxi province demonstrate a slightly varied pattern because Nanning and Qinzhou are much higher than others in the ranking but Yulin, Beihai and Fangchenggang cities are only higher than the rest cities (counties) of Hainan province excluding Haikou city that is ranked the second. The majority of cities and counties in Hainan province are not ranked high, and with high spatial variation (Table 5).

Table 4 The eigenvalue and variance contribution

Component	Initial eigenvalue			Extraction load square sum			Sum of squares of rotational loads		
	Total	% of variance	Cumulative %	Total	% of variance	Cumulative %	Total	% of variance	Cumulative %
1	6.355	45.395	45.395	6.355	45.395	45.395	5.647	40.334	40.334
2	2.796	19.97	65.365	2.796	19.97	65.365	3.454	24.673	65.006
3	1.595	11.392	76.757	1.595	11.392	76.757	1.607	11.48	76.486
4	1.217	8.696	85.453	1.217	8.696	85.453	1.255	8.967	85.453
5	0.731	5.223	90.677						
6	0.435	3.11	93.787						
7	0.308	2.2	95.986						

Table 5 The results of urban mass (M) in the Beibu Gulf urban agglomeration

City or County	*FAC*1	*FAC*2	*FAC*3	*FAC*4	Urban mass *M*	Adjusted urban mass *Mβ*	Ranking
Nanning	2.52	0.52	2.08	0.02	1.39	19.87	1
Haikou	0.07	3.09	−0.83	0.49	0.74	13.42	2
Zhanjiang	1.28	−0.46	−1.30	0.56	0.30	9.02	3
Qinzhou	0.27	−0.50	0.71	0.32	0.10	6.96	4
Maoming	0.95	−0.60	−1.54	−1.21	−0.05	5.50	5
Yangjiang	−0.11	−0.13	0.01	−0.04	−0.08	5.20	6
Chongzuo	−0.44	−0.83	−0.06	3.14	−0.11	4.94	7
Danzhou	−0.67	0.84	−0.64	0.13	−0.12	4.76	8
Yulin	0.66	−1.15	−0.69	−0.45	−0.14	4.61	9
Beihai	−0.16	0.09	−0.35	−0.91	−0.16	4.37	10
Fangchenggang	−0.69	0.07	0.86	−0.34	−0.19	4.09	11
Changjiang county	−0.97	0.09	1.18	−0.69	−0.29	3.06	12
Chengmai county	−0.99	−0.22	1.05	−0.41	−0.37	2.29	13
Dongfang	−0.97	−0.10	−0.42	−0.40	−0.50	1.02	14
Lingao county	−0.77	−0.71	−0.07	−0.22	−0.51	0.88	15

Note For the purpose of easy calculation and comparison, the urban mass calculation is adjusted as: $M\beta = (M + 0.6) * 10$

4.3.2 Interaction and Relationship Between Cities

The interaction intensity of 15 cities (counties) across the Beibu Gulf urban agglomeration is calculated according to the formulas (7) and (8) and further classified into three categories: weak, normal and strong (Fig. 3). The spatial development of inter-city interaction intensity is uneven, which is characterized by the stronger inter-province interaction than the intra-province interaction. At the province level, the interaction intensity between Guangdong and Guangxi provinces is stronger than others. The interaction between Hainan and Guangdong provinces is stronger than that of Hainan and Guangxi. Guangdong and Guangxi provinces, adjacent to each other, have strong transport links. Although Guangdong and Hainan provinces are separated by the sea, Guangdong is the main channel connecting Hainan to the mainland. The following six pairs of cities: Nanning–Qinzhou, Nanning–Chongzuo, Nanning–Fangchenggang, Qinzhou–Fangchenggang, Zhanjiang–Maoming and Haikou–Chengmai, adjacent to each other, have relatively higher interaction intensity than the rest. It can be seen that the interaction intensity between cities is affected by city administrative division, transport accessibility, spatial distance and geographical barrier. All these results have demonstrated a loose spatial structure of BGUA,

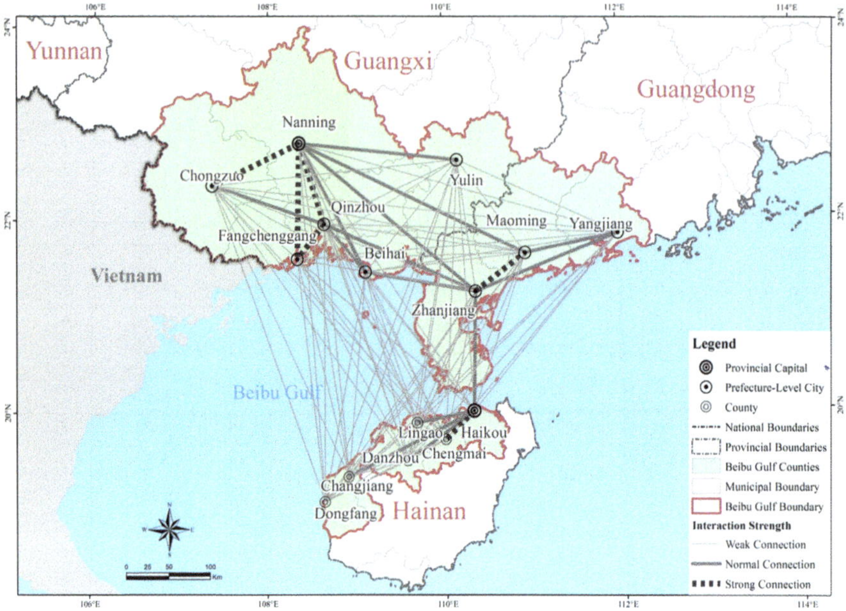

Fig. 3 Urban interactions between cities (counties) within the BGUA

compared with other urban agglomerations across China. Nanning, Zhanjiang and Haikou have shown clear provincial centrality within the urban agglomeration. Nanning city has the largest total intensity of interaction, which has reached 151.97, but it is more connected to the cities within Guangxi province and less to the cities and counties within Hainan province. The Nanning city has been planned as the core city of this urban agglomeration but there will be a long way to achieve this objective. As the central cities of BGUA, the interaction between Zhanjiang and Haikou cities is only limited to the provincial level. Accordingly, it can be seen that the core and central cities within the BGUA are centralized only locally, so it is imperative to strengthen its intensity and distance of radiation.

The interaction intensity between cities in urban agglomeration has demonstrated a large spatial variation. The interaction intensity between the six cities within Guangxi province is relatively stronger, indicating clear regional integration but there is weaker interaction between the three cities (Yulin, Beihai and Chongzuo) and the others. Consequently, the integral development of Nanning, Beihai, Qinzhou, Fangchenggang, Chongzuo and Yulin should be further enhanced in the future. The three cities in Guangdong and seven cities (counties) in Hainan have shown relatively average interaction intensity in their own province, with an emerging trend of regional integration. Thereby, it is necessary to strengthen not only the intra-province but also inter-province regional integration process within the BGUA.

5 Discussion and Conclusions

This chapter has analyzed the spatial organization of the BGUA by using the methods of location quotient, urban flow and interaction intensity models. In general, the development within the BGUA is still weak or in its infancy in terms of spatial connection, total urban flow intensity, and overall economic competitiveness of urban agglomeration, which are not high or strong enough at the current stage of urbanization.

The integral results of urban flow model, the quality of urban economy, and urban interaction model have revealed the "short board" pattern in the process of the regional development in the BGUA as there has been limited cooperative development between these cities and counties within the region, and regional resources have not been properly configured as well.

Spatially, the central city's functions of BGUA have not yet been clearly defined and highlighted. Nanning as the core city and Zhanjiang and Haikou as the two central cities have not played their leading roles in the regional development of this agglomeration as planned due to their weak external radiation capacity.

The main contributing reasons include city administrative divisions, geography barrier and other political and socio-economic factors; for example, the inter-provincial urban interaction is weak. Consequently, the BGUA is characterized by a "weak-nuclear" development pattern. We argue that the spatial linkage of cities and counties within the BGUA should be further enhanced to achieve a coordinated development, particularly from the following four aspects: transport infrastructure, economic development, regional governance, and urban/town development.

By improving regional transport infrastructure, a fast and efficient transport network system, composed of high-speed railway, expressway, aviation, waterway and highway, can enhance the urban interactions and flow intensity across the urban agglomeration, which enables the integration of this region into a "city group" and benefits to the coordinated development between all the cities and counties.

The BGUA serves as a bridge between the Eastern and Western regions, Guangdong, Hong Kong, Macao, Taiwan and China ASEAN Cooperation by actively undertaking industrial transfer and attracting investment. The homogenization of the internal city industry can be solved by strengthening the construction of the central city industrial system and improving the rationality of the industrial structure. The regional competitiveness can be further enhanced as well by properly guiding the inter-city industrial division and cooperation. A strategic fulcrum for the "One-Belt and One-Road" initiative will be developed in the BGUA by promoting its development from a simple collated model of city "addition" to a city "multiplication" model with orderly division of labor and cooperation. All these help the urban agglomeration play a leading role in the regional development.

As the BGUA covers three administrative provinces, there has been lacking consistent development goals and regional governance between them. The administrative division "imprisonment" should be broken to construct an agency to coordinate the integral development of the three provinces (autonomous regions).The disparity of regional policies can be removed or reduced by implementing a joint conference system between the three provinces. By strengthening the propaganda of BGUA development, a common dialogue mechanism to promote the participation of "government organizations, enterprises, social organizations, and residents" can be formed to guide regional integral development.

Although a "weak-nuclear" urban agglomeration has been initially formed within the study area, which is centered by "Nanning, Zhanjiang and Haikou cities", the centrality of these three cities should be further enhanced to achieve the transfer from the weak-nuclear to nuclear pattern on the one hand, and the urban and rural development in particular the development of small and medium-sized cities and towns should be motivated on the other hand. As such, the small-scale development with low density within the urban agglomeration can be changed to promote the regional sustainable development.

Acknowledgements This work was jointly supported with funding given by the National Natural Science Foundation of China (Grant No: 41461007) and the Early Career Foundation from the Guangxi Teachers Education University (Grant No: 8656-QQJJ2013005).

References

Cabus P, Vanhaverbeke W (2006) The territoriality of the network economy and urban networks: evidence from Flanders. Entrep Reg Dev 18:25–33

Castells M (1996) The rise of the network society. Black-well Publishers, MA

Chen QY, Song YX (2011) Analysis on the special relation of Circum-Chang-Zhu-Tan urban agglomeration based on urban flow. Econ Geogr 31:1840–1844

Cheng JQ (2000) Urban systems engineering. Press of Wuhan University, Wuhan

Cheng JQ, Berterloni L (2013) Understanding urban networks: comparing a node-, a density-and an accessibility-based view. Cities Int J Urban Policy Plan 31:165–176

Djankov S, Freund C (2002) Trade flows in the former Soviet Union from 1987 to 1996. J Comp Econ 30:76–90

Fang CL (2014) Progress and the future direction of research into urban agglomeration in China. Acta Geogr Sin 69:1130–1144

Fang CL (2017) Theoretical foundation and patterns of coordinated development of the Beijing-Tianjin-Hebei urban agglomeration. Prog Geogr 36:15–24

Fang CL, Qi WF, Song JT (2008) Researches on comprehensive measurement of compactness of urban agglomerations in China. Acta Geogr Sin 63:1011–1021

Frideman JR (1986) The world city hypothesis: development & change. Urban Stud 23:5–137

Friedman J (1973) The spatial organization of power in the development of urban systems. Dev Change 4:12–50

Gu CL, Zhang M, Zhang C (2006) Prospects of urban agglomeration in the Yangtze river delta. Sci Geogr Sin 15:771–775

Guan XG, Liu L (2014) Spatial linkage analysis of Beijing Tianjin Hebei urban agglomeration based on modified gravity model. Urban Probl 11:21–26

Guan W, Zhou XT (2014) The spatio-temporal evolvement of spatial interaction among cities of south central Liaoning. Econ Geogr 34:48–55

Guo J (2015) Research on the optimization of spatial structure of Beibu Gulf urban agglomeration. Guangxi normal university

Hu Y, Zhang J, Liu ZH, Lin AW (2016) The research of the middle reaches of the Yangtze river's spatial contact-based on the Gravity model and the urban flow. Mod Urban Res 1:52–57
Jiao LM, Tang X, Liu XP (2016) Spatial linkage and urban expansion: an urban agglomeration perspective. Prog Geogr 35:1177–1185
Kunzmann KR, Wegener M (1991) The pattern of urbanization in Western Europe. Ekisties 50:156–178
Leng BR, Yang YC, Tan YM (2014) City network studies: the transformation of research perspective from hierarchy to network. Urban Plan Int 29:1–7
Li SM, Shum YM (2001) Impacts of the national trunk highway system on accessibility in China. J Transp Geogr 9:39–48
Liu JY, Yang HL, Song Q, Fan XX (2014) Spatio-temporal pattern evolution of the interaction among the cities in central plains economic zone. Sci Geogr Sin 34:1060–1068
Lu C, Wang DQ (2016) A comparative analysis of economic connections between cities in Shangdong Peninsula urban agglomeration. J Beijing Jiaotong Univ (social sciences edition) 15:22–32
Meng DY, Zhao WL (2011) Spatial structure of urban economic linkage in Henan province based on travel time cost. Urban Stud 18:104–110
Meywe DR (1980) A dynamic model of the integration of frontier urban places into the United States system of cities. Econ Geogr 56:39–120
Miao CH, Wang HJ (2006) On the direction and intensity of urban economic contacts in Henan province. Geogr Res 25:222–232
Ning Y (2011) Definition of Chinese metropolitan areas and large urban agglomeration—role of large urban agglomeration in regional development. Sci. Geogr Sin 31:257–263
Niu FQ, Liu WD, Song T, Hu ZD (2015) A multi-level spatial structure analysis algorithm for urban agglomeration study in China. Geogr Res 34:1447–1460
Pan JH, Liu WS (2014) Identification of spatial influence sphere of urban agglomerations in China based on urban hinterland delimitation. Adv Earth Sci 29:352–360
Pyrgiotis YN (1991) Urban networking in Europe. Ekistics 50:350–351
Sun J, Xu CZ (2011) The construction and application of the economic gravity model on the perspective of spatial economic-example from the economic data of Yangtze river delta in 2010. Economist 7:37–44
Ullman E (1957) American commodity flow Seattle. University of Washington press, pp 60–73
Wang T, Zeng JX (2014) Spatial patterns and development strategy of competition-integration between cities in urban agglomeration in the middle reaches of the Yangtze river. Trop Geogr 34:390–398
Wang B, Zhen F (2016) China's city hierarchy under internet and its influencing mechanism: an empirical analysis based on Baidu search. Economic geography 36:46–52
Wen YZ, Zhong YX, Huang J (2013) The effects of the traffic network evolution on the Hinterland of central cities—a case of Jiangxi province. Econ Geogr 6:59–65
Wheeler SM, Beebe CW (2011) The rise of the postmodern metropolis: spatial evolution of the Sacramento metropolitan region. J Urban Des 16:307–332
Xi GL, Zhen F, Zhang M, Yin G, Jiang JL (2015) Spatio-temporal evolution and regional connection of online consumption: a case study on Jingdong mall. Scintia Geogr Sin 35:1372–1379
Xu LY, Wu LP (2016) Economic linkages and Geo-economic relations among cities in the Beibu gulf urban agglomeration. Urban problems 10:59–64
Xue JF (2008) Hierarchical structure and distribution pattern of Chinese urban system based on aviation network. Geogr Res 27:23–33
Yao SM, Zhou CS (2016) New perspectives on urban agglomerations in China. Science press, Beijing
Zhang XK, Zhou PF (2012) Research on city agglomeration coordinated development based on urban flow intensity analysis in Beibu gulf. J Guangxi Univ Financ Econ 25:29–35
Zhang HZ, Wu YD, Wang YC (2014) Research of the match state between the economic contact and geo-economics relationship in Yunnan central urban agglomeration. Areal Res Dev 33:16–19

Zhen F, Wang B, Wei ZC (2015) The rise of the internet city in China: production and consumption of internet information. Urban Stud 52:2313–2329
Zhen F, Cao Y, Qin X, Wang B (2017) Delineation of an urban agglomeration boundary based on sina weibo microblog 'check-in' data: a case study of the Yangtze river delta. Cities 60:180–191
Zheng LH, Deng XL, Hou Y (2011) A measurement analysis of economic connection between Guanzhong cities based on Gravity model. Hum Geogr 26:80–84
Zhou YX (1998) Major directions of economic linkages: some theoretical considerations. Urban Plan 2:22–25
Zhou XY, Hua M, Qin YW, Ma XX (2016) Urban agglomeration spatial contacts research in the middle reach of Yangtze river. Resour Environ Yangtze Basin 25:1492–1500
Zhu MM (2013) A comparative study on spatial structure evolution of urban agglomerations in Pearl River delta and Beibu gulf. Guangxi university
Zhu SP, X B, Mao JX (2013) Analysis on the spatial relation of Beibu Bay urban agglomeration. J Northwest Norm Univ (natural science) 6:106–111
Zhu YM, Yu NW (2002) Study on urban flow in Shanghai Hangzhou Ningbo Urban Agglomeration Area. Urban Plan Forum 1(31–33):44

Junneng Wang is a Lecturer of Human Geography in the School of Geography and Planning, Guangxi Teachers Education University, China. His research interests include human geography and regional planning particularly in the less-developed region of China. Wang has published about 10 peer-reviewed papers.

Dr. Jianquan Cheng is a Reader of urban studies at the Manchester Metropolitan University (MMU), UK and a deputy director of MMU Crime and Well-being Big Data Centre. Cheng's research interests include urban and regional studies, urbanization of China, and urban analytics. Cheng has authored or co-authored two books and about 60 peer reviewed papers.

Dr. Nianxiu Qin is an Associate Professor of Geography at the Guangxi Teachers Education University. His research interests include physical geography and climate change. Qin has published about 30 peer reviewed papers.

Qingyi Chen is a postgraduate student at the Guangxi Teachers Education University, China. Chen's research interest includes human geography and urban-rural planning.

Strategies to Promote Green Development in Coal Resource-Based Regions with China's Shanxi Province as an Example

Jianhui Cong, Jun Yang, Qin Zhang and Zhou Zhou

Abstract The economic development in coal resource-based regions of China has been quite difficult. The traditional "black" development pattern based on mineral resources has brought out some negative consequences to economic development, and its transformation towards the green development pattern becomes imminent. In this chapter, we define the essential dimensions of the green development pattern and examine several major obstacles toward the green development in coal resource-based regions with China's Shanxi province as an example. We further discuss some strategies to tackle the dilemma and suggest that coal resource-based regions should promote the coal supply-side structural reform, improve the environmental capacity and resource values, and promote the quality of life for people.

Keywords Green development · Coal resource-based regions
Coal supply-side structural reform · Shanxi province · China

1 Introduction

The bearing capacity of China's ecological environment has currently reached the upper limit, and the need to promote the ecological civilization construction is more urgent than ever before. As an important guideline during the 13th five-year period or a longer time period in China, green development will play an essential role in constructing a moderately prosperous society in all aspects and realizing comprehensive, coordinated and sustainable development. Green development is of great significance because it helps adjust and lead the "Economic new normal", cultivate the new driving force for economic growth, and optimize and upgrade the industrial

J. Cong · J. Yang (✉) · Q. Zhang · Z. Zhou
School of Economics and Management, Shanxi University, Taiyuan 030006, Shanxi, China
e-mail: yangjun@sxu.edu.cn

J. Cong · J. Yang · Q. Zhang · Z. Zhou
Shanxi Research Center for Green Development, Taiyuan 030006, Shanxi, China

X. Yang and S. Jiang (eds.), *Challenges Towards Ecological Sustainability in China*, https://doi.org/10.1007/978-3-030-03484-9_14

structure. However, coal resource-based regions have faced many lasting challenges, such as single industrial structure, coal resource depletion, environment destruction, and so on (Takatsuka et al. 2015; Li and Dewan 2017). These regions cannot meet the current development requirement and achieve a sound and rapid economic development.

Shanxi province is located in central China, with a total area of 156,700 km^2 and population of 36.64 million by 2015. The province is rich in mineral resources, and its economic development heavily depends on coal resource exploitation and utilization. Shanxi province occupies an important share in the national mining economy. As a typical coal resource-based region, there are a large number of coal-based cities in Shanxi, which extensively distribute across the province. According to the comprehensive classification of resource-based cities in "China's National Sustainable Development Plan for Resource-Based Cities (2013–2020)" (The State Council 2013), there are 13 resource-based cities in Shanxi, including all four types of resource-based cities, of which there are 10 prefecture-level cities, accounting for 8% among 126 prefecture-level administrative regions (including prefecture-level city, prefecture district, autonomous prefecture, and autonomous union in the national resource-based cities). The economic development in Shanxi province has been quite difficult. The traditional "black" development pattern based on coal resources can hardly achieve sustainable growth, and the economic improvement and upgrading are facing many difficulties. Therefore, it is difficult to tackle the dilemma in the long run, and the transformation towards green development pattern is urgent (Qin et al. 2015; Yu et al. 2016). At the same time, the Shanxi provincial government has proposed that the economic and social development should be dominated by the concept of ecological civilization. The dual promotion of policy and practice indicates that the green development pattern will become an important direction for socio-economic transformation in the future and a key task at the regional level in Shanxi.

In this chapter, we begin with the traditional understanding of the green development, and then discuss the meaning of green development emphasizing economic development. Taking Shanxi Province as an example, we discuss some possible paths for economic growth towards green development in Shanxi and other provinces with similar types of coal resources. The purpose of our research is to provide guidance for Shanxi province and other coal resource provinces in moving away from heavy mining resources dependency and implementing the transformation from low-value resources to high-value development, from high carbon resources to low-carbon development, and from black coals towards green development.

2 Evolution of the Green Development Concept in China

2.1 *Theoretical Foundation*

The concept of green development has mainly been used in the fields of resources and the environment. Some scholars pointed out that green development is a comprehensive development in such aspects as economy, environmental protection, ecosystem management, and resource conservation. Another group of scholars considered that green development may not be limited to resource and environmental protection and ecosystem management, but incorporate social and governmental factors and residents' lifestyles. There are several representative views on green development by international organizations, China's central administrations, and individual scholars as follows:

(1) Research Center for Energy Environment and Green Development at Tianjin University of Science and Technology (2010): Promote energy, environment, and economy coordination and sustainability.
(2) World Bank and the State Council Development Research Center Joint Task Force (2013): Economic growth should be least dependent upon resource exploitation while minimizing carbon emission and environmental destruction; promoting growth by creating new green product markets, green technologies, green investments, and changing consumer and environmental practices.
(3) Hu (2014): Forming the coupling relationship between green growth, green treasure, and green welfare through the synthesis of economic, social, and natural systems; emphasizing the systematic, integrality, coordination among the system of economy, society, and nature.
(4) Li et al. (2014): Socioeconomic sustainability and eco-environment sustainable development are of equal importance.

2.2 *Redefining the Concept of Green Development*

Based on the above theoretical basis and the realistic development requirement, we do not simply define the green development as afforesting, protecting environment, conserving resources or sustainable development, but reexamine the green development pattern from the development perspective. In general, green development includes four main characteristics. First, green development is a concept of guidance and it is also a realistic requirement. Second, green development emphasizes the development of morality and prevents the behavior of morality losses. Third, green development focuses on the future success, overcoming of the impediment for development, and the details of evolution. Last, the premise of green development is the harmonious coexistence between man and nature.

Based on the above characteristics, we believe that for understanding green development, we should consider green production mode, green lifestyle, and eco-environment protection. As for green production mode, green development requires to use renewable resources to replace the non-renewable resources during the production period while maximizing the utilization efficiency. For lifestyle, green development emphasizes the upgrade of people's living standards, including the improvement of the health level, people's happiness, individual character and social custom. For eco-environment protection, green development should only cast a small influence on natural environment and grow together with environment. It also requires that all activities not only observe ecological functioning assurance, environmental quality safety, and natural resource conservation, but also make sure the disturbed environment can be restored shortly. Based on the above considerations, we define green development pattern with four dimensions: factor composition, negative externality minimization, environmental value, and consumption morality.

(1) Factor composition advancement: Emphasizing the transition from non-renewable resources to renewable resources and the improvement of the efficiency of individual factors.
(2) Negative externality minimization: Development should only have small influences on natural environment and social dimension; and environment and society grow together.
(3) Environmental value maximization: Real values: from green protection to green growth and to green creation; monetary values: from resources to asset and to capital.
(4) Consumption morality promotion: Emphasizing the improvement of the life quality for people and the culture. Meantime, it requires the improvement of nutrition, health, happiness, the quality of people and social custom.

3 The Challenge Towards Green Development in Shanxi

According to the above discussion on the basic characteristics of green development and the situation in Shanxi province, we have identified four major impediments for green development, which will be discussed in the following sections.

3.1 Coal Resources Reliance and the Single Industrial Structure

Due to the impact of resource endowment, Shanxi province has formed a "coal-dominated" industrial structure for a long time. Resource-based industries which are dominated by coal account for over 60% of GDP, and the economic growth relies heavily on the coal industry (Fig. 1).

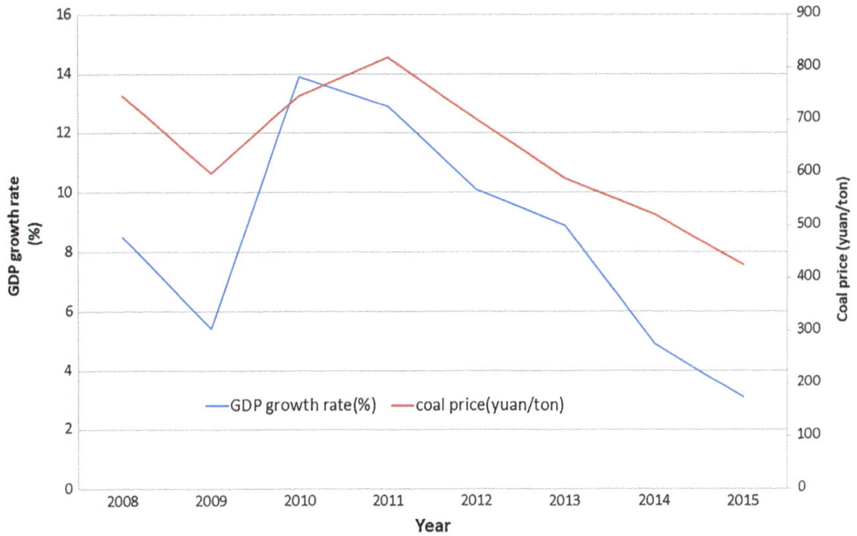

Fig. 1 Trend of economic growth and coal price fitting

In 2016, the added value of the province's mining industry accounted for 62.08% of the total industrial added value (Fig. 2). The persistent dependence upon coal resources seriously restricts the sustainable development. It will be a long way to dilute the overcapacity of coal, although the industrial structure has begun to adjust in recent years. In terms of the overall industrial structure, however, coal and its ancillary industries still account for a large proportion. Industrial growth mainly relies on the consumption of mineral resources and the environmental resources, which does not bring about high economic benefit. After a "golden decade" of coal, the economy grew from 12.9% in 2011 to 4.5% in 2015, and the economic booming and falling were extremely obvious. Economic aggregations, especially structural and quality problems are prominent because of the lack of the capacity for independent innovation and replacing old growth drivers with new ones. The economic growth mainly relies on the coal resources, and the number of employees is large, who are engaged in resource mining, processing industry and other related industries. Currently, the development stamina is lacking with the decline in resource reserves, the increased difficulty in production, and the rise of cost. The economy growth driver of Shanxi province is now weak and is highly dependent upon the economic cycle of resources. The financial income is noticeably reduced, and the social enterprise security capacity is insufficient.

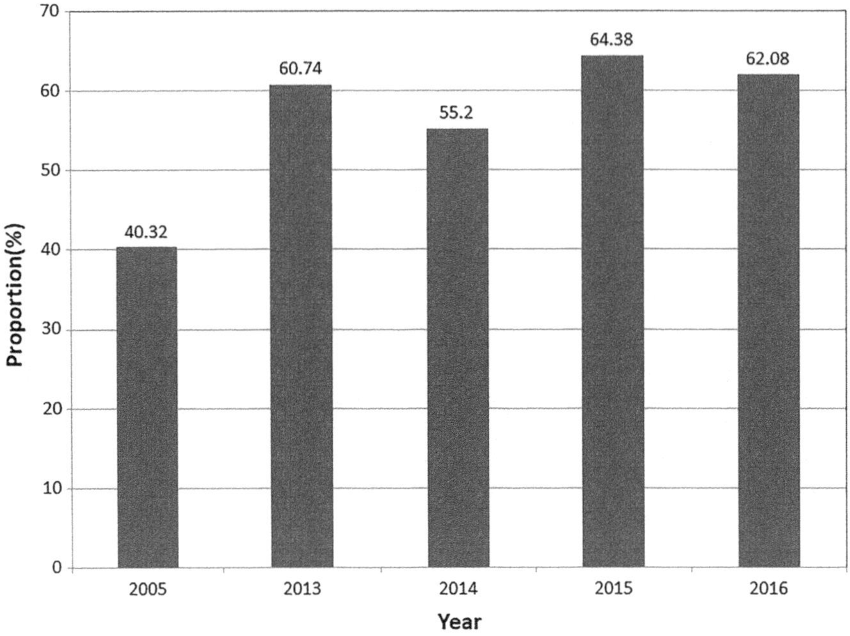

Fig. 2 The proportion of the added value of mining industry in the total industrial added value

3.2 *Negative Externalities in Coal Exploitation*

The negative externalities have caused the resource losses and the ecological degradation, along with many other problems being related with safety production, income gap and employment stabilization. Waste waters, flue gases and waste residues coming from the resource exploitation process over a long time period were not managed well and have caused a series of environmental problems, such as surface subsidence, water resource loss, environmental pollution, and ecological degradation. The social cost of coal mining can be enormous. There are many heavy transformation tasks in coal mining subsidence sites and shanty towns, and large mining subsidence sites should be remediated. The population of poor people in cities is quite large. With the progress of the coal supply-side structural reform and the state-owned enterprise reform, the problem caused by the transfer of workers' position has heightened, and the pressure for re-employment has increased.

3.3 Fragile Ecological Environment and Unreasonable Coal Resources Utilization

Shanxi is located at the loess plateau and belongs to the most vulnerable eco-environmental zone. Its economic growth heavily relies on the coal mining, which has aggravated the ecological environment degradation. Nowadays, Shanxi is among the provinces experiencing serious ecological degradation, poor environmental quality, and low sustainable development capacity. The problem of the ecological environment has become a big constraint to the green development in Shanxi. For example, the lack of vegetation and heavy coal exploitation has led to serious water loss and soil erosion as well as rapid development of land desertification. Forest deterioration has led to the biodiversity loss. The higher utilization intensity has brought out the deterioration of grasslands.

As for the utilization of coal resources revenue, the income from coal development and utilization has been used to support the development of resource departments and improve people's livelihood. The former increased the difficulty of upgrading industrial structure and intensified the influence of "resource curse", and the economic revenue is low and fluctuates easily. The latter can only improve people's livelihood temporarily, but it's difficult to alter their living condition essentially. The coal resources cannot form an excellent support to the local economy and social development because of the unreasonable utilization of coal resource.

3.4 Low Quality of the Life for Residents

In terms of residents' health level, the air pollution caused by coal burning has led to a higher proportion of people in Shanxi suffering from respiratory diseases and lung cancer. The average hospital stay in Shanxi province was 10.8 days in 2015, which was 1.4 days more than the national average. In 2015, the total number of doctors per 10,000 permanent residents was 0.89, and the beds per thousand permanent residents were 1.13, which were lower than the national average, indicating a relative shortage of medical and health resources. The qualified rate of drinking water was only 87.9%. From the residents' living standard, the rural poverty rate of Shanxi province was 14.08% in 2015, which was higher than the national average. From the personal quality of residents, the proportion of people in the primary school, middle school, high school and higher education was 24.51%, 30.73%, 20.38% and 13.74% in 2015, respectively, showing that the population with high education was quite small. From the public awareness, the first survey data "National Survey of Ecological Civilization Research" in 2014 showed that the public overall recognition, the popularity and practice of ecological civilization scoring 74.8, 48.2, 60.1 points, respectively, indicating status with "high recognition, low cognition, and lack of practice". These further indicate that the awareness of ecological environment was weak.

4 Some Strategies for Green Development in Shanxi

4.1 Promoting the Rational Exploitation and Comprehensive Utilization of Coal Resources as Well as the Structural Reform of the Coal Supply Side

4.1.1 Regulating the Exploitation Intensity of Coal Resources and Promoting Conservation and Comprehensive Utilization of Coal Resources

The mineral resources potential evaluation, reserve utilization investigation and mining license verification should be carried out to fully understand the reserve of mineral resources that can be critical for assessing the potential and planning future development. Construction and exploitation of coal resources should be based on the characteristics of regional coal quality, which should also guide the development of industrial clusters and parks, the establishment of a coal recycling concept, and the optimization of the industrial framework. We need to rigorously enforce the mineral development access conditions and prohibit non-eligible enterprises from entering the field of resource development.

We need to improve the recycling level of coal resources, strengthen the comprehensive utilization of industrial solid waste, and increase the added value of mineral resources products. Mining enterprises should promote associated resources, tailings and the important beneficial elements in the solid waste for recycling and reusing. For example, they should steadily promote the comprehensive utilization of coal gangue, and increase the power of coal gangue for reclamation, land reclamation and subsidence area governance. They should support the construction of a demonstration project on the comprehensive utilization of coal resources and encourage mining enterprises, universities and research institutes to form collaborative innovation platforms for industrial solid waste comprehensive utilization.

4.1.2 Promoting the Coal Supply-Side Structural Reform

Efforts should be made to promote the optimization and upgrading of the coal industry. Through the reduction of replacement, and reorganization and other measures in dealing with the excess capacity of coal, we can vigorously develop advanced production capacity and improve the concentration of coal industry. We need to implement the strategic restructuring, and develop large coal groups to promote the coal industry intensive scientific development. Relying upon the three major bases including Jinbei, Jinzhong and Jindong and the three major brands of thermal coal, coking coal and anthracite, through the comprehensive utilization of market, legal systems and economy, with the modern enterprise system as the operational mode, we can implement the strategic reorganization of coal mining enterprises and continue to promote the merger and reorganization of coal enterprises between different regions and different ownerships. We will need to promote the moderate concentration of production factors such as resources, capital, technology, management and equipment in an orderly manner to reduce the number of coal mines and build modern large-scale coal enterprise groups. This can further improve industrial concentration and comprehensively promote the intensive development of coal industry in Shanxi.

4.2 *Reducing the Negative Externalities of Coal Mining and Improving People's Livelihood*

We need to strengthen the ecological restoration in mining areas. Efforts should be made to improve the regional environmental compensation mechanism, strengthen the ecological reconstruction of mining areas, and promote mining vegetation restoration and carbon sequestration project of afforesting. We need to strengthen the prevention of underground exploration and mining activities from destroying the groundwater system, and eliminate high water consumption industry gradually. We need to carry out the treatment of geological and environmental problems (such as mining subsidence areas and open pit mines), promote land reclamation and ecological restoration in mines, strengthen the prevention and control of the key pollutants, and control the pollution of the atmosphere, water and soil. We need to control coal pollution that can further help, prevent the heavy pollution weather, especially in haze treatments, and strengthen the prevention and control of industrial water pollution and the management of urban sewage and the ecological protection of the basin. We need to promote land reclamation, mining geology and ecological environment restoration and control in mining area of mining subsidence area in order to improve soil environmental quality.

Additionally, we need to increase employment and social security. We should improve the basic public services, production and living environment and the people's living standards so that the achievement of the coal mining and economic development can benefit the human well beings. Also, we should pursue industrial structure transformation, upgrading and expansion of employment to benign interaction, encourage and support the resource-based areas to develop employment-driven ability of labor-intensive enterprises and small micro-enterprises. We need to speed up the improvement of education and culture, health care, employment services, social security, environmental protection, and other basic public service system to enhance the quality of public services. We should properly resolve the historical legacy of the closure of retirees of the bankrupt collective enterprise for medical insurance and the old workers' injury insurance. Meanwhile, it is important to promote the work of relocation and comprehensive management of the independent industrial and mining area, coal mining subsidence area and shantytowns, and to speed up the construction of urban infrastructure, carry out underground integrated corridor pilot, and promote the construction of sponge city. Furthermore, it is also important to build a functional, safe and efficient modern urban infrastructure system to enhance the comprehensive functions of urban production and living services, and accelerate the new urbanization to enhance the city's bearing capacity and radiation-driven role. Last, we should speed up the beautiful and livable rural construction to achieve an integrative development between the urban and rural areas.

4.3 Enhancing the Environmental Capacity and Resource Values

There are some ways to strengthen the construction of ecological protection system. First, promote the restoration of ecological fragile areas and the water and soil conservation along the Yellow River. Then, use enclosure, grazing prohibition and grazing rest and other measures to increase forest and grass vegetation, improve forest coverage, control soil erosion, and strengthen sand governance. Meanwhile, promote wetland conservation and restoration emphasizing the national geological parks in the province in order to strengthen the construction of nature reserve management, build an ecological protection system, enhance the overall function of natural ecosystems, and promote the environmental capacity. Such efforts can help digest and absorb more pollutants and can provide a good basis for economic and social development.

On the other hand, promoting resource value upgrading and recycling is also very important. The earnings of the coal exploitation and utilization should not be used directly to improve the lives of residents or continue to support the development of traditional coal resources sector. The government should establish a "reserve fund" or a "regional development fund". The former should be used to upgrade some advantageous traditional industries and developing new industries to promote quality and efficient development of traditional industries and alternative industries, which can enhance the ability to accumulate capital to promote sustainable economic development. The later fund should be used to compensate the social and economic development losses caused by coal mining that can support the development of middle and small-sized enterprises having low market access threshold, flexible operation and a large number. This can help absorb the surplus population outside the resource industry, tap the potential of local labor resources, and promote the update of public infrastructures to create a good living environment.

4.4 Improving the Residents' Quality and Strengthening the Publicity of Green Production, Living and Consumption

In order to improve the residents' quality, we need to optimize their health and education levels. Some specific strategies are discussed here. First, increase the supervision and control of food safety, drinking water and air pollution. Use the resources such as public health, medical treatment, medical insurance and all social forces to improve the mechanism of disease prevention and control system, and to implement comprehensive prevention and control works of chronic diseases, especially respiratory system diseases. Furthermore, increase the investment in education, deepen the comprehensive reform in the field of education, promote the balanced development of education, and establish a unified educational funding guarantee mechanism for urban and rural areas. Lastly, promote the fair development of education, support the development of private education, meet the residents' demand for educational resources, develop the education for all and lifelong education to build a learning society, and improve the education level of residents gradually.

Promoting green production, living and consumption concepts requires the joint effort from the government, enterprises, and the society. The government should strengthen the cultivation of the green consumption consciousness of residents and the green marketing concept of enterprises, strengthen government responsibilities, establish an independent green service sector, take effective measures to regulate the green market order, and strengthen macro management to provide administrative guarantee for green consumption. The enterprise should strengthen the green management and implement green production to provide consumers with green products and create a better support environment for green consumer. The public should try to choose green goods with green product identification, provide green services, and should reject non-green goods that cost resources and pollute the environment.

In short, coal-based provinces should promote green development since it is the only way to achieve long term prosperity. The economic, ecological and social problems caused by relying on coal resource development are not helpful for sustainable development of human society. Therefore, green transition is imperative. In view of the challenges towards green transformation in coal resource-based provinces, it is necessary to change the "coal-dominated" industrial structure to break through the development bottleneck. Meanwhile, we need to pay attention on the ecological environment and social livelihood issues. We need to make efforts to improve the public quality, foster good social fashion, and make development really benefit the livelihood of the people. These can help develop a healthy, sustainable society.

References

Hu A, Zhou S (2014) Green development functional definition, mechanism analysis and development strategies. China Popul Resour Environ 1:14–20

Li B, Dewan H (2017) Efficiency differences among China's resource-based cities and their determinants. Resour Policy 3:31–38

Li X, Liu Y, Song T (2014) Calculation of the green development index. Soc Sci China 6:69–95

Qin S, Wang X, Fu H (2015) The predicament and countermeasures of the green development strategy in China: based on ecological civilization construction integrating into economic construction. Ecol Econ 7:168–171

Takatsuka H, Zeng DZ, Zhao L (2015) Resource-based cities and the Dutch disease. Resour Energy Econ 3:57–84

The State Council (2013) China's national sustainable development plan for resource-based cities (2013–2020). http://www.gov.cn/zwgk/2013-12/03/content_2540070.html. Accessed 3 Dec 2013

World Bank and the State Council Development Research Center Joint Task Force (2013) China in 2030: building a modern, harmonious and creative society. China Financial & Economic Publishing House

Yu C, de Jong M, Cheng B (2016) Getting depleted resource-based cities back on their feet again–the example of Yichun in China. J Clean Prod 10:42–50

Zhu H (2010) Research Center for Energy Environment and Green Development at Tianjin university of Science and Technology was established. http://society.people.com.cn/GB/86800/12882320.html. Accessed 6 Oct 2010

Dr. Jianhui Cong is an Associate Professor of Economics at the Shanxi University, China. His research interests include resource economy and management, green and low-carbon development, and climate change governance. Cong has authored or co-authored three books and over 20 articles in these areas.

Dr. Jun Yang is currently a Full Professor of Economics at the Shanxi University, China. His research current interests include regional economic development, green economy, and industrial investment. He has authored or co-authored over 10 books and 80 articles in these areas. Yang is one of the leading talents in philosophical and social sciences under China's "Ten Thousand People Plan". He is a Vice President of Shanxi University, China.

Qin Zhang is currently a postgraduate student at the Shanxi University, China. Her research interests include the green development of industry and poverty-relief by industrialization.

Zhou Zhou is currently a senior undergraduate student with major in Economics at the Shanxi University, China. Her research interest includes low-carbon development.

Index

X. Yang and S. Jiang (eds.), *Challenges Towards Ecological Sustainability in China*, https://doi.org/10.1007/978-3-030-03484-9

Printed by Printforce, the Netherlands